Elements of Earthquake Engineering and Structural Dynamics

André Filiatrault

POLYTECHNIC INTERNATIONAL PRESS

Elements of Earthquake Engineering and Structural Dynamics, André Filiatrault

Production team
Editorial management and production: Éditions de l'École Polytechnique de Montréal
Editing: Käthe Roth
Illustrations: André Filiatrault and Flavio Mini
Typesetting: Claire Bertin
Cover page: Chantal Fauteux and Daniel Viens

For information on distribution and points of sale, see our Web site: http://www.polymtl.ca/pub
E-mail of Éditions de l'École Polytechnique de Montréal: editions@mail.polymtl.ca

Legal deposit: 3rd quarter 1998   ISBN 2-553-00629-2
Bibliothèque nationale du Québec   Printed in Canada
National Library of Canada   1 2 3 4 5   02 01 00 99 98

To Louise, Lou-Anne, and Sydney

PREFACE

Major earthquakes can occur in Canada. The earthquake hazard is, however, variable across the country. Therefore, the earthquake-engineering background of Canadian engineers may vary greatly from region to region. Engineers educated in regions of low seismic hazard, such as the Canadian prairies, may be faced with new earthquake problems if they are involved in projects on the west coast of British Columbia, for example. Current literature on earthquake engineering can be difficult for an engineer who has not had previous training in seismic design to grasp fully. Furthermore, no formal textbook on earthquake engineering is adapted to Canadian standards and practice. This latter observation sparked the writing of this book.

This book can be used by practising engineers, senior undergraduate and junior graduate structural-engineering students, and university educators. Its main goal is to provide basic knowledge to structural engineers who have no previous knowledge about earthquake engineering and structural dynamics. Problem sets offered at the end of most chapters allow the reader to appreciate the practical aspects of the subject. After assimilating the material contained in this book, readers will have the necessary background in earthquake engineering to be able to consult more specialized references on particular issues related to earthquakes or seismic design.

Earthquake engineering is a multidisciplinary science. This book is not limited to structural analysis and design. The basics of other relevant topics (such as geology, seismology, and geotechnical engineering) are also covered to ensure that structural engineers can interact efficiently with other specialists during a construction project in a seismic zone.

In chapter 1, basic concepts of earthquake engineering are presented. Chapter 2 discusses seismological aspects relevant to structural engineering. Chapter 3 gives a description of the seismic hazard and how to model it in order to construct seismic-zoning maps. In chapter 4, the cornerstone of the book, the fundamental notions of linear and nonlinear dynamic analyses of structures are discussed. The concepts of dynamic analysis are used in chapter 5 to evaluate the influence of a soil deposit on the movement at the base of a structure; this chapter also addresses the problem of liquefaction of saturated granular soils and the soil-structure interaction phenomenon. The last four chapters are aimed at practising structural engineers who must consider the effects of earthquakes in their designs. Chapter 6 gives a detailed presentation of the seismic provisions included in the 1995 edition of the National Building Code of Canada. Chapters 7 and 8 describe specific seismic detailing requirements that must be incorporated into steel and concrete structures. Finally, chapter 9 presents some energy concepts relevant to earthquake engineering, which lead naturally to innovative earthquake resistant systems based on energy dissipation. A computer program (RESAS) is described in the appendix and is available on the web site : http://www.polymtl.ca/pub. This program computes the response spectra of earthquake accelerograms.

The material covered in this book originates from class notes of an undergraduate structural dynamics course and a graduate earthquake engineering course I have taught, first at the University of British Columbia in Vancouver from 1988 to 1990, then at École Polytechnique in Montreal since 1991. Chapter 4 can be used as a stand-alone reference for a senior undergraduate structural dynamics course. The entire book can be used as a reference for a basic graduate earthquake engineering course.

I would like to thank les Éditions de l'École Polytechnique de Montréal for publishing this book, particularly Lucien Foisy and Diane Ratel. I would also like to thank my wife, Louise Daoust, for translating into English the original French version of this book.

In addition, I would like to thank my colleagues from the structures group of the civil engineering department at École Polytechnique, in particular, Professor René Tinawi who has graciously provided me with class notes from the structural dynamics course he has been teaching for many years. Several examples taken from these notes have been incorporated into chapter 4 of this book. Several research results generated by graduate students I had the immense pleasure of supervising in the last nine years are also included in the book. In particular, I would like to acknowledge the following former graduate students: Marc Sarrazin for recompiling the computer program RESAS, as well as Danilo D'Aronco, Sylvain Pineau, and Mario Hernandez-Cervantes.

The final revision of the manuscript was done during a sabbatical leave at the University of Canterbury in New Zealand, from September 1994 to June 1995. Many thanks to Professors Thomas Paulay, Robert Park, Athol Carr, and Jose Restrepo, all faculty members at Canterbury, for the long discussions and exchanges, mostly over tea, that positively influenced the final outcome of this book. Of course, I thank also École Polytechnique in Montreal for awarding me this sabbatical leave.

Finally, I would like to express my deepest gratitude to my doctoral supervisor, Shel Cherry, Professor Emeritus, University of British Columbia, in Vancouver, for introducing me to earthquake-engineering research. His professionalism, enthusiasm, and broad-mindedness continue to be a source of inspiration exceeding the bounds of my technical activities.

André Filiatrault, Ph.D., Eng.
Professor of Civil Engineering
École Polytechnique de Montréal
June, 1997

FOREWORD

Earthquake Engineering is concerned with the development of effective and efficient procedures for mitigating hazards due to earthquakes – hazards that are essentially related to human works. Engineers and geoscientists have been the prime movers in the advancement of this relatively new multidisciplinary activity, and they have been joined in this endeavour by planners, architects, building officials, and social scientists. While the record of earthquake damage and disaster in Canada has, fortunately, been relatively low, this has been more the result of luck than the work of benevolent favour. Various regions of the country are situated in active seismic zones, and the probability of occurrence of a major earthquake disaster is very real and must be addressed.

The National Building Code of Canada (NBCC) recognizes this risk and offers a clear warning of the serious prospect of major earthquakes occurring in regions in which five of the largest Canadian cities are located. Due to population increase, economic and industrial growth, and urban area development, the potential hazard from future seismic events has become much more acute in the past quarter-century. Attempts to mitigate such dangers are clearly in the national interest, and our society presently devotes a share of its total resources to providing a reasonable degree of safety from the earthquake hazard.

Although the existing catalogue of Canadian earthquakes begins around 1534, perhaps the earliest descriptive account of this hazard in Canada concerns the great St. Lawrence River earthquake of 5 February 1663, which was centred in the Trois-Rivières region of Quebec. Based on historical documentation, it has been assigned the modern equivalent of a Richter magnitude 7.5–8.0. Reportedly, "it did a great deal of good to souls" – presumably those of the sinners in the area! It was not until two centuries after this event that the earthquake phenomenon was freed from explanations based on superstition and folklore and became rooted in studies that had a scientific basis.

The modern origins of earthquake engineering are founded in the observations and investigations of a number of catastrophic events that took place early in this century. But the most significant development and advances in this subject were initiated over the past four decades, during which time we have seen a quantum increase in our knowledge and understanding of the earthquake problem. This trend can be attributed to a number of factors, including the strong impetus for earthquake-engineering research and technical development fostered by the International Association for Earthquake Engineering (IAEE), whose formation, following the successful first World Conference on Earthquake Engineering (Berkeley, 1956), was initiated at the second World Conference (Tokyo, 1960), and became a reality in 1963. The IAEE is now responsible for quadrennially sponsoring the World Conferences on Earthquake Engineering and has been the catalyst for the development of national and regional conferences on a regular basis in many parts of the world, including Canada.

Canada was a founding member of IAEE, and the first milestone heralding the development of earthquake engineering in this country dates from the formation by the National Research Council (NRC), also in 1963, of the Canadian National Committee for Earthquake Engineering (CANCEE). CANCEE serves as an advisory body to the NBCC, and in that capacity has provided, and continues to provide, distinguished service to the profession by ensuring that the seismic provisions of the NBCC continuously incorporate the latest advances in earthquake engineering. The earthquake load requirements that appeared in the first (1941) NBCC were based on rigid body dynamics and essentially duplicated U.S. building code practice. CANCEE's initial imprint on the NBCC occurred in the 1965 edition of that document when, for the first time, the quasi-static treatment of its seismic load provisions was formulated on the basis of rational analysis.

As originally conceived, CANCEE represented Canada with the world organization and played a fundamental role in encouraging the involvement of researchers and the education of practitioners in earthquake engineering matters. In this latter capacity, CANCEE was instrumental in initiating two very important developmental landmarks, a pair of earthquake engineering symposia (the first in 1965 in Vancouver, closely followed by a corresponding event in Montreal in 1966) and the subsequent quadrennial series of Canadian Conferences on Earthquake Engineering (the first in 1971 consisting of 24 papers, the seventh in 1995 containing 118 papers).

In the mid-1960s, publication of the first probabilistic seismic risk maps of Canada, by the then Seismological Branch of the Dominion Observatory, represented a major achievement. These were incorporated by CANCEE into the NBCC (1970), which, at the time, was one of the first modern building codes, worldwide, that used this scientific approach for seismic zonation. Around 1970, the Seismological Branch began its development of the Canadian strong-ground-motion network, which has since captured the characteristics of a number of strong-motion earthquake records in different regions of the country. Both of these milestones must be recognized as significant and identifiable contributions to the Canadian earthquake hazard-reduction program. Also beginning in the mid-1960s, the commitment of our universities, government agencies, and professional societies – by their offerings of credit and continuing education courses and/or sponsoring of seminars and workshops – played a major role in ensuring the education and development of a knowledgeable body of earthquake engineers and scientists in this country.

CANCEE was reorganized by the NRC in late 1991, and its new mandate is basically focused entirely on code-related matters. Given this development, and recognizing the very noticeable and growing number of individuals interested and engaged in earthquake engineering, it was decided that the time was ripe to move beyond the committee stage and form a truly national organization, which would be separate and distinct from CANCEE and whose membership would be open to all interested individuals. As a result, the Canadian Association for Earthquake Engineering/Association canadienne du génie parasismique (CAEE/ACGP) was federally incorporated in January 1993 as an independent, multidisciplinary, technical society with a broad mandate to foster earthquake engineering practice and research in Canada and to represent Canada on the IAEE. This formation of a network of collegial earthquake engineers throughout the nation represents a most important and exciting milestone in the history of earthquake engineering in Canada.

The preceding brief journey through that history has, of necessity, drawn attention only to some of its highlights. But, before concluding, I would like to suggest that the book that gave rise to this foreword itself represents a bright new event, in the sense that it is the first modern text published in Canada that is devoted to the subject of earthquake engineering. Professor André Filiatrault has used his considerable talent, scholarship, and organizational ability to produce a valuable primer on earthquake engineering that will be of particular interest to both students and practitioners. The material is clearly and concisely presented and follows a logical progression that incorporates the fundamental principles of the major disciplines contributing to earthquake engineering. Following an initial commentary, the reader will find a good, basic introduction to seismology and seismicity (chapter 2); analysis of seismic risk (chapter 3); dynamic analysis of structures (chapter 4); soil dynamics, soil liquefaction and soil-structure interaction (chapter 5); the design of earthquake-resistant building structures (chapter 6); detailing for the seismic resistance of concrete structures (chapter 7) and steel structures (chapter 8); and an introduction to some innovative systems for control of seismic structural response (chapter 9). Exercises are offered to illustrate the application of the theory covered, and the design section is based on Canadian code requirements.

This book is ideally suited for use by senior undergraduates, first-year graduate students, and professional engineers. It offers a solid basis for the self-study of more advanced topics and current literature on earthquake engineering. Professor Filiatrault, one of Canada's leading young earthquake-engineering experts, is to be congratulated for his efforts, which admirably fill a niche and which represent a real contribution to the profession.

Shel Cherry
Professor Emeritus
University of British Columbia
June, 1998

TABLE OF CONTENTS

CHAPTER 1

Preliminary Notions

1.1   BRIEF HISTORY OF EARTHQUAKE ENGINEERING

Earthquake engineering is a young science, with most of its major developments occurring in the twentieth century, so making an accurate outline of its history is difficult. At the Eighth World Conference on Earthquake Engineering held in San Francisco in 1984, Housner (1984) presented a historical review of earthquake engineering. The following is inspired by his paper.

The foundations of earthquake engineering were laid in the eighteenth and nineteenth centuries by scientists most of whom were British – not because England has high seismicity, but because the industrial revolution (1700–1900) strongly influenced developments in a variety of sciences. Robert Hooke (1635–1703), known for his law of elasticity, was one of the first scientists to become interested in the earthquake phenomenon, giving several talks on earthquakes and volcanoes at the Royal Society of London in 1667 and 1668.

In the nineteenth century, no distinction was made between seismology and earthquake engineering. Seismology is derived from the Greek word *sismo* (vibration), and was used for the first time by British engineer Robert Mallet (1810–81). Mallet also introduced the now well-known terms "epicentre" and "focal point." In 1848, he published plans for the first electromagnetic seismograph, but the instrument was never built. Italian-born Luigi Palmieri (1867–96) modified Mallet's concept, built the first automatic seismograph, and obtained the first modern earthquake records.

Meanwhile, on the other side of the globe, John Milne, Robert Ewig, and Thomas Gray, then teaching at the Imperial College of Engineering in Tokyo, founded the Seismological Society of Japan in 1880. It was the first scientific association entirely dedicated to the study of seismic phenomena and the forerunner of the International Association for Earthquake Engineering.

Around the turn of the century, three major earthquakes provided a strong impetus for developing further knowledge related to seismic phenomena: Mino-Awari, Japan (1891), San Francisco (1906), and Messina, Italy (1908). This last earthquake, which occurred on December 28, was particularly important: because of its seriousness – more than 83 000 people died – it led to the first earthquake-resistant construction methods and, therefore, the science of earthquake engineering. The Italian government set up a special committee composed of nine practising engineers and five engineering professors, which was asked to study the earthquake and formulate recommendations. In the committee's report, M. Panetti, a professor of applied mechanics at the University of Turin, proposed that for active seismic zones, civil-engineering structures be calculated with a uniform static lateral load represented by a "seismic coefficient" expressed as a fraction of the structure's weight. He also noted that the effects of earthquakes on structures are in fact a "structural dynamics" problem; however, this problem was considered "much too complicated" to address, and an equivalent static approach was preferred for the design of earthquake-resistant structures. This approach remained in use until engineers finally gained access to powerful computers. In 1909, A. Danusso, professor of structural engineering at the University of Milan, published a paper that explained in detail the application of the static approach proposed by Panetti. This method was slowly integrated into building codes with few modifications until the 1940s. After Tokyo's great earthquake in 1923, Japanese provisions adopted a seismic coefficient of 10%. The city of Los Angeles, however, adopted a seismic coefficient of 8% following the 1933 Long Beach earthquake. A seismic coefficient that varies with the height of the structure was introduced only in 1943. This event marked the birth of the static approach method, as found in modern building codes.

1.2 DEVELOPMENT OF EARTHQUAKE ENGINEERING IN CANADA

Every year, 200 to 300 earthquakes are recorded in Canada. On average, 15% of these earthquakes occur in the eastern part of the country, 25% in western Canada, 60% in the north, and very few in the Prairies. Over the last century, the eastern part of the country has been hit, on average, by one earthquake exceeding magnitude 6 every 10 years, while an average of two earthquakes exceeding magnitude 6.5 has occurred every 10 years in western Canada.

In Canada, the development of earthquake engineering is closely linked with implementation of the seismic provisions of the National Building Code of Canada (NBCC). The first edition of the NBCC was published in 1941. Uzumeri, Otani, and Collins (1978) published an overview of the evolution of earthquake provisions included in the NBCC until its 1977 edition. The first Canadian seismic provisions, included in the Code's appendices, were strongly influenced by the procedures presented in the 1937 edition of the Uniform Building Code in the United States. The static method was used with a uniform lateral load based on a seismic coefficient of 2% to 5%, depending on the soil-bearing capacities.

The earthquake provisions were modified and integrated into the main body of the Code for its second edition in 1953 and its third edition in 1960. Two major changes were introduced: the first seismic zoning map for Canada and consideration of structural flexibility in the calculation of the seismic coefficient. Until 1970, the seismic zoning map divided Canada into four zones of "seismic intensity."

Developments in earthquake engineering in California strongly influenced the 1965 edition of the NBCC. For the first time in Canada, the importance of structural ductility in seismic response was recognized in a design "base shear" formula, which varied along the height of the building and was proportional to the weight of each floor. Furthermore, the notion of "torsional couples" appeared in this edition of the NBCC; it had already been included in the Mexican building code. Finally, the 1965 edition of the NBCC recognized that a dynamic analysis undertaken by a "competent authority" was an alternative to the static design procedure.

The 1970 edition of the NBCC included a new seismic zoning map for Canada. For the first time, the map of the entire country presented contours of peak horizontal accelerations (PHAs) based on a return period of one hundred years, although the design base shear formula did not include the 100-year acceleration values. To consider the higher modes effect, the 1970 edition of the NBCC suggested that a portion of the lateral static load be applied to the roof of the structure. For the same reason, the code allowed a reduction of the overturning moment at the base of the structure.

In the 1975 edition of the NBCC, the base shear formula was calibrated to introduce the 100-year acceleration values directly in the lateral load calculation, ensuring that the design base shear level remained comparable to the previous edition of the code. For the first time, the steps for carrying out the dynamic analysis of a "complex or irregular" structure were described in the Code's commentary. This dynamic analysis was based on an "elastic design response spectrum."

No major changes were included in the 1977 and 1980 editions of the NBCC. However, the Code required that the base shear obtained by a dynamic analysis be at least 90% of the base shear calculated with the Code's static method.

Since 1980, the NBCC has been revised every five years; a fair number of changes appeared in its 1985 and 1990 editions. The 1985 edition presented new seismic zoning maps for PHAs and peak horizontal velocities (PHVs), drawn up based on a uniform return period of 475 years for the entire Canadian territory. The base shear was recalibrated to consider the 475-year values without, however, drastically altering the level of protection suggested in previous NBCC editions. The 1990 edition differed by presenting a base shear equation that emphasized the influence of ductility on reduced lateral design forces. Finally, the 1995 edition of the NBCC presented minor modifications and will be discussed in chapter 6.

1.3   COMPUTERS AND EARTHQUAKE ENGINEERING

The evolution of computers from analog to digital was paramount for earthquake engineering. For example, computers allowed earthquake records to be analyzed in a reasonable amount of time and were the cornerstone of practical application of the seismic response spectrum method, proposed by George Housner (1910 –) at the beginning of the 1950s. Computers also allowed for dynamic analysis of structures subjected to seismic excitations. This type of analysis, which Panetti in 1908 had called "much too difficult," was now readily accessible to engineers. Numerical procedures such as the finite element method were also used in dynamic analyses of complex structures. Although powerful computers were beneficial to earthquake engineering, machines cannot replace the judgment of structural engineers. The design and construction of an earthquake-resistant structure is both a science and an art.

1.4   EXPERIMENTAL EARTHQUAKE ENGINEERING

1.4.1 Importance of Experimental Earthquake Engineering Research

Although powerful computers allow for dynamic analysis of complex structures, earthquake engineering is not beyond experimental testing methods. Laboratory experimentation remains a necessary tool for maintaining an adequate level of seismic protection for civil-engineering structures.

Around the world, experimental earthquake engineering research is used to study the behaviour of elements, assemblies, and complete structural systems in a seismic environment. Three types of tests may be performed: quasi-static; pseudo-dynamic; and shake table.

1.4.2 Quasi-Static Tests

Quasi-static tests replace the inertia forces generated by an earthquake on a structure with equivalent static loads. Hydraulic actuators produce static forces between a reaction frame and a specimen. Generally, quasi-static tests are performed on large-scale structural elements. Basic information, such as strength, stiffness, and ductility, is obtained to predict the behaviour of a structure. The results can then be used to validate and develop numerical models. Unlike dynamic tests, quasi-static tests may be interrupted at any time to assess the condition of the specimen. The main weakness of quasi-static tests resides in the comparison between the specimen's ability to dissipate energy and the energy-dissipation capacity required to ensure seismic safety. A constant problem during a quasi-static test is to know whether the specimen is overloaded or not. Most universities in Canada are able to perform quasi-static tests. In 1992, McGill University in Montreal acquired a tension-compression universal testing machine with a capacity of 11 MN. École Polytechnique in Montreal has a testing frame with a capacity of 1,8 MN, which is used mainly to test full-scale steel bracing elements.

1.4.3 Pseudo-Dynamic Tests

Pseudo-dynamic tests combine quasi-static tests with a numerical analysis to simulate the seismic response of a structure. However, the speed at which the tests are performed is much slower than the speed of a real earthquake. Pseudo-dynamic tests directly measure the internal forces at specific points on the structure within a given time interval. These forces are used to solve numerically the system's equations of motion and to calculate the resulting displacements. Hydraulic actuators then apply these displacements to the specimen. Although pseudo-dynamic tests are relatively simple and more realistic than quasi-static tests, they still have some limitations:

- Materials sensitive to strain-rate effects may produce results that are invalid in a real earthquake.
- Because of its low speed, the pseudo-dynamic test cannot be performed on viscoelastic damping systems.

- Because of cost, only a few hydraulic actuators are used, thereby limiting the number of degrees of freedom. The complexity of the experiment is thus greatly affected.

The University of Ottawa has the only pseudo-dynamic testing facility in Canada. The equipment comprises three hydraulic actuators with a capacity of 1 MN each.

1.4.4 Shake-Table Tests

The shake-table test is the only experimental technique allowing direct simulation of inertia forces on a structure with a distributed mass. A shake table can easily reproduce different types of ground motions, such as recorded earthquakes, synthetic accelerograms, or simple signals such as sine waves. In this controlled environment, the ground-motion intensity can be increased to induce an inelastic response in the specimen. The results allow evaluation of the nonlinear response and the failure modes of structures in a realistic seismic environment. No other testing methods produce these types of results. The major weakness of the shake-table test is related to the load-bearing capacity of the testing platform. In most cases, small-scale testing is used. In Canada, only two universities have shake-table facilities with a capacity exceeding 10 tons: the University of British Columbia in Vancouver and École Polytechnique in Montreal. The larger one, at École Polytechnique, has a testing platform of 3,4 m × 3,4 m and a load-bearing capacity of 15 tonnes.

1.5   REFERENCES

Housner, G.W. (1984). "Historical View of Earthquake Engineering." *Proceedings of the Eighth World Conference on Earthquake Engineering*, San Francisco, CA, July 21– 28. Englewood Cliffs, NJ: Prentice-Hall, pp. 764-77.

National Building Code of Canada and Commentaries. 1st ed. (1941); 2nd ed. (1953); 3rd ed. (1960); 4th ed. (1965); 5th ed. (1970); 6th ed. and supplement (1975); 7th ed. (1977); 8th ed. (1980); 9th ed. (1985); 10th ed. (1990); 11th ed. (1995). Ottawa: National Research Council of Canada.

Uzumeri, S.M., Otani, S., and Collins, M.P. (1978). "An Overview of Canadian Code Requirements for Earthquake Resistant Concrete Buildings." *Canadian Journal of Civil Engineering*, 5(3): 427–41.

CHAPTER 2

Elements of Seismology and Seismicity

2.1 INTRODUCTION

The earthquake design of a structure always depends on the degree of the seismic activity in the region. Many seismological factors directly influence the work of structural engineers. Consider the following:

- the distribution of earthquake sources affecting the construction site;
- the fault mechanisms of the various sources;
- the seismic activity of the various sources in terms of the recurrence of magnitudes;
- the ground-motion intensity;
- the attenuation of the ground motion with distance.

This chapter presents an overview of the fundamental properties of these various seismological aspects, in order to acquaint structural engineers with seismologists' "language."

2.2 CAUSES OF EARTHQUAKES

2.2.1 Natural Earthquakes

Most natural earthquakes occur in the earth's crust. The crust measures between 60 and 100 km in depth and is made of different segments that are continuously in motion. Because of this motion, deformations occur in the rock, which cause a build-up of elastic energy. A rupture or slip along a fault line causes a sudden release of this energy into a seismic shock (an earthquake), which, in turn, causes propagation of seismic waves and ground shaking. Most earthquakes occur in two specific zones on the planet (fig. 2.1):

- the Circum-Pacific Belt: South America, the California coast, Alaska, Japan, Formosa, the Philippines, New Zealand;
- the Alpine Belt: the Mediterranean, North India, Indonesia.

Damage from an earthquake can be caused by:

- *fault movement* (one-dimensional);
- *ground shaking* (three-dimensional).

The vast majority of structural damage is caused by ground shaking.

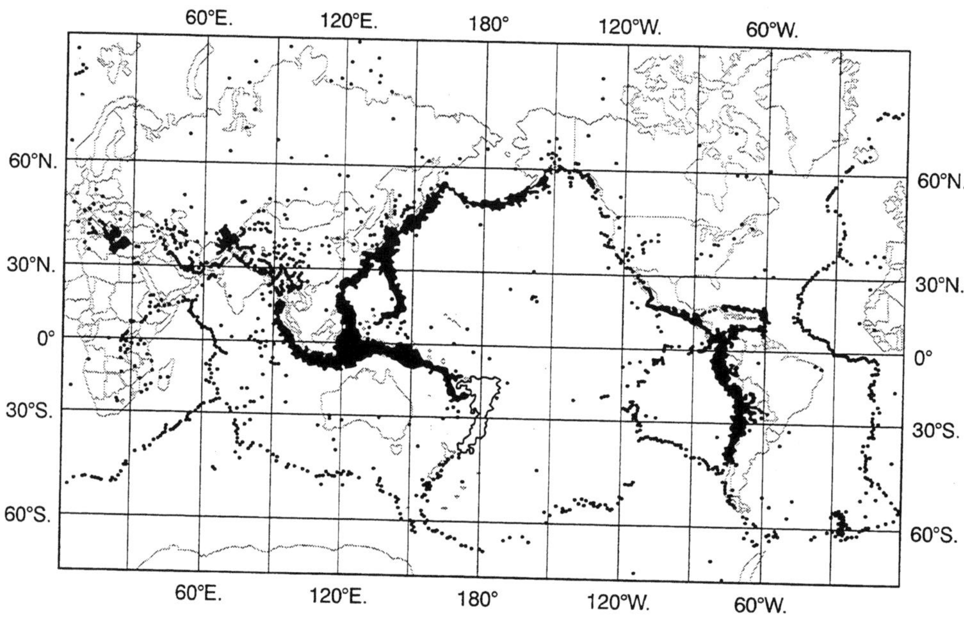

Figure 2.1   World seismicity from 1963 to 1988, magnitude > 5 (from Bolt, 1993).

2.2.2 Induced Earthquakes

Some human interventions influence the amplitude and distribution of strains in the earth's crust. Interventions such as the filling of a water reservoir, mining, excavation of huge quarries, high-pressure injection of fluids to generate geothermic energy, oil wells, and underground nuclear explosions may cause major induced earthquakes.

Among these human activities, the filling of water reservoirs usually causes the most severe induced earthquakes, which may reach magnitude 6 on the Richter scale (section 2.9). More than 70 earthquakes of this type have been recorded worldwide. For further reading, Gupta (1992) is an excellent reference on earthquakes induced by the filling of water reservoirs. This book will be limited to naturally occurring earthquakes.

2.3   THEORY OF PLATE TECTONICS

Plate tectonics is generally accepted as the explanation for the occurrence of most earthquakes. According to this theory, proposed in the 1960s, the earth's crust is composed of several large plates that float on a viscous medium and support continents and oceans. All continents were originally linked together, but they started drifting apart 200 million years ago. The plates move from 1 to 15 cm every year. Three types of plate motion can be observed:

- *transform motion:* the plates slide past each other;
- *diverge motion:* the plates diverge from each other, forming ocean ridges;

- *subducted motion:* the plates converge on each other, causing the subduction of one plate under another.

According to the theory of plate tectonics, earthquakes arise at the boundaries of adjacent plates (fig. 2.2) and occur when the resistance of the rock is exceeded. These conditions create a fracture. The energy build-up between two plates (a fault) can be estimated, but it is still impossible to predict when this energy will be released. The embryonic science of earthquake prediction has also investigated other factors that might predict seismic events: variations in water levels in wells; changes in wave-propagation velocities; changes in electric resistivity and magnetic distortion; gas emissions; abnormal animal behaviour; and so on.

For example, for more than 10 years, Japanese researchers studied catfish behaviour in relation to the occurrence of earthquakes. Their research concluded that catfish had the ability to predict the occurrence of earthquakes. Unfortunately, their behaviour remained the same regardless of earthquake intensity. This illustrates the great complexity of earthquake prediction: scientists are faced with predicting not only time of the occurrence, but also intensity of the ground motion.

In 1975, the Chinese government evacuated close to three million people from many cities of Manchuria. This decision came following an elaborate study and an actual earthquake of magnitude 4.8. Several days after the evacuation, a 7.3 magnitude earthquake destroyed 90% of the cities, but only several hundred people lost their lives. In 1976, however, people evacuated from the city of Kuantung, near Canton, waited two months, but the predicted earthquake never came. The same year, the Tangshan earthquake was not predicted but an estimated 250 000 people died.

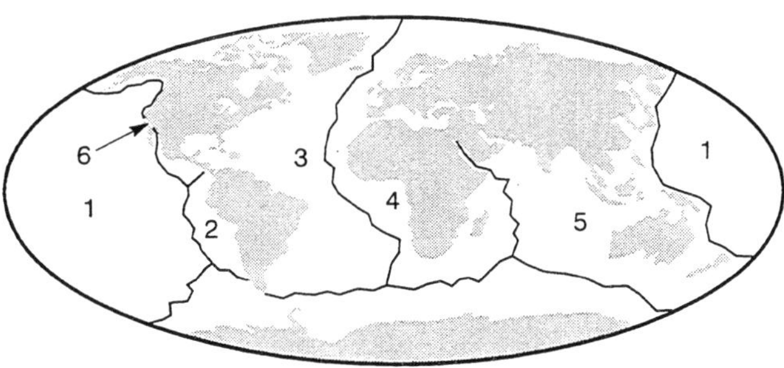

Legend
1- Pacific plate
2- Nazca's plate
3- American plate
4- African plate
5- Eurasian plate
6- San Andreas fault

Figure 2.2   Theory of plate tectonics.

2.4   REID'S ELASTIC REBOUND THEORY

Following the 1906 San Francisco earthquake, H.F. Reid proposed an explanation for the immediate cause of an earthquake. According to Reid's elastic rebound theory (fig. 2.3), a fault is incapable of movement until strain has accumulated in the rock on either side due to gradual shifting of the earth's crust. The rock becomes distorted but holds its original position. When the accumulated stress finally overcomes the resistance of the rock, the earth snaps back into an unstrained position, releasing a large quantity of energy, producing waves that travel through the earth in every direction, and causing what are known as earthquakes.

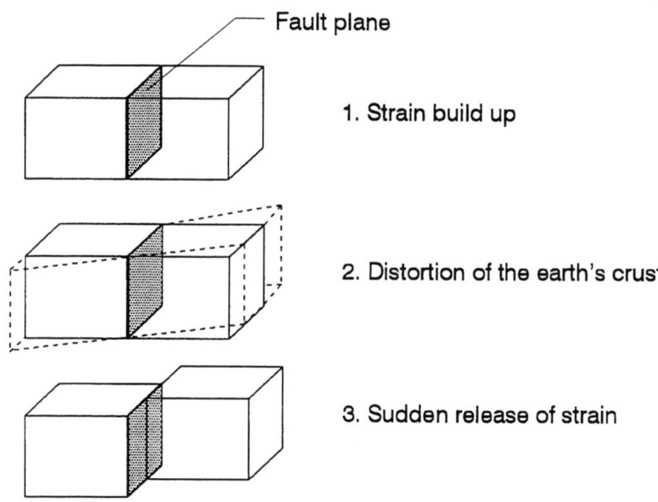

Figure 2.3   Reid's elastic rebound theory.

2.5   FAULT MECHANISMS

There are three types of fault mechanisms (fig. 2.4).

- The *strike-slip fault* is caused by the transform motion (lateral motion) of the plates.
- The *normal fault* is caused by the diverge motion of the plates. The top wall of the fault plane slides downward. As the fault plane is generally inclined, it slides on an inclination angle.

Sometimes normal faults are called rift faults when they are associated with the structure of a tectonic graben. These faults are responsible for the formation of valleys when the earth's crust collapses between parallel normal faults.

- The *underthrust fault* is caused by the subducted motion of the plates. In this case, the top wall of the fault plane is pushed upward. A particular type of underthrust fault, called the thrust fault, is characterized by a small inclination ($< 45\%$) of the fault plane.

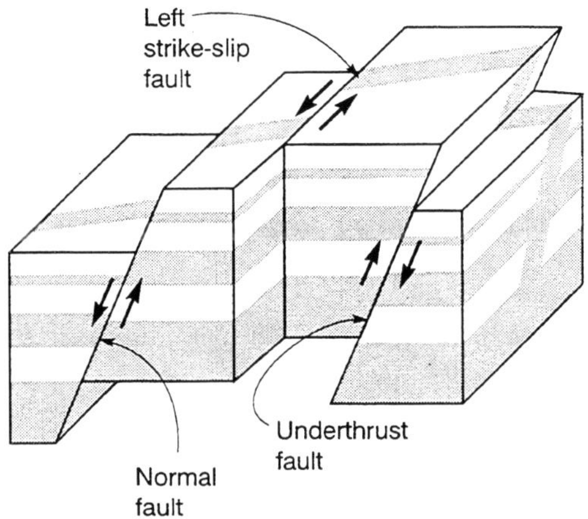

Figure 2.4   Fault mechanisms.

2.6   DEFINITION OF SEISMIC WAVES

Figure 2.5 shows a schematic representation of seismic waves.

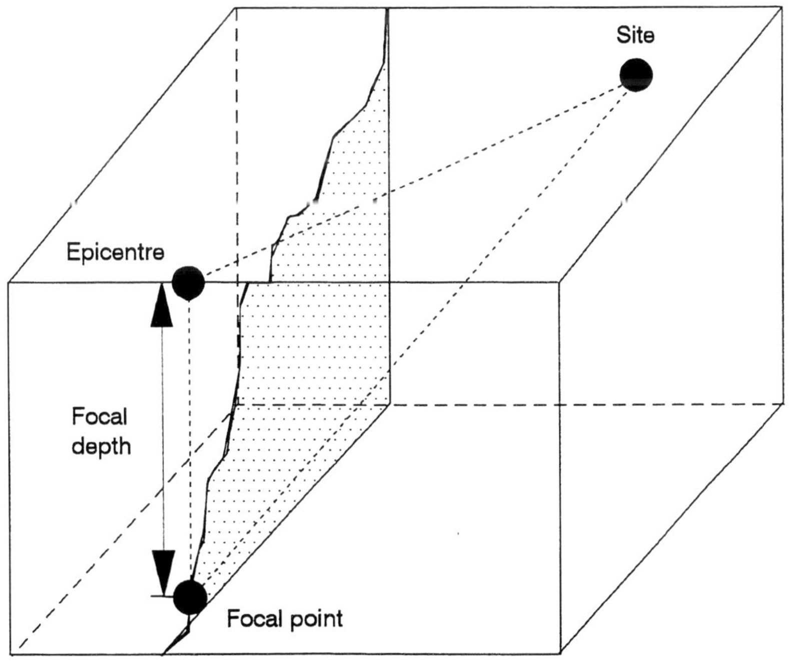

Figure 2.5   Seismic waves.

Consider the following definitions:

- The *focal point* (focus) or hypocentre is where the earthquake occurs, where the fault line originates. This point is located at a depth varying from a few kilometres to 100 km.
- The *epicentre* is the vertical projection of the focal point to the ground surface.
- The *focal depth* is the vertical distance between the focal point and the epicentre.

Various kinds of waves are produced during an earthquake. The waves travelling within the solid earth are called body waves; those travelling near the ground surface are called surface waves.

There are two kinds of body waves:

- Primary waves (P-waves): horizontal tension and compression waves, which travel in the direction of the wave front.
- Secondary waves (S-waves): shear waves, which travel perpendicularly to the wave front.

There are two kinds of surface waves:

- Rayleigh waves: vertical waves travelling on the ground surface.
- Love waves: horizontal waves travelling on the ground surface.

Primary waves are generally high-frequency and are the first to reach a structure. Secondary waves have a lower frequency but greater amplitude and produce, from an engineering point of view, the most destructive vibrations. The amplitude of body waves (primary and secondary) decreases as $1/d$, where d is the distance to the hypocentre (hypocentral distance). The amplitude of the surface waves (Rayleigh and Love) decreases as $1/\sqrt{d}$. So, surface waves travel over a greater distance.

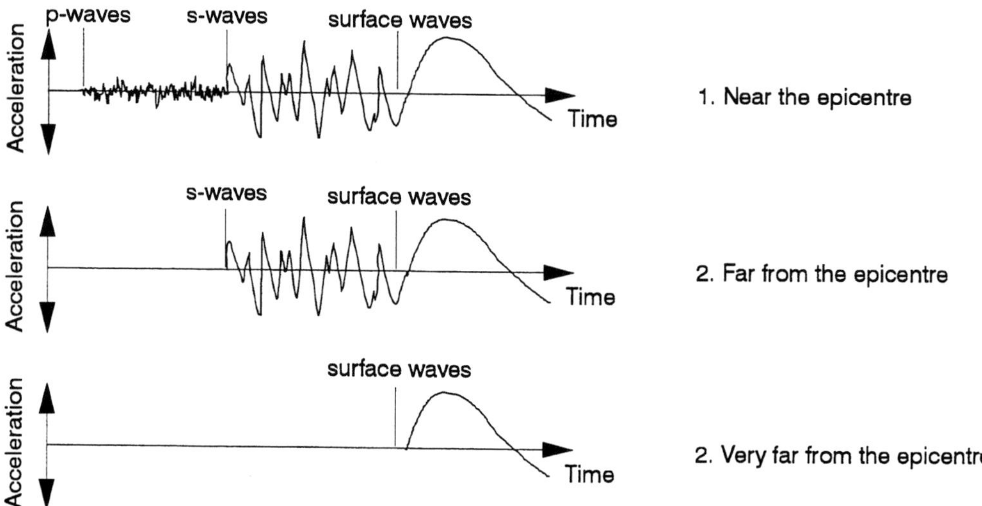

Figure 2.6   Typical earthquake records.

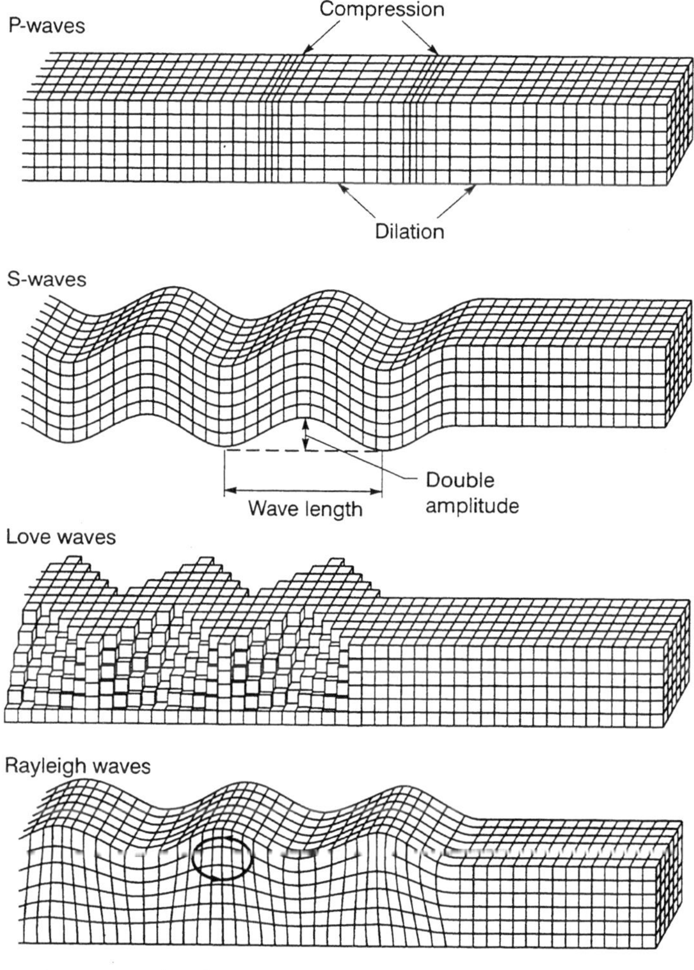

Figure 2.7 Deformations near the ground surface caused by seismic waves.

Typical earthquake records for sites near and far from the epicentre are shown in figure 2.6. Figure 2.7 shows the deformations near the ground surface caused by the four different types of seismic waves.

2.7   LOCATION OF THE EPICENTRE OF AN EARTHQUAKE

At least three geological stations are required to find the location of the epicentre (fig. 2.8).

For each station i, we apply the differential equations between distance, velocity, and time to the primary and secondary waves.

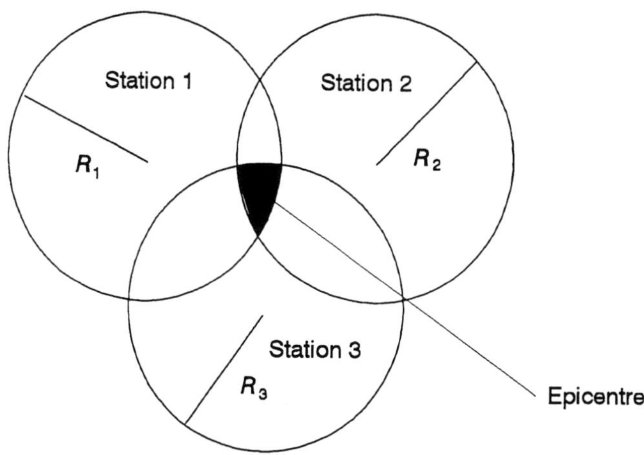

Figure 2.8 Locating the epicentre of an earthquake.

To find the distance of the epicentre (epicentral distance) R_i for station i:

$$R_i = \frac{(t_s - t_p)}{\left[\dfrac{1}{v_s} - \dfrac{1}{v_p}\right]}$$

(2.1)

where $t_s - t_p$ = time difference, taken from the earthquake record at the station, between the arrival of the P- and S-waves
v_p = velocity of the P-waves
v_s = velocity of the S-waves

From the theory of propagation of elastic seismic waves (section 2.14), it can be shown that the velocities of propagation of the P- and S-waves are given by:

$$v_p = \sqrt{\frac{(\Gamma + 2G)}{\rho}}$$

$$v_s = \sqrt{\frac{G}{\rho}}$$

(2.2)

where

$$\Gamma = \frac{\nu E}{(1 + \nu)(1 - 2\nu)} = \text{Lamé constant}$$

G = shear modulus
E = Young's modulus
ν = Poisson's ratio
ρ = density of rock (mass/volume)

(2.3)

2.8   THE MODIFIED MERCALLI INTENSITY SCALE

In 1902, an Italian seismologist, G. Mercalli, introduced an intensity scale before any recording instruments were invented. Later, two Californian seismologists (Wood and Neumann, 1931) adapted this scale to modern construction methods. Known as the modified Mercalli intensity scale, I_{MM}, it is based on subjective feelings and observations of local damage, and is not a scientific measurement of ground motion. It is a function of the epicentral distance and of the quality of the construction at the site, and its scale goes from I to XII (table 2.1). The data necessary to determine the intensity of an earthquake based on the modified Mercalli intensity scale are obtained by interviews and surveys. The results are usually expressed in terms of intensity contours called isoseismal lines. Figure 2.9 shows isoseismal lines obtained during the St. Lawrence Valley earthquake on March 1, 1925. Although the Mercalli scale is far from reliable, it is still used today for areas lacking instrumentation.

Other intensity scales are used around the world. The MSK scale (Meddeved, Sponheuer, and Karnik, 1964), used in Europe, has 12 levels and is similar to the I_{MM} scale. The JMA scale (Japan Meteorological Agency) has only eight levels.

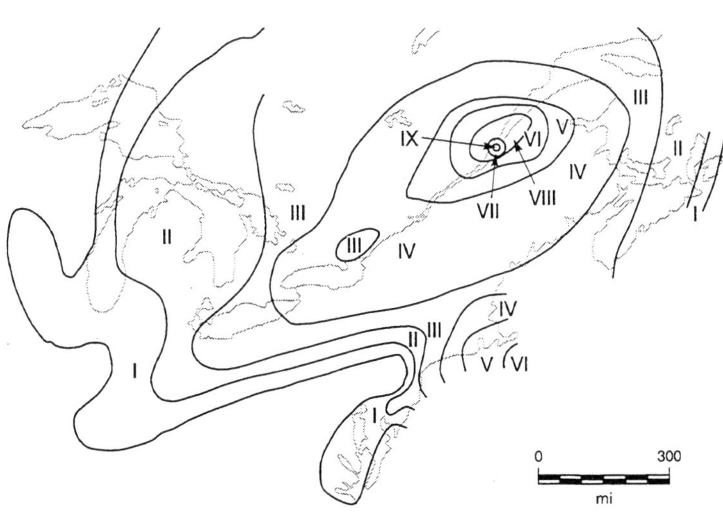

Figure 2.9   Isoseismal lines from the 1925 St. Lawrence Valley earthquake (from Milne and Davenport, 1969).

2.9   THE RICHTER MAGNITUDE SCALE

In the early 1930s, C.F. Richter, then a geophysics professor at the California Institute of Technology, sought a rational way to describe the size of a seismic event (earthquake). Up to that time, only the intensity scales had been used.

Richter adopted the torsion seismograph, developed by Wood and Anderson, to examine his concept. This seismograph has a natural period of 0,8 s, 80% of critical viscous damping, and a static magnification of 2800. This magnification is the ratio of the amplitude read on the seismograph (the trace amplitude) to the actual amplitude of the ground motion.

To construct his magnitude scale, Richter (1935) considered the relationship between the maximum trace amplitude A and the distance R from the epicentre (epicentral distance). He observed that the curves of $\log_{10}A$ vs R were essentially parallel for different earthquakes (fig. 2.10).

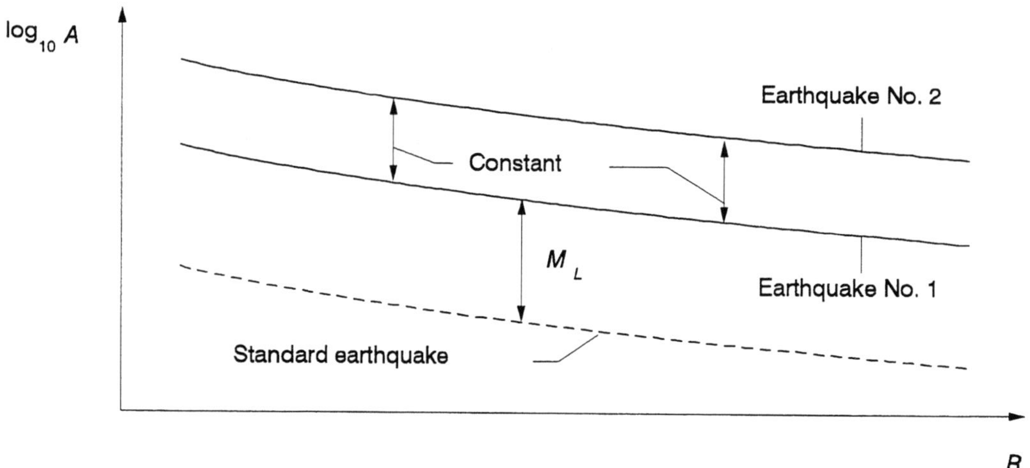

Figure 2.10   Relationship between $\log_{10}A$ and R for the Wood-Anderson seismograph.

Richter first chose a particular curve, which he defined as the "standard event." He then defined the magnitude of a given earthquake, M_L, by calculating the difference between its amplitude and the amplitude of the standard event (Richter, 1958).

$$M_L = \log_{10} A - \log_{10} A_o \qquad (2.4)$$

where A = Wood-Anderson seismograph amplitude of the event in mm
    A_o = Amplitude of the standard event for the same epicentral distance

Richter used the term "magnitude" by analogy to the same term used in astronomy to define the brightness of a star, even though the magnitude-brightness relationship is reversed in astronomy. Table 2.2 shows the amplitude of the standard event (as a function of $-\log_{10}A_o$) for different epicentral distances, R, as defined by Richter (1958).

Table 2.1 Modified Mercalli intensity scale (I_{MM}) (adapted from Wood and Neumann, 1931).

Intensity I_{MM}	Description	Approximate peak ground horizontal acceleration (g)
I	Detected with sensitive instrumentation	< 0,003
II	Felt by a few persons on upper levels; suspended objects may swing	
III	Felt noticeably indoors, but not always recognized as an earthquake; parked cars rock slightly	0,003–0,007
IV	Felt indoors by many, some people awaken; parked cars rock noticeably	0,007–0,015
V	Felt by most people; some breakage of dishes, windows, and plaster	0,015–0,030
VI	Felt by all; many are frightened; falling plaster and chimneys; minor damage	0,030–0,070
VII	Everybody runs outdoors; damage to buildings varies, depending on quality of construction	0,070–0,150
VIII	Panel walls thrown out of frames; walls, monuments, chimneys fall; drivers disturbed	0,150–0,300
IX	Buildings shifted off foundations, cracked, thrown off plumb; ground cracked; underground pipes broken	0,300–0,700
X	Landslides; rails bent; most masonry and framed structures destroyed; ground cracked	0,700–1,500
XI	Bridges destroyed; new structures remain standing but are greatly damaged	1,500–3,000
XII	Total destruction	3,000–7,000

The epicentral distance R is obtained by the difference in arrival times between the primary and secondary waves (equation 2.1). For $R = 100$ km, the amplitude on the seismograph of the standard event is equal to $A_o = 0,001$ mm ($-\log_{10} A_o = 3$). Richter's logarithmic scale means that an earthquake of $M_L = 7$ produces waves that are 10 times the amplitude of an earthquake with $M_L = 6$ and 100 times the amplitude of one with $M_L = 5$.

Hutton and Boore (1987) analyzed almost 10 000 records from 972 earthquakes recorded in southern California and proposed a new expression for the Richter scale's standard event.

$$-\log_{10} A_o = 1,110 \log_{10}\left[\frac{d}{100}\right] + 0,00189 (d - 100) + 3,0 \qquad (2.5)$$

where d is the distance to the hypocentre (hypocentral distance) in km. Equation 2.5 is valid for distances between 10 and 700 km from the epicentre. Figure 2.11 compares equation 2.5 with the original values proposed by Richter. The two equations are similar, and almost identical for epicentral distances of between 50 and 200 km.

Table 2.2 Amplitude of Richter scale's standard event (1958).

R (km)	$-\log_{10}A_o$	R (km)	$-\log_{10}A_o$	R (km)	$-\log A_o$
0	1,4	150	3,30	390	4,4
5	1,4	160	3,30	400	4,5
10	1,5	170	3,40	410	4,5
15	1,6	180	3,40	420	4,5
20	1,7	190	3,50	430	4,6
25	1,9	200	3,50	440	4,6
30	2,1	210	3,60	450	4,6
35	2,3	220	3,65	460	4,6
40	2,4	230	3,70	470	4,7
45	2,5	240	3,70	480	4,7
50	2,6	250	3,80	490	4,7
55	2,7	260	3,80	500	4,7
60	2,8	270	3,90	510	4,8
65	2,8	280	3,90	520	4,8
70	2,8	290	4,00	530	4,8
80	2,9	300	4,00	540	4,8
85	2,9	310	4,10	550	4,8
90	3,0	320	4,10	560	4,9
95	3,0	330	4,20	570	4,9
100	3,0	340	4,20	580	4,9
110	3,1	350	4,30	590	4,9
120	3,1	360	4,30	600	4,9
130	3,2	370	4,30		
140	3,2	380	4,40		

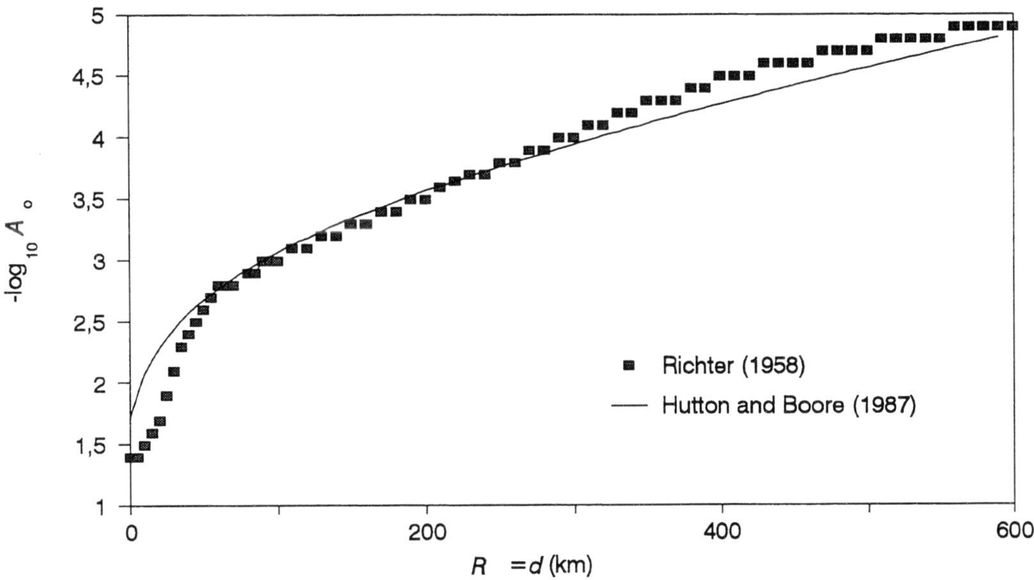

Figure 2.11   Amplitudes of standard event according to Richter (1958)
and Hutton and Boore (1987).

The following assumptions and limitations of the local magnitude scale must be noted:

- The scale was developed for southern California. For other regions, a correction factor must be introduced to reflect the different structures of the earth's crust.
- The scale does not take the focal depth into account because California earthquakes are generally shallow. To compensate, the hypocentral distance can be used, as proposed in equation 2.5.
- The scale is restricted to the Wood-Anderson seismograph. A correction to the amplification must be introduced when using other instruments.
- The scale does not take into account the characteristics of the recording sites. A station correction is usually introduced based on the systematic deviations from the mean obtained from a large number of records.

Every year, around the world, thousands of earthquakes of $M_L < 4$ are recorded, as are about a dozen of magnitude $M_L = 7$, and, fortunately, very few of magnitude $M_L > 8$. The Richter scale has no upper or lower limits. It is therefore possible for an earthquake to exhibit an amplitude lower than the amplitude of the standard event, creating a negative magnitude. Physically, the maximum value for M_L is about 9. This maximum value depends on the rupture length and depth of the fault and on local site conditions. For example, table 2.3 shows M_L values and the rupture length for recent recorded earthquakes.

Table 2.3 Local magnitude and length of rupture
for recent earthquakes.

	M_L	L (km)
San Francisco (1906)	8,3	430
El Centro (1940)	7,1	65
Chile (1960)	8,6	1 000
Alaska (1964)	8,4	700
San Fernando (1971)	6,6	20
Loma Prieta (1989)	7,0	40

2.10 EVOLUTION OF MAGNITUDE SCALES

Local magnitude, M_L, measures the spectral amplitude of the ground motion on a frequency band of 1 to 5 Hz. Following Richter's original work, many modifications were introduced, leading to a variety of magnitude scales depending on different types of waves and frequency bands. As a result, the value of an earthquake's magnitude is not absolute but depends on the type of waves considered. Nowadays, in addition to the local magnitude, which is used mostly in California, four other magnitude scales are used: surface wave magnitude (M_s), body wave magnitude (m_b), moment magnitude (M_w), and Nuttli's magnitude (m_N), used by the Geological Survey of Canada.

2.10.1 Surface Wave Magnitude (M_s)

Gutenberg and Richter (1936) proposed the surface wave magnitude, M_s, for earthquakes recorded at great distances (teleseismic distances) from the epicentre. M_s measures the spectral amplitude of the ground motion at a frequency of 0,05 Hz.

The following empirical equation approximates the relationship between the energy released by an earthquake, E, and the surface wave magnitude.

$$\log_{10} E = 11,8 + 1,5\, M_s \tag{2.6}$$

where E = energy released in ergs (1 erg = $99,9 \times 10^{-9}$ N-m = $73,7 \times 10^{-9}$ ft-pound)

2.10.2 Body-Wave Magnitude (m_b)

As body-wave amplitudes are relatively insensitive to depth, Gutenberg (1945) proposed the body-wave magnitude scale, m_b, for deep-focus earthquakes. This scale is based on the amplitude of the primary waves, for a frequency of 1 Hz, recorded at hypocentral distances greater than 2 000 km.

In certain regions of the world, such as the west coast of North America, primary waves are significantly attenuated at a depth of 75 to 200 km. As a result, for these regions, the value of m_b is greatly diminished. To overcome this problem, Rayleigh's surface waves are used to determine the body-wave magnitude. In this case, the body-wave magnitude is written $m_b L_g$. Herrmann and Nuttli (1981) demonstrated that $m_b L_g$ and M_s values are practically identical for earthquakes in California.

2.10.3 Moment Magnitude (M_W)

This measure of the size (dissipated energy) of an earthquake is much more recent. The seismic moment is defined as the rigidity of the rock times the fault area times the length of slip.

$$M_o = G \times s \times A \tag{2.7}$$

where M_o = seismic moment
$\quad\quad G$ = shear modulus of rock
$\quad\quad s$ = average slip
$\quad\quad A$ = fault area

The seismic moment has units of work or energy.

Many researchers believe that the seismic moment is the most fundamental parameter that can be used to measure the strength of an earthquake caused by fault slip. Table 2.4 presents the order of magnitude for seismic moments.

Kanamori (1977) defined the moment magnitude, M_w, from the seismic moment, M_o, expressed in dyn-cm.

$$M_w = \frac{2}{3} \log_{10} M_o - 10,7 \tag{2.8}$$

Table 2.4   Values of seismic moments.

Earthquake	M_o (dyn-cm)
Chile, 1960	1 030
Micro earthquakes	1 012
Micro-fractures (laboratory)	105

Note: 1 dyn = 1 g-cm/s^2.

2.10.4 Nuttli's Magnitude (m_N)

In the late 1960s, the Geological Survey of Canada started using the magnitude scale proposed by Nuttli (1973). This scale is based on the maximum amplitude of the Rayleigh's surface waves for a frequency of 1 Hz.

$$m_N = -0,1 + 1,66 \log_{10} R + \log_{10} \frac{A}{KT} \tag{2.9}$$

where R = epicentral distance (km)
$\quad\quad A$ = amplitude read on the seismograph (mm)
$\quad\quad K$ = amplification of the seismograph (x 1000)
$\quad\quad T$ = natural period of the seismograph (s)

Nuttli's magnitude is used for epicentral distances greater than 50 km, and for instruments with a natural period smaller than 1,3 s. The scales m_N and m_bL_g result in practically the same numerical values.

2.11 RELATIONSHIPS BETWEEN MAGNITUDE SCALES

The use of different scales to define the magnitude of an earthquake may cause confusion. Usually, the news media will refer to Richter's magnitude scale without specifying the type of scale used. When Richter and Gutenberg proposed their magnitude scales, M_L, M_s and m_b, their hope was that all three would produce the same numerical values for any earthquake. This hypothesis proved wrong, as each scale measures the amplitude in a specific frequency band, and the amplitude is not constant throughout the frequency spectrum of any typical earthquake. So, the interpretation of the magnitude of an earthquake depends on the scale used. For example, in California, Richter's magnitude is associated with M_L when the magnitude is smaller than 6,5 and with M_s for larger earthquakes.

Nuttli (1981) demonstrated that the relationships between magnitude scales are a function of the seismic environment. He proposed two types of relationships: one for earthquakes along the boundaries of the tectonic plates (interplate zones) and one for earthquakes within continental regions (intraplate zones). Tables 2.5 and 2.6, inspired by Nuttli and Herrmann (1982), show the relationships between magnitude scales.

Table 2.5 Relationships between magnitude scales for interplate earthquakes.

M_L or m_bL_g or m_N	m_b	M_s	M_w
5,0	4,6	4,15	4,6
5,2	4,8	4,35	4,8
5,4	5,0	4,60	5,0
5,6	5,2	5,00	5,2
5,8	5,4	5,80	5,8
6,0	5,6	6,55	6,3
6,2	5,8	7,00	6,9
6,4	6,0	7,45	7,4
6,6	6,2	7,90	7,9
6,8	6,4	8,15	8,2
7,0	6,6	8,40	8,6
7,2	6,8	8,70	9,2

Table 2.6 Relationships between magnitude scales
for intraplate earthquakes.

M_L or $m_b L_g$ or m_N	m_b	M_s	M_w
4,0	4,0	2,85	3,8
4,2	4,2	3,10	3,9
4,4	4,4	3,30	4,1
4,6	4,6	3,60	4,3
4,8	4,8	4,00	4,6
5,0	5,0	4,40	4,8
5,2	5,2	4,80	5,1
5,4	5,4	5,20	5,4
5,6	5,6	5,60	5,6
5,8	5,8	6,00	5,9
6,0	6,0	6,40	6,2
6,2	6,2	6,80	6,4
6,4	6,4	7,20	6,7
6,6	6,6	7,60	7,0
6,8	6,8	8,00	7,2
7,0	7,0	8,40	7,5
7,2	7,2	8,70	7,8

2.12 SEISMIC PARAMETERS INFLUENCING STRUCTURAL RESPONSE

As we will see in chapter 4, to determine the response of a structure subjected to a particular earthquake the complete acceleration time-history (the accelerogram) is required. Typical accelerograms of real earthquakes are shown in figure 2.12. For design purposes, it is unlikely that any one record will be adequate. It is usually necessary to examine a group of records.

Structural damage is caused mainly by three important seismic parameters: the amplitude of the ground motion, the frequency contents of earthquake accelerograms, and earthquake duration.

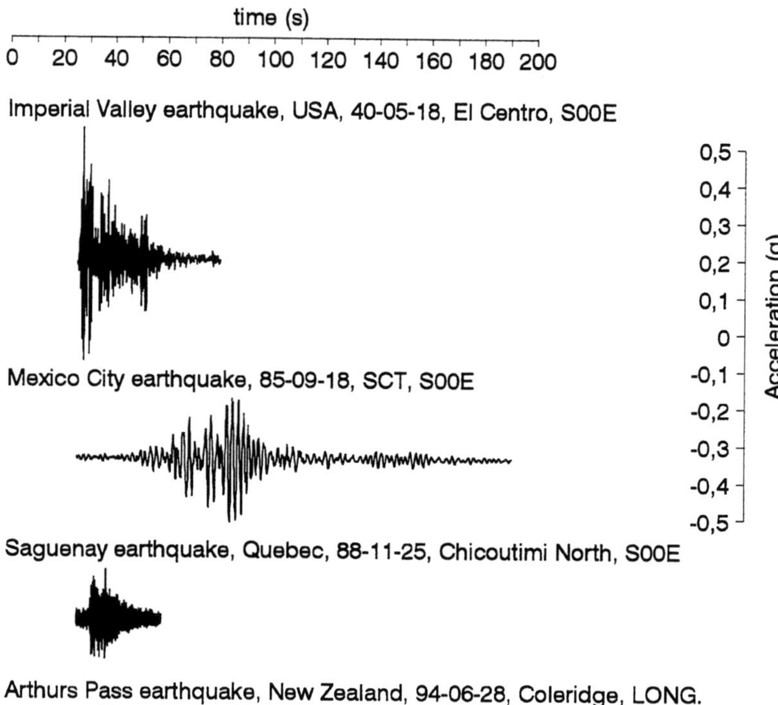

Figure 2.12 Typical earthquake accelerograms.

2.12.1 Amplitude of the Ground Motion

The amplitude of the ground motion can be expressed in various ways:

- the peak ground acceleration a_{max}
- the root mean square (RMS) acceleration a_{RMS}:

$$a_{RMS} = \sqrt{\frac{1}{t_o} \int_0^{t_o} a(t)^2 \, dt} \qquad (2.10)$$

where t_o is the total duration of the record.

Generally, the peak ground acceleration is used to define the amplitude of a specific earthquake accelerogram. However, this parameter does not take into account the complete ground acceleration time-history; therefore, it is not necessarily a good indicator of the earthquake's damage potential. In contrast, the RMS acceleration is a factored mean amplitude for the entire accelerogram.

2.12.2   Frequency Contents of Earthquake Accelerograms

The frequency contents of earthquake accelerograms are evaluated in various ways:

- by counting the total number of cycles (or zero crossings) and dividing by the total duration of the record;
- by plotting a response spectrum of the accelerogram representing the maximum responses of a series of single-degree-of-freedom oscillators with different natural periods and damping ratios (earthquake response spectra, chapter 4);
- by plotting a Fourier spectrum, $|F(\omega)|$, representing the frequency distribution, ω, of the energy contained in the accelerogram.

$$|F(\omega)| = \sqrt{\left(\left[\int_0^{t_o} a(\tau)\cos\omega\tau\,d\tau\right]^2 + \left[\int_0^{t_o} a(\tau)\sin\omega\tau\,d\tau\right]^2\right)} \qquad (2.11)$$

2.12.3   Earthquake Duration

For engineering purposes, the duration of an earthquake is usually described by its "strong motion duration," or the duration of the central segment of the earthquake, which generally causes structural damage. Various methods are used to calculate the strong motion duration:

- the direct approach, which takes the time interval between the first and last peak greater than a given value on the accelerogram (usually 0.05 g);
- the method proposed by Dobry, Idriss, and Ng (1978), which defines the duration as the time interval required to accumulate between 5% and 95% of the accelerogram's total energy. This energy is represented by a measure of intensity, I_A (Arias, 1969).

$$I_A = \frac{\pi}{g}\int_0^t a^2(t)\,dt \qquad (2.12)$$

2.13   ATTENUATION RELATIONS

An attenuation relation is a mathematical expression describing the variation (attenuation) of the ground motion as a function of the distance from the focal point and of the earthquake's magnitude. Obviously, this equation is a simplification of the attenuation phenomenon related to complex physical and tectonic characteristics. In many modern building codes, attenuation relations for peak ground horizontal acceleration and/or for peak ground horizontal velocity are used to construct seismic zoning maps. Many authors have developed different attenuation relations. Campbell (1985) presented an excellent paper on the development of attenuation relations. In the following sections, some of those relations, relevant to Canada, will be discussed.

2.13.1 Milne and Davenport (1969)

The first seismic zoning map of Canada based on attenuation relations appeared in the 1970 edition of the National Building Code of Canada. Milne and Davenport (1969) had proposed an attenuation relation for earthquakes occurring in western Canada.

$$a_{max} = \frac{0,0069\,e^{1,64\,M_L}}{1,1\,e^{1,10\,M_L} + d^2} \tag{2.13}$$

where a_{max} = peak ground horizontal acceleration (g)
M_L = Richter's local magnitude
d = hypocentral distance (km)

Seismologists soon noticed that this relationship was not applicable to earthquakes in eastern Canada, where attenuation characteristics of seismic waves are different. Taking into account the lack of available earthquake records for eastern Canada, Milne and Davenport decided to use an attenuation relation based on the modified Mercalli intensity scale, I_{MM}. Using isoseismal lines of five eastern Canadian earthquakes, they proposed the following empirical relation for eastern Canada:

$$I_{MM} = I_7 - 9,66 - 0,00370\,d + 1,38\,M_L + 0,00528\,dM_L \tag{2.14}$$

where I_{MM} = modified Mercalli intensity
M_L = Richter's local magnitude
d = hypocentral distance (km)
I_7 = intensity at a distance d of an earthquake of local magnitude 7

The term I_7 was calibrated to minimize the difference between equation 2.14 and the intensities of the five earthquakes used as references. Then, Richter's (1958) empirical relationship was used to link the intensity with the peak ground horizontal acceleration, a_{max}.

$$\log_{10}\left(a_{max}\right) = \frac{I_{MM}}{3} - 3,5 \tag{2.15}$$

2.13.2 Milne (1977)

The following attenuation relations, proposed by Milne (1977), originated from 200 earthquake records from the western United States and were calibrated for earthquakes in western Canada.

$$a_{max} = 0{,}04 \; e^{1{,}00\,M_L} \; d^{-1{,}4}$$

$$v_{max} = 0{,}58 \; e^{1{,}17\,M_L} \; d^{-1{,}2}$$

(2.16)

where a_{max} = peak ground horizontal acceleration (g)
$\;\;v_{max}$ = peak ground horizontal velocity (cm/s)
$\;\;M_L$ = Richter's local magnitude
$\;\;d$ = hypocentral distance (km)

The Milne relations can be used for sites located between 1 and 380 km from the hypocentre, and for magnitudes, M_L, between 3.5 and 7.7.

2.13.3 Hasegawa, Basham, and Berry (1981)

In 1981, Hasegawa, Basham, and Berry proposed attenuation relations for eastern and western Canada. Known as the HBB relations, they were used to construct seismic zoning maps that were included in the 1985, 1990, and 1995 editions of the National Building Code of Canada.

For western Canada:

$$a_{max} = 0{,}01019 \; e^{1{,}3\,M_L} \; d^{-1{,}5}$$

$$v_{max} = 0{,}00040 \; e^{2{,}3\,M_L} \; d^{-1{,}3}$$

(2.17)

For eastern Canada:

$$a_{max} = 0{,}00347 \; e^{1{,}3\,M_L} \; d^{-1{,}1}$$

$$v_{max} = 0{,}00018 \; e^{2{,}3\,M_L} \; d^{-1{,}0}$$

(2.18)

where a_{max} = peak ground horizontal acceleration (g)
$\;\;v_{max}$ = peak ground horizontal velocity (cm/s)
$\;\;M_L$ = Richter's local magnitude
$\;\;d$ = hypocentral distance (km)

The HBB relations are used for rock sites located between 10 and 200 km from the hypocentre, and for magnitudes, M_L, between 4,0 and 7,0. The authors also estimated the maximum vertical acceleration and velocity by multiplying equations 2.17 and 2.18 by 2/3 for western and eastern Canada, respectively.

2.13.4 Atkinson (1984)

Atkinson (1984) proposed new attenuation relations for rock sites in eastern and western Canada. These relations were developed from a shear waves radiation model.

For western Canada:

$$a_{max} = \frac{0,084\, e^{\,0,573\, M_w}\, e^{\,-0,00587\, d}}{d}$$

$$v_{max} = \frac{0,188\, e^{\,1,12\, M_w}\, e^{\,-0,00589\, d}}{d}$$

(2.19)

For eastern Canada:

$$a_{max} = \frac{0,063\, e^{\,0,673\, m_b}\, e^{\,-0,001\, d}}{d}$$

$$v_{max} = \frac{0,0349\, e^{\,1,35\, m_b}}{d}$$

(2.20)

where a_{max} = peak ground horizontal acceleration (g)
v_{max} = peak ground horizontal velocity (cm/s)
M_w = moment magnitude
m_b = body wave magnitude
d = hypocentral distance (km)

The Atkinson relations are used for rock sites located between 5 and 1 000 km from the hypocentre and for magnitudes, m_b or M_w, between 5 and 7.

2.13.5 Nuttli-Herrmann (1984–87)

Between 1984 and 1987, Herrmann and Nuttli used a new theoretical model of the intraplate slip process to devise attenuation relationships for eastern North America. They also used micro-earthquake records from the east to calibrate their model for hypocentral distances greater than 50 km. These attenuation relations (Herrmann and Nuttli, 1984) are:

$$\log_{10}\left[a_{max}\right] = -2,42 + 0,50\, m_b - 0,83 \log_{10}\left[\left(R^2 + h^2\right)^{1/2}\right] - 0,0069\, R$$

$$\log_{10}\left[v_{max}\right] = -3,60 + 1,00\, m_b - 0,83 \log_{10}\left[\left(R^2 + h^2\right)^{1/2}\right] - 0,00033\, R$$

(2.21)

$$\log_{10}\left[u_{max}\right] = -6,81 + 1,50\, m_b - 0,83 \log_{10}\left[\left(R^2 + h^2\right)^{1/2}\right] - 0,00017\, R$$

where a_{max} = peak ground horizontal acceleration (g)
v_{max} = peak ground horizontal velocity (cm/s)
u_{max} = peak ground horizontal displacement (cm)
m_b = body wave magnitude
R = epicentral distance (km)
h = focal depth (km)

These three relations are solely used for magnitudes, m_b, between 4,0 and 5,0 and for hypocentral distances greater than 50 km.

In 1987, the same authors proposed a slightly different version of these attenuation relations. The following relations are used for hypocentral distances less than 20 km (Nuttli and Herrmann, 1987):

$$\log_{10}\left[a_{\max}\right] = -2,42 + 0,50\, m_b - 0,83 \log_{10}\left[\left(R^2 + h^2\right)^{1/2}\right] - 0,00120\, R$$

$$\log_{10}\left[v_{\max}\right] = -3,60 + 1,00\, m_b - 0,83 \log_{10}\left[\left(R^2 + h^2\right)^{1/2}\right] - 0,00052\, R \tag{2.22}$$

$$\log_{10}\left[u_{\max}\right] = -6,81 + 1,50\, m_b - 0,83 \log_{10}\left[\left(R^2 + h^2\right)^{1/2}\right] - 0,00024\, R$$

2.13.6   Boore and Atkinson (1987)

Boore and Atkinson (1987) proposed attenuation relations for rock sites in eastern North America. Based on random vibration theory, their model did not require the use of western earthquake records. The authors proposed relations for the peak ground horizontal acceleration, $a_{\max}$, and for the pseudo-spectral velocity, PS_v , with a damping ratio of 5%. The response spectra (including PS_v) will be defined in chapter 4.

The relationship for $a_{\max}$ is:

$$\log_{10}\left[a_{\max}\right] = c_0 + c_1 d - \log_{10} d \tag{2.23}$$

where the coefficients c_0 and c_1 are defined by:

$$c_0 = 0,771 + 0,3354\left(M_w - 6\right) - 0,02473\left(M_w - 6\right)^2$$

$$c_1 = -0,003885 + 0,001042\left(M_w - 6\right) - 0,00009169\left(M_w - 6\right)^2 \tag{2.24}$$

where $a_{\max}$ = peak ground horizontal acceleration (g)
   M_w = moment magnitude
   d = hypocentral distance (km)

The relation for PS_v in cm/s is written the same way as the acceleration relation:

$$\log_{10}\left[PS_v\right] = c_0 + c_1 d - \log_{10} d \tag{2.25}$$

with the coefficients c_0 and c_1 defined by:

$$c_i = x_0^i + x_1 i\left(M_w - 6\right) + x_2^i\left(M_w - 6\right)^2 + x_3^i\left(M_w - 6\right)^3 \qquad i = 0 \text{ ou } 1 \tag{2.26}$$

The regression coefficients x_0^i, x_1^i, x_2^i and x_3^i, for different vibration frequencies, are shown in table 2.7.

In order to compare the various proposed attenuation relations with eastern earthquake records, even though these records are not expressed in terms of the moment magnitude, the authors developed an empirical relation based on a least square regression to relate the M_w and $m_b L_g$ scales.

$$M_w = 2,715 - 0,277\, m_b L_g + 0,127\, (m_b L_g)^2 \tag{2.27}$$

The Boore and Atkinson relations are used for rock sites located at a hypocentral distance of less than 200 km, and at magnitudes, M_w, between 4,5 and 7,5.

Table 2.7   Regression coefficients of the Boore and Atkinson (1987) attenuation relations for PS_v with 5% damping.

Vibration frequency (Hz)	Coefficient	x_0	x_1	x_2	x_3
0,2	c_0	1,743	1,064	-0,04293	-0,05364
	c_1	-0,0003130	0,0014150	-0,00102800	0
0,5	c_0	2,1410000	0,8521000	-0,16700000	0
	c_1	-0,0002504	0	-0,00026120	0
1,0	c_0	2,3000000	0,6655000	-0,15380000	0
	c_1	-0,0010240	-0,0001144	0,00011090	0
2,0	c_0	2,3170000	0,5070000	-0,09317000	0
	c_1	-0,0016830	0,0001492	0,00012030	0
5,0	c_0	2,2390000	0,3976000	-0,04564000	0
	c_1	-0,0025370	0,0005468	0,00007091	0
10,0	c_0	2,1440000	0,3617000	-0,03163000	0
	c_1	-0,0030940	0,0007640	0	0
20,0	c_0	2,0320000	0,3438000	-0,02559000	0
	c_1	-0,0036720	0,0008956	-0,00004219	0

2.13.7   Toro and McGuire (1987)

Toro and McGuire (1987) also used the random vibration theory to determine attenuation relations for rock sites in eastern North America. They proposed relations for the peak ground horizontal acceleration, a_{max}, and for the pseudo-spectral velocity, PS_v, at 5% damping.

The attenuation relation for the peak ground horizontal acceleration is:

$$\ln\left[a_{max}\right] = 0{,}982\, m_b L_g - 1{,}004\,\ln d - 0{,}00468\, d - 4{,}465 \qquad (2.28)$$

where a_{max} = peak ground horizontal acceleration (g)
$m_b Lg$ = body wave magnitude measured from Rayleigh's surface waves
d = hypocentral distance (km)

The attenuation relation for the pseudo-spectral velocity (PS_v in cm/s) is:

$$\ln\left[PS_v\right] = c_0 + c_1 + c_2\,\ln(R) + c_3\,R \qquad (2.29)$$

where the coefficients c_0, c_1, c_2 and c_3 are defined in terms of the vibration frequencies (table 2.8).

Table 2.8   Regression coefficients of the Toro and McGuire (1987) attenuation relation for PS_v with 5% damping.

Vibration frequency (Hz)	c_0	c_1	c_2	c_3
1	-9,283	2,289	-1,000	-0,00183
5	-2,757	1,265	-1,000	-0,00310
10	-1,717	1,069	-1,000	-0,00391

The Toro and McGuire relations are used for rock sites in eastern North America located between 10 and 100 km of the hypocentre and with magnitudes, $m_b Lg$, between 5,0 and 7,0.

2.13.8   Atkinson and Boore (1990)

Atkinson and Boore (1990) studied different attenuation relations proposed for western and eastern North America. They presented new equations that closely replicated available earthquake records. They proposed relations for the peak ground horizontal acceleration, a_{max}, for the peak ground horizontal velocity, v_{max}, and for the pseudo-spectral velocity, PS_v, at 5% damping.

For western North America:

$$\log_{10} y = x_0 + x_1\left(M_w - 6\right) + x_2\left(M_w - 6\right)^2$$
$$+ x_3\,\log_{10}\left[\sqrt{R^2 + h^2}\right] + x_4\,\sqrt{R^2 + h^2} + s \qquad (2.30)$$

where y = seismic parameter (a_{max}, v_{max} or PS_v)
 M_w = moment magnitude
 R = epicentral distance (km)
 h = focal depth (km)

The different regression coefficients in equation 2.30 are shown in table 2.9. Note that the coefficient s, in equation 2.30, is a correction factor for soil deposits greater than 5 m in depth. Also, as the focal depth, h, is approximate, standard values are proposed to represent a more reliable correlation with available earthquake records (table 2.9).

Table 2.9 Regression coefficients for the Atkinson and Boore (1990) relation for western North America.

y		x_0	x_1	x_2	h (km)	x_3	x_4	s
	0,25 Hz	1,96	0,88	-0,24	4,7	-0,95	0	0,29
	0,50 Hz	2,12	0,79	-0,20	4,7	-1,00	-0,0015	0,32
PS_v (cm/s)	1,00 Hz	2,28	0,67	-0,17	4,7	-1,00	-0,0039	0,27
	2,00 Hz	2,41	0,52	-0,14	5,1	-1,00	-0,0051	0,14
	5,00 Hz	2,46	0,35	-0,09	9,6	-1,00	-0,0063	-0,01
	10,00 Hz	2,16	0,25	-0,06	11,3	-1,00	-0,0073	-0,02
a_{max} (g)		0,43	0,23	0	8,0	-1,00	-0,0027	0,00
v_{max} (cm/s)		2,09	0,49	0	4,0	-1,00	-0,0026	0,17

The Atkinson and Boore relations for western North America are used for magnitudes, M_w, between 5,0 and 7,7.

For eastern North America:

$$\log_{10} y = x_0 + x_1(M - 6) + x_2(M - 6)^2 - \log_{10} d + x_3 d \qquad (2.31)$$

where y = seismic parameter (a_{max}, v_{max} or PS_v)
 M = magnitude (M_w or m_N)
 d = hypocentral distance (km)

The different regression coefficients in equation 2.31 are shown in table 2.10 as a function of M_w or m_N.

Table 2.10 Regression coefficients of the Atkinson and Boore (1990) relation for eastern North America.

M	y	Hz	x_0	x_1	x_2	x_3
M_w	PS_v (cm/s)	0,2	1,73	0,96	-0,03	-0,000340
		0,5	2,16	0,85	-0,18	-0,000370
		1,0	2,30	0,67	-0,15	-0,000640
		2,0	2,30	0,53	-0,09	-0,001020
		5,0	2,21	0,44	-0,04	-0,001700
		10,0	2,11	0,42	-0,03	-0,002500
		20,0	1,97	0,41	-0,03	-0,003500
	a_{max} (g)		0,66	0,42	-0,03	-0,002810
	v_{max} (cm/s)		2,16	0,66	-0,03	-0,001310
m_N	PS_v (cm/s)	0,2	1,36	1,21	+0,09	-0,000340
		0,5	1,83	1,17	-0,18	-0,000370
		1,0	2,04	0,93	-0,16	-0,000640
		2,0	2,10	0,71	-0,08	-0,001020
		5,0	2,04	0,58	+0,01	-0,001700
		10,0	1,95	0,54	+0,01	-0,002500
		20,0	1,81	0,53	+0,01	-0,003500
	a_{max} (g)		0,50	0,54	0	-0,002810
	v_{max} (cm/s)		1,91	0,85	-0,04	-0,001131

The Atkinson and Boore relations for eastern North America are used for rock sites located between 10 and 400 km from the hypocentre, with magnitudes, M_w or m_N, between 4,5 and 7,5.

2.13.9 Boore and Joyner (1991)

Until Boore and Joyner's (1991) work, the attenuation relations proposed for eastern North America were applicable only to rock sites. Boore and Joyner modified the equations to take into account the presence of deep soil deposits. They developed attenuation relations for peak ground horizontal acceleration, a_{max}, for pseudo-spectral relative velocity, PS_v, and for maximum values of the pseudo-spectral relative velocity, PS_{vmax}, and absolute acceleration, PS_{amax}.

These new equations for eastern North America are:

$$\log_{10} y = x_0 + x_1(M - 6) + x_2(M - 6)^2 + x_3(M - 6)^3 - \log_{10} d + x_4 d \qquad (2.32)$$

where y = seismic parameter (a_{max}, PS_v, PS_{vmax} or PS_{amax})
$\quad\ M$ = magnitude (M_w or m_N)
$\quad\ d$ = hypocentral distance (km)

The regression coefficients in equation 2.32 are shown in table 2.11 as a function of M_w, and in table 2.12 as a function of m_N. The Boore and Joyner relations for eastern North America are used for deep soil sites, located between 10 and 400 km from the hypocentre, with magnitudes, M_w, between 5,0 and 8,5. These relations may be used when detailed geotechnical studies are not available.

Table 2.11 Regression coefficients of the Boore and Joyner (1991) relations for eastern North America, as a function of M_w.

y	Hz	x_0	x_1	x_2	x_3	x_4
PS_v (cm/s)	0,25	2,059	1,039	-0,145	-0,022	-0,00020
	0,33	2,258	0,973	-0,176	-0,008	-0,00027
	0,50	2,432	0,851	-0,180	0,002	-0,00039
	0,67	2,511	0,763	-0,165	0,004	-0,00050
	1,00	2,567	0,655	-0,135	0,002	-0,00058
	1,33	2,586	0,592	-0,111	-0,001	-0,00072
	2,00	2,588	0,526	-0,080	-0,007	-0,00095
	2,50	2,575	0,499	-0,066	-0,009	-0,00110
	3,33	2,543	0,472	-0,051	-0,012	-0,00140
	5,00	2,461	0,447	-0,037	-0,016	-0,00168
	6,67	2,377	0,437	-0,031	-0,017	-0,00190
	10,00	2,267	0,429	-0,026	-0,018	-0,00240
	20,00	1,946	0,431	-0,028	-0,018	-0,00350
a_{max} (g)		0,671	0,448	-0,037	-0,016	-0,00220
PS_{vmax} (cm/s)		2,596	0,608	-0,038	-0,022	-0,00055
PS_{amax} (g)		1,050	0,433	-0,029	-0,017	-0,00180

Table 2.12 Regression coefficients of the Boore and Joyner (1991) relations for eastern North America, as a function of m_N.

y	Hz	x_0	x_1	x_2	x_3	x_4
PS_v (cm/s)	0,25	1,784	1,425	-0,044	-0,135	-0,00020
	0,33	2,000	1,356	-0,129	-0,094	-0,00027
	0,50	2,207	1,199	-0,174	-0,056	-0,00039
	0,67	2,310	1,077	-0,169	-0,044	-0,00050
	1,00	2,395	0,920	-0,133	-0,039	-0,00058
	1,33	2,431	0,827	-0,100	-0,041	-0,00072
	2,00	2,452	0,724	-0,050	-0,050	-0,00095
	2,50	2,446	0,682	-0,026	-0,055	-0,00110
	3,33	2,421	0,640	-0,001	-0,061	-0,00140
	5,00	2,345	0,600	0,025	-0,068	-0,00168
	6,67	2,264	0,583	0,036	-0,071	-0,00190
	10,00	2,156	0,571	0,045	-0,074	-0,00240
	20,00	1,835	0,575	0,043	-0,075	-0,00350
a_{max} (g)		0,555	0,602	0,028	-0,071	-0,00220
PS_{vmax} (cm/s)		2,439	0,808	0,057	-0,093	-0,00055
PS_{amax} (g)		0,938	0,578	0,039	-0,072	-0,00180

2.13.10 Discussion

In this chapter, attenuation relations have been presented in a deterministic manner for eastern and western Canada. These relations are to be considered mean values. Because of the lack of earthquake records in Canada, there is great uncertainty regarding the predictions of these simple equations. For example, figure 2.13 compares the predictions for a_{max}, based on the attenuation relations for eastern Canada, with the recorded values at rock sites during the November 25, 1988, Saguenay earthquake ($M_s = 6$). This earthquake is the largest recorded seismic event in eastern Canada (Tinawi et al., 1990). Except for two sites, all proposed relations underestimated the values of a_{max}.

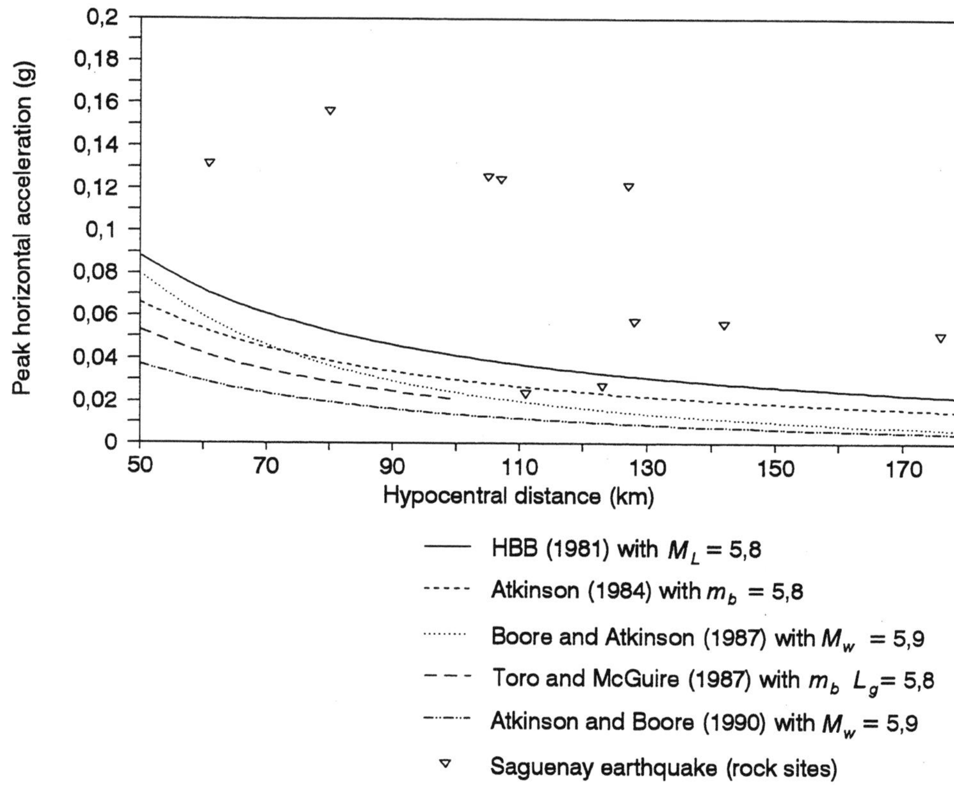

Figure 2.13   Comparison between the eastern Canada attenuation relations and the recorded values of the Saguenay earthquake ($M_s = 6$).

To take into account the great uncertainty in evaluation of a region's seismic hazard, attenuation relations are considered random variables. In doing so, the general equation for an attenuation relation is written as:

$$\log_{10} y = x_0 + f_1(M) + f_2(d) + f_3(s) \pm \epsilon \tag{2.33}$$

where y = seismic parameter (a_{max}, v_{max}, etc.)
  x_0 = constant
  f_1 = mathematical function of magnitude *(M)*
  f_2 = mathematical function of the hypocentral distance *(d)*
  f_3 = mathematical function of the soil conditions at the site *(s)*

The term ϵ in equation 2.33 is a random variable with zero mean and non-zero standard deviation.

2.14 THE ELASTIC WAVE MODEL

The simplest seismological model considers the earth's crust to be a perfectly elastic medium in which the theory of elasticity is adopted. This model represents an ideal situation and neglects the following effects:

- large strains near the earthquake source;
- geological anisotropy;
- discontinuities near the source due to fractures and faults in the earth's crust.

Although this model is quite simple, it proves the existence of seismic waves. The following two sections will review the mathematical background and provide the basic equations for this simple model.

2.14.1 Mathematical Background

The elastic wave model is developed in Cartesian coordinates. The coordinates of a point A in the earth's crust are given by x_i ($i = 1,2,3$). All vectors originating from the earth's crust are shown using unit vectors $\vec{i}$, $\vec{j}$ and $\vec{k}$ and the index notation is used. The basic rules of this notation are:

two repeated indices indicate the summation of all indices:

$$\epsilon_{ii} = \epsilon_{11} + \epsilon_{22} + \epsilon_{33} \tag{2.34}$$

and a comma indicates a derivative:

$$\epsilon_{11,2} = \frac{\partial \epsilon_{11}}{\partial x_2} \tag{2.35}$$

The gradient "grad (or ∇)" of a function $u_i(x_1,x_2,x_3)$ produces a vector given by:

$$\nabla u_i = \frac{\partial u_i}{\partial x_1}\vec{i} + \frac{\partial u_i}{\partial x_2}\vec{j} + \frac{\partial u_i}{\partial x_3}\vec{k} \tag{2.36}$$

The divergent "div (or $\nabla \bullet$)" of a vector $\vec{n} = n_1\,\vec{i} + n_2\,\vec{j} + n_3\,\vec{k}$ produces a scalar given by:

$$\nabla \bullet \vec{n} = \frac{\partial n_1}{\partial x_1} + \frac{\partial n_2}{\partial x_2} + \frac{\partial n_3}{\partial x_3} \tag{2.37}$$

The curl "rot (or $\nabla \times$)" of a vector $\vec{n} = n_1\,\vec{i} + n_2\,\vec{j} + n_3\,\vec{k}$ produces a vector given by:

$$\nabla \, X \, \vec{n} = \begin{vmatrix} \vec{i} & \vec{j} & \vec{k} \\ \dfrac{\partial}{\partial x_1} & \dfrac{\partial}{\partial x_2} & \dfrac{\partial}{\partial x_3} \\ n_1 & n_2 & n_3 \end{vmatrix} \tag{2.38}$$

$$\nabla \, X \, \vec{n} = \left(\frac{\partial n_3}{\partial x_2} - \frac{\partial n_2}{\partial x_3} \right) \vec{i} + \left(\frac{\partial n_1}{\partial x_3} - \frac{\partial n_3}{\partial x_1} \right) \vec{j} + \left(\frac{\partial n_2}{\partial x_1} - \frac{\partial n_1}{\partial x_2} \right) \vec{k}$$

Note that the curl of the gradient of a function is equal to zero.

$$\nabla \times \nabla u = \vec{0} \tag{2.39}$$

We are now ready to develop the basic equations for the elastic wave model, combining different elasticity relations.

2.14.2 Strain-Displacement Equations

The six equations relating strains, ϵ_{ij}, to displacements, u_{ij}, are written:

$$\epsilon_{ij} = \frac{1}{2}\left(u_{i,j} + u_{j,i} \right) \tag{2.40}$$

The dilation (or negative compression), D, is a scalar defined as the increase in volume of a unit cube of the earth, and is given by:

$$D = \epsilon_{11} + \epsilon_{22} + \epsilon_{33} = \epsilon_{ii} \tag{2.41}$$

or:

$$D = \frac{\partial u_1}{\partial x_1} + \frac{\partial u_2}{\partial x_2} + \frac{\partial u_3}{\partial x_3} = \nabla \bullet \vec{u} \tag{2.42}$$

2.14.3 Stress-Strain Equations

The six equations relating stress, σ_{ij}, to strain, ϵ_{ij}, are written:

$$\sigma_{ij} = \Gamma D \, \delta_{ij} + 2G \, \epsilon_{ij} \tag{2.43}$$

where

$$\Gamma = \frac{vE}{(1 + v)(1 - 2v)} = \text{Lamé constant}$$

G = shear modulus

E = Young's modulus

v = Poisson's ratio

ρ = density of rock (mass/volume)

$\delta_{ij} = 1$ if $i = j$

$\delta_{ij} = 0$ if $i \neq j$

(2.44)

2.14.4 Equations of Motion

Let us assume that the static equilibrium of the earth's crust is disturbed by an earthquake that changes the stress field by σ_{ij}. Each element will experience an acceleration, so that, according to Newton's second law, the dynamic equilibrium equations will be:

$$\sigma_{ij,j} = \rho\frac{\partial^2 u_i}{\partial t^2}$$

(2.45)

where ρ represents the density (mass per unit volume) of the earth's crust and t represents time.

The static gravitational forces (body forces) are not shown in equation 2.45, because these forces produce insignificant effects on seismic wave propagation.

Substituting equation 2.43 into equation 2.45 yields:

$$\left[\Gamma D \delta_{ij} + 2G\epsilon_{ij}\right]_{,j} = \rho\frac{\partial^2 u_i}{\partial t^2}$$

(2.46)

Now, substituting equation 2.40 into equation 2.46 leads to:

$$\left[\Gamma D \delta_{ij} + G\left(u_{i,j} + u_{j,i}\right)\right]_{,j} = \rho\frac{\partial^2 u_i}{\partial t^2}$$

(2.47)

Using the chain rule, the terms on the left-hand side of equation 2.47 can be solved separately.

$$\left(\Gamma D \delta_{ij}\right)_{,j} = \Gamma D_{,j}\,\delta_{ij} + \Gamma D \delta_{i,j} = \Gamma\frac{\partial D}{\partial x_i}$$

(2.48)

$$Gu_{i,jj} = G\left[\frac{\partial^2 u_i}{\partial x_1^2} + \frac{\partial^2 u_i}{\partial x_2^2} + \frac{\partial^2 u_i}{\partial x_3^2}\right] = G\nabla \bullet (\nabla u_i) = G\nabla^2 u_i$$

(2.49)

where ∇^2 is called the Laplacian.

$$G u_{j,ij} = G \left[\frac{\partial^2 u_1}{\partial x_i \partial x_1} + \frac{\partial^2 u_2}{\partial x_i \partial x_2} + \frac{\partial^2 u_3}{\partial x_i \partial x_3} \right]$$

$$= G \frac{\partial}{\partial x_i} \left[\frac{\partial u_1}{\partial x_1} + \frac{\partial u_2}{\partial x_2} + \frac{\partial u_3}{\partial x_3} \right] \tag{2.50}$$

$$= G \frac{\partial D}{\partial x_i}$$

Therefore, equation 2.47 is written:

$$(\Gamma + G)\frac{\partial D}{\partial x_i} + G \nabla^2 u_i = \rho \frac{\partial^2 u_i}{\partial t^2} \tag{2.51}$$

Equation 2.51 represents three equations of motion. These equations, also called Navier's equations, can be written in vector notation:

$$(\Gamma + G)\nabla D + G \nabla^2 \vec{u} = \rho \frac{\partial^2 \vec{u}}{\partial t^2} \tag{2.52}$$

where

$$\nabla^2 \vec{u} = \nabla^2 u_1 \, \vec{i} + \nabla^2 u_2 \, \vec{j} + \nabla^2 u_3 \, \vec{k} \tag{2.53}$$

2.14.5   P-Wave Equation

The divergent is used on both sides of equation 2.49 to obtain:

$$\left(\frac{\Gamma + 2G}{\rho} \right) \nabla^2 D = \frac{\partial^2 D}{\partial t^2} \tag{2.54}$$

which is the standard wave equation governing the propagation of dilatational waves (p-waves).

The p-wave velocity is given by:

$$v_p = \sqrt{\frac{(\Gamma + 2G)}{\rho}} \tag{2.55}$$

2.14.6   S-Wave Equation

An alternative to equation 2.52 is also possible. Waves that involve a distortion of the earth's crust can be expected to produce a relative rotation of neighbouring particles causing shear stresses. In vector notation, rotation is expressed by taking the curl of the displacement vector. Operating with the curl on both sides of equation 2.52 yields:

$$(\Gamma + G)(\nabla X \nabla D) + G\nabla^2(\nabla X \vec{u}) = \rho\frac{\partial^2(\nabla X \vec{u})}{\partial t^2} \tag{2.56}$$

but recalling that the curl of a gradient is equal to zero, equation 2.56 is simplified to:

$$\frac{G}{\rho}\nabla^2(\nabla X \vec{u}) = \frac{\partial^2(\nabla X \vec{u})}{\partial t^2} \tag{2.57}$$

Equation 2.57 is, again, the standard wave equation, but this time, governing the propagation of a pure rotational disturbance or shear waves (s-waves). The s-wave velocity is given by:

$$v_s = \sqrt{\frac{G}{\rho}} \tag{2.58}$$

The elastic wave model proves mathematically that an earthquake produces two different types of body waves: purely dilatational waves (tension-compression) called p-waves, and purely rotational waves (shear) called s-waves. Similar equations can be derived for surface waves (Aki and Richards, 1980).

2.15 SEISMICITY AND SEISMOTECTONICS IN CANADA

In Canada, British Columbia and Quebec are two provinces with high population densities in urban centres that are very susceptible to earthquakes. The seismic risk associated with each region is the product of the seismic hazard, which is the probability of occurrence of an earthquake causing a certain level of ground shaking, multiplied by the structural vulnerability of built structures.

$$\textbf{Risk = Hazard} \times \textbf{Vulnerability} \tag{2.59}$$

Although the seismic hazard in western Canada seems greater, large cities in the east are also at risk due to the presence of a large number of unreinforced masonry buildings, which generally behave poorly during earthquakes.

2.15.1 Situation in Eastern Canada

Quebec is located on a very large, relatively stable continental plate. However, some zones in Quebec have very high seismicity. Figure 2.14 shows three zones in this province where the seismic activity is significant: western Quebec, the Charlevoix region, and the lower St. Lawrence region. Table 2.13 shows significant earthquakes, with magnitude greater than 5, that occurred in these three regions during the last 350 years. The hypocentre, or focus, for most earthquakes in these regions is located at a depth of between 5 and 25 km in the earth's crust dating from the pre-Cambrian era, the geological period dating back 400 to 600 million years. For each of these seismic regions, many mechanisms and explanations for the origin of the earthquakes have been proposed. Generally, it seems that the occurrence of earthquakes is linked to the reactivation of a system of rift faults that are between 250 and 800 million years old. It is believed that earthquakes east of the Cordillera, including Quebec, occur within a regional compression field dominated by a compression in the east-northeast direction (Adams and Basham, 1989). Major earthquakes observed in these regions have occurred near recent faults and fractures that wear away the integrity of the North American tectonic plate. But, even though Quebec's seismotectonics seem understood, many aspects remain vague and without valid explanation. Examining each seismic zone in more detail will highlight some of those uncertainties.

Western Quebec. The Ottawa River region has a high concentration of seismic activity, which seems to be distributed among two distinct strips that can be considered two separate zones. The first strip, oriented west-northwest, mostly follows the Ottawa River from Lake Temiscamingue to Ottawa. Then, it expands southeastward to Cornwall and eastward to Montreal. The second strip, which produces more earthquakes but of less intensity, is oriented north-northwest and covers an area from Montreal to the Baskatong Reservoir. These two strips, although quite distinct in the north, where their tectonic characteristics are considered to be different, are relatively similar around the Montreal area.

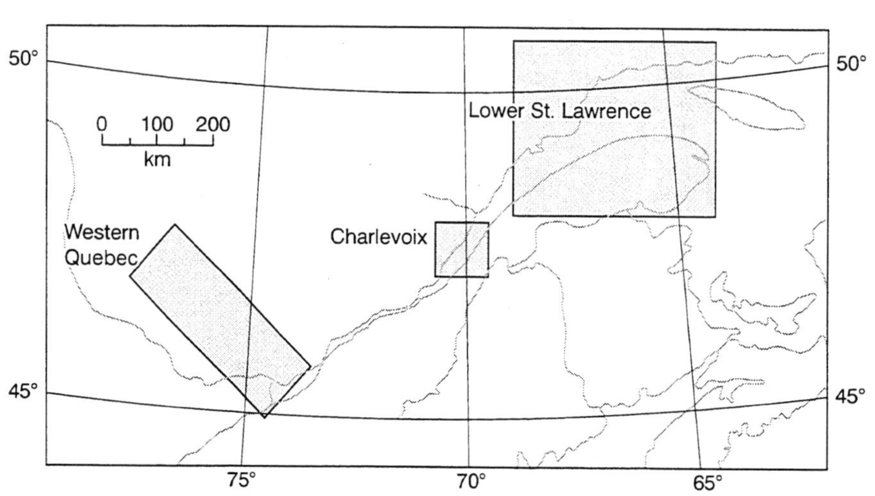

Figure 2.14 Seismicity of eastern Canada before 1987: western Quebec, Charlevoix, and the lower St. Lawrence (adapted from Adams and Basham, 1989).

Forsyth, cited by Adams and Basham (1989), demonstrated that earthquakes along the first strip, including important historical earthquakes, are associated with a fault zone along the Ottawa River that was active 250 to 500 million years ago. Recent studies seem to indicate that the fault network, or rift faults, of the Ottawa Valley is in fact active, explaining earthquakes observed in this region.

For the second strip, the explanation for the observed earthquakes is not as clear. The presence of a hot spot under North America 120 to 140 million years ago seems to be at the origin of the geological structures explaining the presence of earthquakes in this region. In fact, it is believed that during the hot spot's passage, the pre-Cambrian crust was thermally loaded and was cracked due to a differential uplift. This recent cracking of the earth's crust could explain the high seismicity of the region. However, evidence of the hot spot's presence under other regions of Quebec and Labrador remains to be found. Although this theory has not been validated, it remains the most plausible explanation.

Table 2.13   Major earthquakes in eastern Canada from 1663 to 1994 (from Tinawi et al., 1990).

Region	Location	Year	Magnitude
Western Quebec	Montreal	1732	~ 6,0
		1816	~ 5,5
		1897 (2)	~ 5,5
	Ottawa	1861	~ 5,5
	Temiscamingue	1935	$M_L = 6,2$
	Cornwall-Massena	1944	$M_L = 5,7$
Charlevoix	Malbaie region	1663	~ 7,0
		1665	~ 5,5
		1791	~ 6,0
		1831 (2)	~ 5,0
		1860	~ 6,0
		1870	~ 6,5
		1924	~ 5,5
		1925	$M_L = 7,0$
	Saguenay	1988	$M_s = 6,0$
Lower St. Lawrence	-	-	-

Charlevoix region. Historically, the Charlevoix region is the most seismically active region of eastern Canada. Since 1663, the region experienced six earthquakes with magnitude greater than or equal to 6. As for western Quebec, the seismic activity occurs inside a pre-Cambrian layer. Earthquakes seem concentrated between a network of Paleozoic rift faults on the north shore of the St. Lawrence River and a bathymetric boundary close to the south shore. No clear rupture mechanism (inverse, normal faulting, etc.) can be specifically identified. Moreover, the influence of the crash of a large meteorite on this region 350 million years ago is controversial when it comes to explaining the region's seismicity (Adams and Basham, 1989; Lamontagne, 1987). The crash caused annular faults to appear in the earth's crust and a peculiar fault distribution; it is believed that these faults may affect current earthquake distribution. However, according to Adams and Basham (1989), the strain and damage in the rock surrounding the meteorite's impact site do not seem to be the decisive factors in the region's seismicity.

Lower St. Lawrence region. Like the Charlevoix region, the lower St. Lawrence region has its seismicity concentrated mostly under the St. Lawrence River. The findings of rupture mechanisms of many earthquakes showed slips along normal faults, slips that are the results of uniform compression of eastern Canada in the northeast and southwest directions.

Discussion. Most earthquakes that occurred in Quebec were caused by the reactivation of a system of rift faults along the St. Lawrence and Ottawa rivers, except perhaps for the northern strip of western Quebec. This fault system, continuous over thousands of kilometres but only sporadically active, shows seismic activity only in specific regions, such as Charlevoix and the lower St. Lawrence. The high seismic activity in the Charlevoix region, compared to other high-risk areas, is difficult to explain mainly to a lack of geological indices along its faults. It is believed that the activity observed in Charlevoix has a return period of about ten thousand years with an irregular reoccurrence.

2.15.2   Situation in Western Canada

In contrast with eastern Canada, the relationship between seismotectonics and seismicity on the west coast of Canada is well defined (figure 2.15). It is common knowledge today that most earthquakes in western Canada are caused by the movements of three tectonic plates: the Pacific, the North American, and the Juan de Fuca. Three distinct tectonic regions can be identified: the Queen Charlotte region (lateral movement between the Pacific and North American plates), the offshore region of the west coast (interaction between the Pacific and Juan de Fuca plates), and the continental region (subduction of the Juan de Fuca plate beneath the North American plate). Table 2.14 shows major earthquakes that occurred in western Canada over the last 125 years.

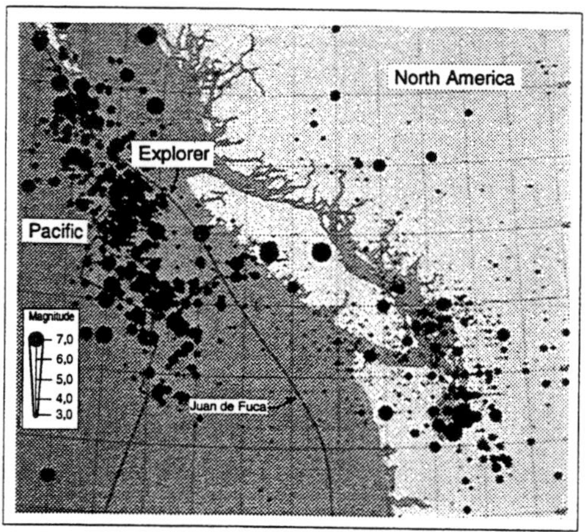

Figure 2.15 Seismicity and seismotectonics of western Canada.

Queen Charlotte region. North of the 51st parallel, between the Pacific and North American plates, a right strike-slip movement occurs at a rate of 5 cm per year. This movement is almost purely horizontal until it reaches the oceanic Aleutians ridge. To the north, the plate boundaries produce a convergent movement that complicates the tectonic mechanism. In addition to the Queen Charlotte main fault, other parallel secondary faults have been identified. Westward, it is believed that part of the continental plate is coupled with the Pacific plate, therefore creating major compression zones. Between the 55th and 51st parallels, seismic activity is limited to a belt on the border of the continent that follows the Queen Charlotte fault. Even though most recorded earthquakes show a series of strike-slip faults, it is believed that underthrust faults are also present.

Table 2.14 Major earthquakes in western Canada, 1872-1978 (from Milne et al., 1978).

Region	Year	Magnitude (M_L)
Washington State	1872	~ 7,0
Yakutat Bay, Alaska	1899	~ 8,6
Vancouver Island	1918	7,0
Queen Charlotte Islands	1929	7,0
Vancouver Island	1946	7,3
Puget Sound, Washington	1949	7,1
Queen Charlotte Islands	1949	8,0
Vancouver Island	1957	6,0
Puget Sound, Washington	1965	6,5
Vancouver Island	1972	5,7

Offshore region of the Pacific coast. South of the 51st parallel, the Pacific and North American plates separate to form two smaller oceanic plates: the Juan de Fuca plate to the south, and the Explorer plate to the north. Off the Pacific coast, the boundary between these oceanic plates and the Pacific plate contains several secondary fracture zones, all the way north to the Queen Charlotte plate. The relation between this triple tectonic system and observed seismicity is not well defined. Most observed epicentres are to the south of the 51st parallel, near the northern tip of the Juan de Fuca plate and east of the Explorer plate.

Continental region. The continental region south of the 51th parallel is a subduction zone. The oceanic Juan de Fuca plate is sinking under the North American plate. The convergence rate between these two plates is 2 to 3 cm per year. On the other hand, the convergence rate between the Explorer plate and the North American plate is only 1 to 1,5 cm per year (Milne et al., 1978). Normally, such a continental subduction zone should generate intense seismic activity at great depth. It is the opposite for the subduction zone of western Canada. Only a few shallow (< 70 km) earthquakes have been observed in this region. The focal depth of similar subduction systems around the world can reach 700 km. The reason for this apparent seismic quietness is not known; many seismologists think it is possible that very large (megathrust) earthquakes could originate from this subduction zone.

2.15.3 Differences and Similarities between Eastern and Western Canadian Earthquakes

Although the number of earthquake records in Canada is limited, particularly for the eastern part of the country, the overall seismicity of North America can still be used to pinpoint the major differences and similarities between earthquakes in eastern and western Canada. The following discussion on these similarities and differences is inspired by the work of Nuttli (1988).

Return period. Based on the modified Mercalli intensity scale, I_{MM}, we can observe that the return period of earthquakes in western Canada with $I_{MM} >$ VII is half the return period of similar earthquakes occurring in eastern Canada. Obviously, this result is influenced by the intense seismic activity in California, and does not reflect completely the situation in western Canada.

By combining the historical seismicity of the Eastern United States and Canada, we can observe that since the arrival of the first European settlers, 2 earthquakes with $I_{MM} >$ VIII occurred in the sixteenth century, 4 in the seventeenth century, 16 in the nineteenth century, and none in the twentieth century. The seismic quietness during the twentieth century in eastern North America, and in eastern Canada, explains in part the lack of interest in seismic hazard. Nuttli (1988) thinks that if the seismic activity of the nineteenth century in eastern North America had occurred in the twentieth century, with the current population density and number of buildings and bridges, the interest in earthquake engineering would be much higher than it is right now.

Attenuation of seismic waves. The seismic wave attenuation characteristics of eastern Canada counterbalance the higher rate of seismic activity observed in western Canada. Seismic waves propagate much farther in eastern Canada than in western Canada. The 1988 Saguenay earthquake in Quebec is the most vivid example of this low attenuation. This earthquake was felt as far south as Washington, D.C., and as far west as Thunder Bay, Ontario. In contrast, earthquakes in western Canada attenuate very quickly, usually within 100 km of the epicentre. Nuttli (1988) believes that if a magnitude 7 were to strike the eastern United States, acceleration levels higher than 0,25 g would be felt over a distance ten times larger than if the same earthquake occurred in California. A similar conclusion could apply for eastern and western Canada.

Fault mechanisms. Earthquakes in western North America, particularly California, are associated with fault systems that extend to the surface. In the east, no earthquake has, since the arrival of the first European settlers, produced surface faulting. The consequence of this major difference in fault mechanisms is that only in western North America can evaluation of the seismic hazard be based on concrete geological evidence. In the east, on the contrary, the distribution of seismic sources is based on an ill-defined, and mostly uniform, seismicity.

Focal depth offers an important similarity between eastern and western earthquakes in North America. In both regions, earthquakes originate at focal depths varying between 5 and 15 km. However, the continental region of western Canada is an exception; the subduction mechanism of this region could produce focal points much deeper than 15 km.

2.16 PROBLEMS

2.1 Figure 16 shows an earthquake record obtained from a station in southern California. The record was obtained with an instrument with a static amplification factor of 7180. The nominal properties of the rock in the region can be estimated as:

Young's modulus:	140,000 MPa
Poisson's ratio:	0,28
Specific weight:	27 kN/m³

a)   Compute the distance between the station and the focal point of the earthquake.
b)   Determine the local magnitude, M_L, of this earthquake.

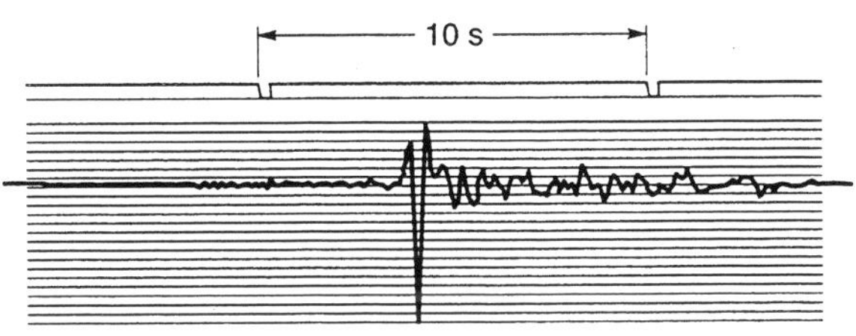

Figure 2.16 Earthquake record from a station in southern California.

Note: The interval between two horizontal lines is 1 mm.

2.2 Read verses 14:3 to 14:5 of the Book of Zachariah in the Bible and interpret the geological phenomenon described.

2.3 Compute the maximum energy transferable to seismic waves from the rock fall on Mount Huascaran, Peru, during the May 31, 1970, earthquake. Fifty million cubic metres of rock fell from a height of 1 km. What is the equivalent surface wave magnitude, M_s, of this event?

2.4 An important structure in eastern Canada is to be located 30 km from a known seismic fault. This fault can produce earthquakes at a focal depth of about 15 km. The design specifications require that the structure be able to resist a magnitude 6 earthquake on the Richter local magnitude scale. Based on the applicable attenuation relations discussed in this chapter, recommend a peak ground horizontal acceleration value to be used for the seismic design of this structure.

2.5 Figure 2.17 shows the locations of the 38 seismological stations triggered by the October 17, 1989, Loma Prieta earthquake that occurred south of San Francisco. Table 2.15 lists the peak ground horizontal acceleration values recorded at these stations.

The Richter local magnitude scale of this seismic event is $M_L = 7,1$. The earthquake was centred 16 km northeast of Santa Cruz at a focal depth of 18,5 km.

a) Using only the peak ground horizontal acceleration values recorded by the 38 stations, construct an attenuation relation for the region of the form:

$$a_{max} = 0,01\, e^{AM_L}\, d^{-B}$$

where a_{max} = peak ground horizontal acceleration (g)
       M_L = Richter's local magnitude
       d = hypocentral distance (km)
       A, B = constants obtained from a least square regression on the logarithmic form of the attenuation relation

b) Compare graphically the attenuation relation obtained in a) with the applicable attenuation relations discussed in this chapter and comment on the results.

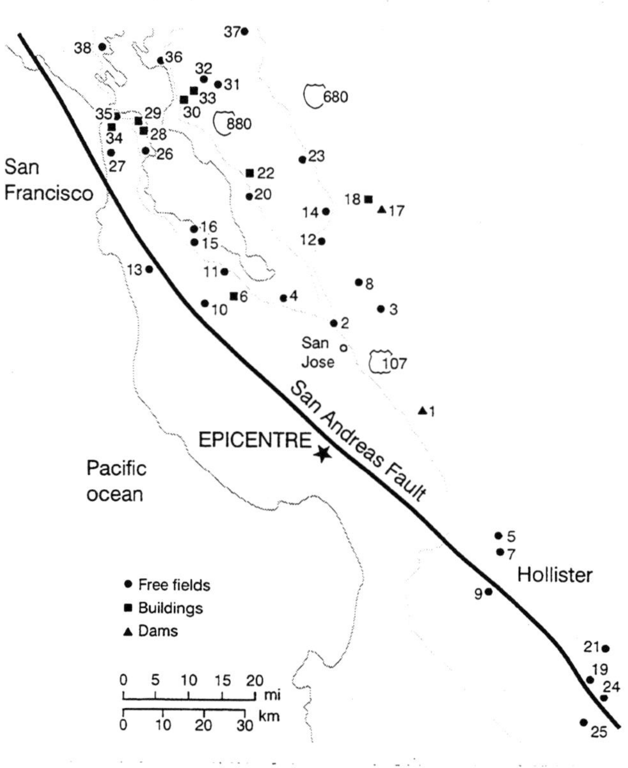

Figure 2.17 Seismological stations triggered by the October 17, 1989, Loma Prieta earthquake.

Table 2.15   Index of seismological stations.

No. on fig. 2.17	Name of station	a_{max} (g)
1	Anderson Dam	0,26
2	San Jose Interchange	0,18
3	Cherry Flat Reservoir	0,09
4	Sunnyvale	0,22
5	Hollister Airport	0,29
6	Palo Alto VA	0,38
7	Hollister City Hall	0,25
8	Calaveras Reservoir	0,13
9	Hollister, SAGO	0,06
10	Stanford, SLAC	0,29
11	Menlo Park VA	0,27
12	Fremont	0,20
13	Crystal Springs Reservoir	0,12
14	Sunol	0,10
15	Redwood City	0,28
16	Foster City	0,12
17	Del Valle Dam	0,06
18	Livermore VA	0,06
19	Bear Valley No. 12	0,17
20	APEEL 2E, Hayward	0,16
21	Bear Valley No. 5	0,07
22	Hayward City Hall	0,10
23	Dublin	0,09
24	Bear Valley No. 10	0,13
25	Bear Valley No. 7	0,06
26	S.F., 1295 Shafter	0,11
27	S.F. State University	0,14
28	S.F., 575 Market	0,13
29	S.F., 600 Montgomery	0,18
30	Emeryville	0,26
31	Berkeley, Strawberry Canyon	0,08
32	Berkeley, Haviland Hall	0,06
33	Berkeley, 2168 Shattuck	0,11
34	San Francisco VA	0,16
35	S.F., Golden Gate Bridge	0,24
36	Richmond	0,11
37	Martinez VA	0,07
38	Larkspur	0,14

2.6   Table 2.16 presents the maximum trace amplitudes obtained on a Wood-Anderson seismograph during an earthquake in southern California. Evaluate the local magnitude based on these recorded values.

Table 2.16   Trace amplitudes on a Wood-Anderson seismograph.

Station	Amplitude on the sismograph (mm)		Epicentral distance (km)
	North-South Component	**East-West Component**	
1	8,4	6,0	114
2	7,9	8,5	179
3	24,5	30,0	90
4	8,1	7,0	246

2.17 REFERENCES

Adams, J., and Basham, P. (1989). "The Seismicity and Seismotectonics of Canada East of the Cordillera." *Geoscience Canada*, 16(1): 3–16.

Aki, K., and Richards, R.G. (1980). *Quantitative Seismology*. San Francisco: W.H. Freeman and Co.

Arias, A. (1969). *A Measure of Earthquake Intensity in Seismic Design for Nuclear Power Plants*. Ed. by R. Hansen. Cambridge: Massachusetts Institute of Technology.

Atkinson, G.M. (1984. "Attenuation of Strong Ground Motion in Canada from a Random Vibrations Approach." *Bulletin of the Seismological Society of America*, 74(6): 2629–53.

Atkinson, G.M., and Boore, B.M. (1990). "Recent Trends in Ground Motion and Spectral Response Relations for North America." *Earthquake Spectra*, 6(1): 15–35.

Bolt, B.A. (1993). *Earthquake*. New York: W.H. Freeman and Co.

Boore, D.M., and Atkinson, G.M. (1987.) "Stochastic Prediction of Ground Motion and Spectral Response Parameters at Hard-Rock Sites in Eastern North America." *Bulletin of the Seismological Society of America*, 77(2): 440–67.

Boore, D.M., and Joyner, W.B. (1991). "Estimation of Ground Motion at Deep-Soil Sites in Eastern North America." *Bulletin of the Seismological Society of America*, 81(6): 2167–85.

Campbell, K.W. (1985). "Strong Ground Attenuation Relation: A Ten Year Perspective." *Earthquake Spectra*, 1(4): 759–805.

Dobry, R., Idriss, I.M., and Ng, E. (1978). "Duration Characteristics of Horizontal Components of Strong Motion Earthquake Records." *Bulletin of the Seismological Society of America*, 68(5): 1487–1520.

Gupta, H.K. (1992). "Reservoir-Induced Earthquake." *Developments in Geotechnical Engineering*, 64. Amsterdam, Elsevier.

Gutenberg, B. (1945). "Magnitude Determination for Deep-Focus Earthquakes." *Bulletin of the Seismological Society of America*, 35(1): 117–30.

Gutenberg, B., and Richter, C.F. (1936). "On Seismic Waves (Third Paper)." *Gerlands Beitraege zur Geophysik*, 47(1–2): 73–131.

Hasegawa, H.S., Basham, P.W., and Berry, M.J. (1981). "Attenuation Relations for Strong Seismic Ground Motion in Canada." *Bulletin of the Seismological Society of America*, 71(6): 1943–62.

Herrmann, R.B., and Nuttli, O.W. (1981). "A Methodology for Automatic Estimation of Magnitudes for Regional Earthquakes." *Abstracts of 21st General Assembly of the International Association of Seismology and Physics of the Earth's Interior*, London, Ontario, p. C4.4.

— (1984). "Scaling and Attenuation Relations for Strong Motion in Eastern North America." *Proceedings of the Eighth World Conference on Earthquake Engineering*, San Francisco, CA. Vol. 2, pp. 305–09.

Hutton, L.K., and Boore, D.M. (1987). "The M_L Scale in Southern California." *Bulletin of the Seismological Society of America*, 77(6): 2074–94.

Kanamori, H. (1977). "The Energy Release in Great Earthquakes." *Journal of Geophysics Research*, 82(20): 2981–87.

Lamontagne, M. (1987). "Seismic Activity and Structural Features in the Charlevoix Region, Québec." *Canadian Journal of Earth Science*, 24(11): 2118–29.

Milne, W.G. (1977). "Seismic Risk Maps for Canada." *Proceedings of the Sixth World Conference on Earthquake Engineering*, New Delhi, India. Vol. 1, p. 930.

Milne, W.G., and Davenport, A.G. (1969). "Distribution of Earthquake Risk in Canada." *Bulletin of the Seismological Society of America*, 59(2): 729–54.

Milne, W.G., et al. (1978). "Seismicity of Western Canada." *Canadian Journal of Earth Science*, 15(5): 1170–93.

Nuttli, O.W. (1973). "Seismic Wave Attenuation and Magnitude Relations for Eastern North America." *Journal of the Geophysical Research*, 78: 867–85.

— (1981). "Empirical Scaling Relations for Midplate and Inter-Plate Earthquakes." *Abstracts of the 21st General Assembly of the International Association of Seismology and Physics of the Earth's Interior*, London, Ontario, p. C4.10.

— (1988). "Similarities and Differences Between Western and Eastern United States Earthquakes, and Their Consequences for Earthquake Engineering." *Earthquake Engineering and Soil Dynamics II: Recent Advances in Ground Motion Evaluation*. ASCE Geotechnical Special Bulletin, no. 20, pp. 25–51.

Nuttli, O.W., and Herrmann, R.B. (1982). "Earthquake Magnitude Scales." ASCE, *Journal of the Geotechnical Engineering Division*, 108(GT5): 783–86.

— (1987). "Ground Motion Relations for Eastern North American Earthquakes." *Proceedings of the Third International Conference on Soil Dynamics and Earthquake Engineering*, Princeton, NJ, Vol. 2, pp. 231–41.

Richter, C.F. (1935). "An International Earthquake Scale." *Bulletin of the Seismological Society of America*, 25(1): 1–23.

— (1958). *Elementary Seismology*. San Francisco and London: W.H. Freeman and Co.

Tinawi, R., Mitchell, D., and Law, T. (1990). "Les dommages dus au tremblement du Saguenay du 25 novembre 1988." *Canadian Journal of Civil Engineering*, 17(3): 366–94.

Toro, G.R., and McGuire, R.B. (1987). "An Investigation into Earthquake Ground Motion Characteristics in Eastern North America." *Bulletin of the Seismological Society of America*, 77(2): 468–89.

Wood, H.O., and Neumann, F. (1931). "Modified Mercalli Intensity Scale of 1931." *Bulletin of the Seismological Society of America*, 21(2): 277–83.

Chapter 3

Elements of Probabilistic Seismic Hazard Analysis

3.1 DEFINITION OF A DESIGN EARTHQUAKE

As seen in chapter 2, the seismic risk associated with a region depends on the probability of occurrence of an earthquake causing a certain level of ground shaking (seismic hazard) and on the structural vulnerability of the built structures. An important part of designing a structure in a seismic zone is the definition of the design earthquake at the specific location. The design earthquake is defined as "the ground movement that will be used in the calculation of the earthquake resistant design" (EERI Committee on Seismic Risk, 1984). Three parameters are used to define the design earthquake: 1) the level of seismic activity of the region; 2) the specific conditions (geotechnic or geological) at the site; and 3) the acceptable risk level. The third criterion (acceptable risk level) is very subjective and is defined as "a probability of social and economic consequences due to earthquakes that is low enough to be judged by appropriate authorities to represent a realistic basis for determining design requirements for engineered structures" (EERI Committee on Seismic Risk, 1984).

The probabilistic method, proposed by Cornell (1968), is used to determine the design earthquake for all locations in Canada. This method gives a relative quantification of the design earthquake for a given site in terms of seismic parameters (acceleration, velocity, etc.), and takes into account uncertainties associated with the occurrence and effects of important earthquakes. This probabilistic seismic hazard analysis cannot predict the date, magnitude, location, and effects of future earthquakes in active seismic zones. However, it provides a basis for the seismic design of structures for a given location.

This method was used to construct the seismic zoning maps included in the 1985, 1990, and 1995 editions of the National Building Code of Canada. It combines two separate models: a seismicity model and an attenuation model.

- Seismicity model

The seismicity model describes the geographical distribution of potential active source zones (seismotectonic sources) and the distribution of magnitudes in each source.

- Attenuation model

The attenuation model describes the effect of an earthquake originating from a specific seismotectonic source, at any given site, as a function of the magnitude and the source-to-site distance.

3.2 SEISMICITY MODEL

3.2.1 Description

On any given fault (source) within a given region, earthquakes occur at irregular times. One basic activity in seismology has long been the search for meaningful patterns of earthquake occurrences. The longer the historical record, the better the picture that can be obtained. In most regions, the useful historical record is short, often showing only a few decades, or sometimes one or two centuries, the great exceptions being China and the eastern Mediterranean, which have useful records going back around 2 000 years.

During any given time along a fault line, the historical distribution of events can be represented by an empirical recurrence relation proposed by Gutenberg and Richter (1965).

$$\log N = A - bM \qquad (3.1)$$

where $\log$ = logarithm ($\log_e = \ln$ or $\log_{10}$)
  N = number of earthquakes of magnitude greater or equal to M per units of time and units of area (also called the return frequency or recurrence rate)
  M = magnitude (M_L, m_b, M_s, M_w or m_N)
  A, b = seismic constants for the region

Table 3.1 Typical values of seismic constants A and b for various regions (from Kaila and Narain, 1971).

Region	Boundaries (latitudes and longitudes)				A	b
Japan	26 N	40 N	132 E	150 E	6,86	1,22
New Guinea	13 S	1 N	132 E	148 E	7,83	1,35
Western Canada	47 N	65 N	142 E	115 W	5,05	1,09
Western United States	25 N	47 N	135 W	105 W	5,94	1,14
Eastern United States	25 N	47 N	105 W	51 W	5,79	1,38
Central America	10 N	25 N	120 W	85 W	7,36	1,45
Colombia, Peru	18 S	6 N	85 W	60 W	5,60	1,11
Northern Chile	37 N	18 S	78 W	60 W	4,78	0,88
Southern Chile	63 S	37 S	78 W	60 W	4,46	0,92
Mediterranean	30 N	50 N	20 W	48 E	5,45	1,10
Iran, Turkmenia	15 N	42 N	48 W	65 W	6,02	1,18
Java	13 S	5 S	90 W	118 E	5,37	0,94
East Africa	40 S	30 N	20 E	48 E	3,80	0,87

Table 3.1 shows typical values of the seismic constants A and b for different regions of the world. These values, presented for comparison purposes only, have been obtained from very large areas. Usually, the seismic constants are defined for much smaller seismotectonic sources. When seismic data exist for a given region, a plot of M against log N can be made and a linear regression analysis is used to estimate the seismic constants A and b. Figure 3.1 is an example of this type of calculation.

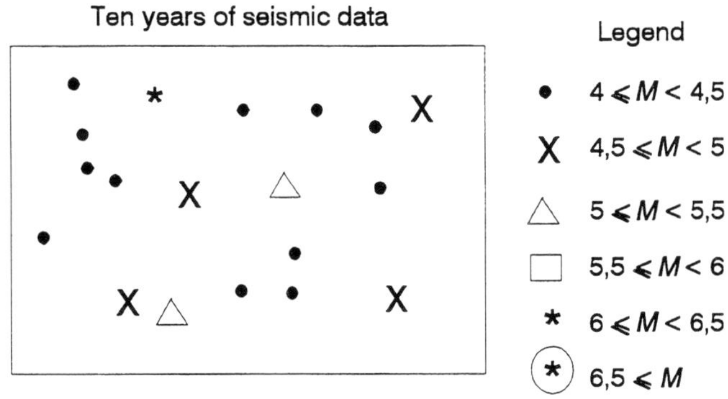

Interval - M	Number of earthquakes	M	Number of quakes / year $\geq M = N$	$\log_{10} N$
4 - 4,5	12	4	19/10 = 1,9	0,279
4,5 - 5	4	4,5	7/10 = 0,7	-0,155
5 - 5,5	2	5	3/10 = 0,3	-0,523
5,5 - 6	0	5,5	1/10 = 0,1	-1,000
6 - 6,5	1	6	1/10 = 0,1	-1,000

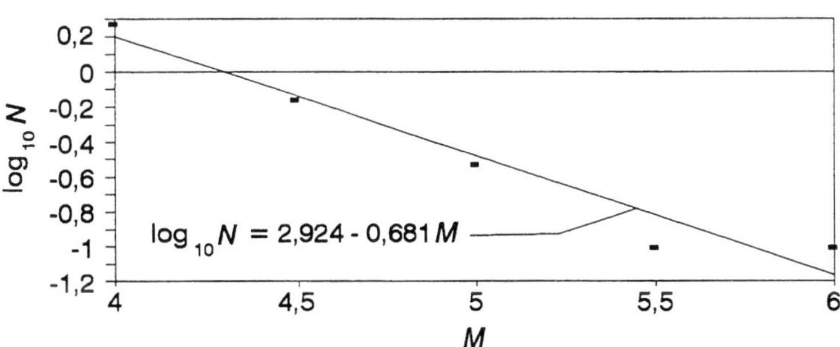

Figure 3.1 Calculation for the magnitude-recurrence relation.

Records of events with magnitude of less than four ($M < 4$), although quite numerous, go back only a few years. These records are usually excluded when calculating the magnitude-recurrence relation, as they may give a misleading bias. As shown in figure 3.2, it would make sense to use a quadratic function for the magnitude-recurrence relation to improve the correlation at high magnitudes. Nevertheless, the linear relation is usually retained, because of its simplicity, to construct seismic zoning maps, although it is unsatisfactory at high magnitudes because, for any fault line, there is a maximum achievable magnitude, M_{max}. This maximum magnitude arises because a given fault has physical constraints on the maximum size of an event it can generate. Despite difficulties in reliably estimating the values of M_{max}, the linear relation for the magnitude distribution has to be interrupted for greater magnitudes. Various expressions have been developed to evaluate the cutoff magnitude (Anderson et al., 1983).

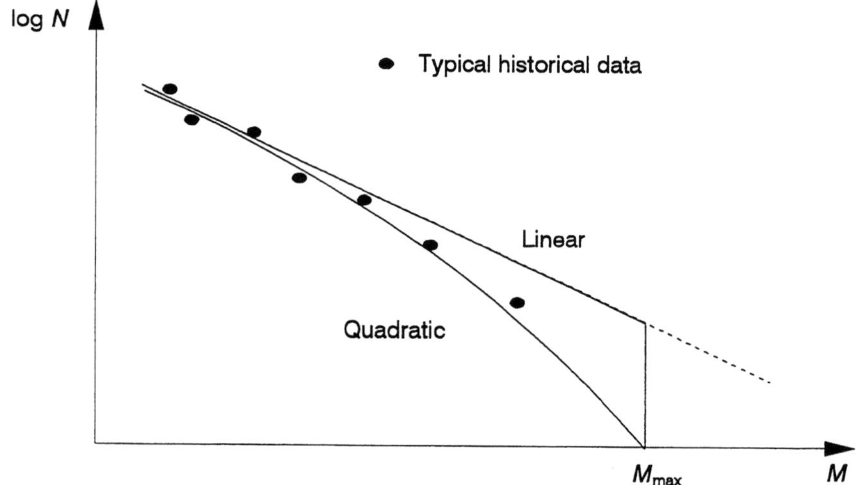

Figure 3.2   Linear and quadratic representation of the magnitude-recurrence relation.

3.2.2 Application for Canada

To construct the seismic zoning maps included in the 1985, 1990, and 1995 editions of the National Building Code of Canada, the seismicity of Canada was modelled into 32 seismotectonic sources (see figure 3.3) (Basham et al., 1982, 1985). The focal depth of all sources was set to 20 km, except for the Puget Sound (PGT) source in western Canada, which was set at 40 km. Seven of the 32 sources are in Quebec, and eleven are in British Columbia. The recurrence relation for each of these sources uses the natural logarithm.

$$\ln N = A - bM_L \tag{3.2}$$

Table 3.2 shows the seismic constants, A and b, the cutoff magnitude, M_{max}, and the surface area for each seismotectonic source in Canada. The M_{max} value was set for each source by adding 0,5 to the magnitude of the largest known earthquake. Note that the coefficients in table 3.2 are related to the recurrence rate, N, for the total area of each source.

To obtain the recurrence rate per unit area, the natural logarithm of the total area of the source (km^2) is subtracted from the constant A. This geographic distribution for the seismotectonic sources of Canada represents the seismic knowledge acquired at the beginning of the 1980s. The hypothesis was made that major earthquakes would occur again in regions with a history of seismic activity. Since then, however, many large-magnitude earthquakes have occurred in regions that were of low seismicity. The 1988 Saguenay earthquake in Quebec is a good example.

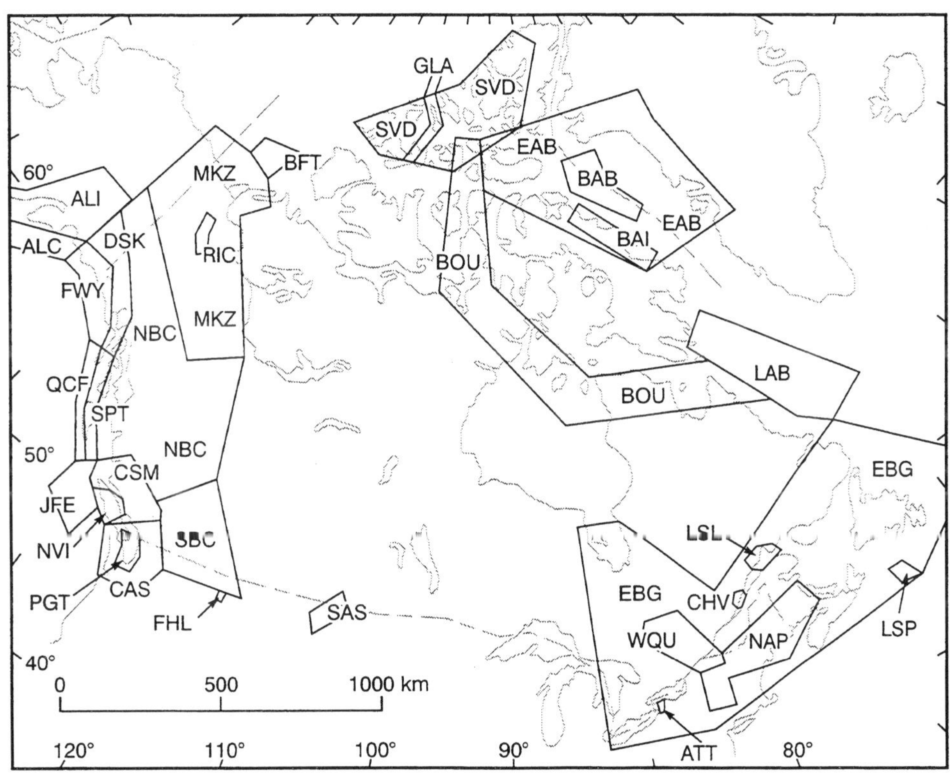

Figure 3.3 Seismotectonic sources in Canada (from Basham et al., 1985).

Because of these observations and pertinent geological evidence, the Geological Survey of Canada is considering modifying the shape and spread of the seismotectonic sources of eastern Canada for the 2000 edition of the National Building Code of Canada (Adams et al., 1995). Among other things, sources found in the Charlevoix (CHV) and lower St. Lawrence (LSL) regions would be replaced by a larger source along the St. Lawrence River.

Table 3.2 Properties of seismotectonic sources in Canada (adapted from Basham et al., 1982).

Region	Abbreviation	Seismic constants		Cutoff magnitude M_{max}	Surface area (km^2)
		A	b		
Western Canada	PGT	6,08	1,58	7,5	28 400
	CAS	6,97	1,87	7,5	145 000
	NVI	3,04	1,04	7,5	27 000
	CSM	5,61	1,77	6,5	139 000
	JFE	8,90	1,72	7,0	84 800
	QCF	7,38	1,50	8,5	46 400
	SPT	7,12	1,87	7,0	53 700
	SBC	8,08	2,28	6,5	255 000
	NBC	7,51	2,28	5,0	875 000
	FHL	10,55	2,58	6,5	2 100
	SAS	5,24	2,07	6,0	--
Northwestern Canada	FWY	8,43	1,66	8,5	111 000
	DSK	7,94	1,96	7,0	110 000
	RIC	7,35	1,76	7,0	20 000
	BFT	6,52	1,76	6,5	39 000
	MKZ	11,43	2,67	6,0	698 000
	ALC	8,25	1,43	8,5	132 000
	ALI	10,95	1,73	8,5	321 000
Eastern Canada	CHV	5,74	1,66	7,5	6 880
	WQU	6,94	1,85	7,0	121 000
	LSL	6,28	1,85	6,0	24 500
	NAP	6,46	1,87	6,0	241 000
	LSP	3,71	1,30	7,5	15 200
	ATT	2,40	1,32	6,0	2 620
	EBG	9,69	2,78	5,0	2 670 000
Northeastern Canada	BAB	6,42	1,64	7,5	100 000
	BAI	10,86	2,54	7,0	85 000
	LAB	7,59	1,95	6,5	352 000
	EAB	6,74	1,81	6,0	1 067 000
	GLA	9,85	2,19	6,5	42 000
	SVD	7,73	2,19	6,0	480 000
	BOU	8,24	2,02	6,5	830 000

3.3 ATTENUATION MODEL

3.3.1 Description

Attenuation relations have been discussed in chapter 2 (sect. 2.13). In general, attenuation models relate the effect of an earthquake, y, at a site, to the magnitude of the earthquake, M, and to the site-to-source distance, d.

$$y = y\,(M,\, d) \tag{3.3}$$

The parameters of an attenuation relation depend mainly on three seismological factors:

1. The characteristics at the hypocentre
   Factors such as magnitude, earthquake depth, geological characteristics at the hypocentre, strain distribution before the earthquake, and strain release during the earthquake will influence the wave propagation in the earth's crust.
2. The characteristics of the propagation trajectory
   Topography and mechanical properties of the earth's crust will influence the reflection and refraction of seismic waves.
3. Conditions at the site
   Stratification and heterogeneity of the rock, as well as topography and soil conditions, will influence the ground shaking at the site.

An attenuation relation like that in equation 3.3 attempts to make a simple model out of a very complex phenomenon. Many variables are involved, and they are difficult to evaluate correctly. Therefore, a high level of uncertainty is associated with existing attenuation relations. Equation 3.3 can also be inverted to obtain an expression for the magnitude necessary to cause a certain effect at a given distance from the hypocentre.

$$M = M\,(y,\, d) \tag{3.4}$$

3.3.2 Application for Canada

As seen in chapter 2, HBB attenuation relations (Hasegawa, Basham, and Berry, 1981) were used to construct the seismic zoning maps of Canada. Equation 3.3 represents a general form of these relations and is commonly expressed as:

$$\log y = b_1 + b_2 M - b_3 \log\!\left(d + b_4\right) \tag{3.5}$$

The parameters of this equation are given in table 3.3.

Table 3.3 Parameters of equation 3.5 for HBB attenuation relations.

Region	log	M	d	y	b_1	b_2	b_3	b_4
Eastern Canada			Hypocentral distance (km)	a_{max} (g)	-5,66	1,30	1,10	0
	ln	M_L		v_{max} (cm/s)	-8,62	2,30	1,00	0
Western Canada				a_{max}(g)	-4,59	1,30	1,50	0
				v_{max}(cm/s)	-7,82	2,30	1,30	0

3.4   PROBABILISTIC SEISMIC HAZARD ANALYSIS PROCEDURE

3.4.1 Description

The probabilistic seismic hazard analysis method, originally proposed by Cornell (1968), attempts to answer the question, "What is the probability per annum of exceeding a given ground motion (peak ground acceleration, velocity, etc.) at a given site?" Usually, the application of the probability seismic hazard analysis requires six separate steps:

- *Step 1. Definition of the seismogenic sources*
  For a given site, geographic zones representing seismotectonic sources are drawn. For each zone, it is assumed that the probability of earthquake occurrence is the same for the entire surface area (a seismogenic source).

- *Step 2. Definition of a seismicity model*
  For each source, a magnitude-recurrence relation of the type $\log N = A - bM$ is defined. Historical seismicity is used to establish the parameters of this relation. A cutoff magnitude, M_{max}, is also defined for each source.

- *Step 3. Definition of an attenuation model*
  An attenuation relation is determined between each source and a given site. Often the same attenuation relation is used for the entire region.

- *Step 4. Calculation of the recurrence rate of a given ground motion level*
  The attenuation relations are combined with the magnitude-recurrence relations to find the annual number of earthquakes causing a chosen ground motion level at any given site.

- *Step 5. Calculation of the annual probability of exceeding a chosen ground-motion level*
  A theoretical probability distribution is used to calculate the annual probability of exceeding a chosen ground-motion level at any given site. The calculation is based on the recurrence rate calculated in step 4.

- *Step 6. Plotting of contour lines having a constant annual probability of excedance*
  Steps 1 to 5 are repeated for each given site. Contour lines are then plotted for the chosen ground-motion parameter. These lines are based on the same annual probability of excedance for the entire surface area.

3.4.2 Illustration of a Large Seismogenic Source Zone

To illustrate the probabilistic seismic hazard analysis procedure, let us consider a site at the centre of a very large seismogenic source (see figure 3.4) and assume that the magnitude-recurrence relation per unit area of the source is known.

$$\log N = A - bM \tag{3.6}$$

Equation 3.6 is inverted to obtain:

$$N = N_o \exp(-bM) \tag{3.7}$$

where

$$N_o = \exp(A) \tag{3.8}$$

The symbol "exp" is considered the general form of the inverse of a logarithm.

$$
\begin{aligned}
\exp &= e & \text{if} \quad \log &= \log_e = \ln \\
\exp &= 10 & \text{if} \quad \log &= \log_{10}
\end{aligned}
\tag{3.9}
$$

Note that N is the number of earthquakes with magnitude greater or equal to M, per annum, and per unit area.

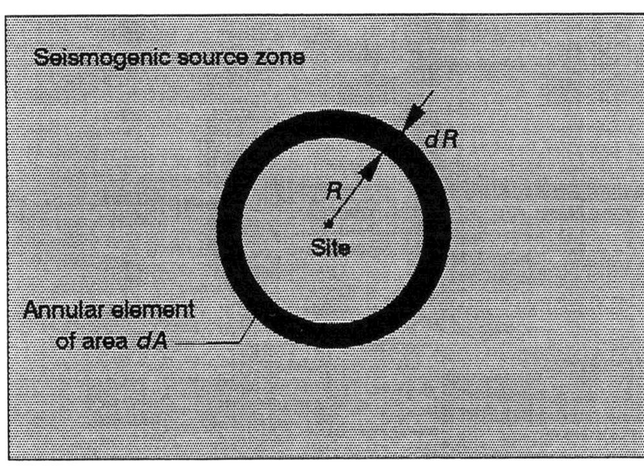

Figure 3.4　Site located at the centre of a large seismogenic zone.

Suppose that the parameter of interest at the site is the peak ground horizontal acceleration, a_{max}. A general attenuation relation, for the region, is given in equation 3.5.

$$\log a_{max} = b_1 + b_2 M - b_3 \log(d + b_4) \tag{3.10}$$

This equation is inverted to isolate a_{max}.

$$a_{max} = \exp(b_1) \exp(b_2 M)(d + b_4)^{-b_3} \tag{3.11}$$

where d represents the hypocentral distance.

$$d = \sqrt{R^2 + h^2} \tag{3.12}$$

where R = epicentral distance
h = focal depth considered constant for the entire source.

To simplify, the parameters of equation 3.11 are rewritten.

$$\exp(b_1) = c_1$$
$$b_4 = 0 \tag{3.13}$$

Equation 3.11 is rewritten as follows:

$$a_{max} = c_1 \exp(b_2 M)(R^2 + h^2)^{-\frac{b_3}{2}} \tag{3.14}$$

Now the equation is inverted in terms of M.

$$M = \frac{\log\left[\left(\dfrac{a_{max}}{c_1}\right)(R^2 + h^2)^{\frac{b_3}{2}}\right]}{b_2} \tag{3.15}$$

Now, consider an infinitesimal annular element of source area dA (figure 3.4). If a unit source area generates N earthquakes per year of magnitude greater or equal to M, then an infinitesimal source area dA generates dN earthquakes per annum of magnitude greater or equal to M. Equation 3.7 is written for dN earthquakes:

$$dN = N_o \exp(-bM)dA \tag{3.16}$$

In figure 3.4, the surface dA can be written as a function of the radius, R, from the site.

$$dA = 2\pi R\, dR \tag{3.17}$$

Equation 3.17 is substituted in equation 3.16.

$$dN = 2\pi N_o \exp(-bM)R\, dR \tag{3.18}$$

Substituting equation 3.15 into equation 3.16 leads to the number of annual earthquakes from surface dA causing, at the site, a peak ground acceleration greater or equal to a_{max} (the recurrence rate).

$$dN = 2\pi N_o \exp\left[\dfrac{-b\log\left[\left(\dfrac{a_{max}}{c_1}\right)\left(R^2 + h^2\right)^{\frac{b_3}{2}}\right]}{b_2}\right] R\, dR \tag{3.19}$$

This equation can be simplified as:

$$dN = 2\pi N_o\left[\left(\dfrac{a_{max}}{c_1}\right)\left(R^2 + h^2\right)^{\frac{b_3}{2}}\right]^{\frac{-b}{b_2}} R\, dR \tag{3.20}$$

The total recurrence rate per annum, N, caused by the entire area of the source is obtained by integrating dN.

$$N = \int_0^{R_o} dN \tag{3.21}$$

where R_o = epicentral distance for which an earthquake with a cutoff magnitude M_{max} causes an acceleration equal to a_{max} at the site.

Equation 3.20 is substituted in equation 3.21.

$$N = 2\pi N_o\left(\dfrac{a_{max}}{c_1}\right)^{\frac{-b}{b_2}} \int_0^{R_o}\left(R^2 + h^2\right)^{\frac{-bb_3}{2b_2}} R\, dR \tag{3.22}$$

The solution is obtained by integrating this equation.

$$N = \frac{\pi K \left[h^{-F} - d_o^{-F} \right]}{F} \tag{3.23}$$

where

$$K = 2N_o \left(\frac{a_{max}}{c_1} \right)^{\frac{-b}{b_2}}$$

$$F = \left(\frac{b_3 b}{b_2} \right) - 2 \tag{3.24}$$

$$d_o = \sqrt{R_o^2 + h^2} = \left(\frac{a_{max}}{c_1 \exp\left(b_2 M_{max}\right)} \right)^{-\frac{1}{b_3}}$$

Equations 3.23 and 3.24 are used for values of a_{max} varying from zero to the possible peak ground horizontal acceleration, $a_{max\,possible}$, at the site. This maximum possible value depends on the attenuation relation and is obtained by replacing $M = M_{max}$ and $R = 0$ in equation 3.14.

$$a_{max\,possible} = c_1 \exp\left(b_2 M_{max}\right) h^{-b_3} \tag{3.25}$$

When the recurrence rate, N, is known, the return period, T_r, of an earthquake with a peak ground acceleration greater or equal to a_{max} at the site is also known.

$$T_r = \frac{1}{N} \tag{3.26}$$

Let us assume that in the previous example, the seismogenic source can be represented by the seismicity model of western Quebec (called WQU). The seismic constants in table 3.2 are modified to reflect the magnitude recurrence per unit area.

$$\begin{aligned} A &= 6,94 - \ln\left(121\ 000\ \text{km}^2\right) = -4,77 \\ b &= 1,85 \end{aligned} \tag{3.27}$$

Table 3.4 shows the numerical results obtained from equations 3.23 to 3.26 using HBB attenuation relations for eastern and western Canada, respectively. Figure 3.5 shows the return period at the site, T_r, as a function of the peak horizontal acceleration, a_{max}, for both attenuation relations. If a return period of 475 years is considered, as adopted by the NBCC, a value of $a_{max} = 0{,}165$ g is obtained when using the attenuation relation for eastern Canada. The corresponding value, when using the attenuation relation for western Canada, is $0{,}095$ g. Although the attenuation model exhibits large uncertainties, it nevertheless plays an important role in evaluating the seismic hazard for a given area. These uncertainties can be included in the probabilistic analysis by considering the attenuation relations as random variables (equation 2.33).

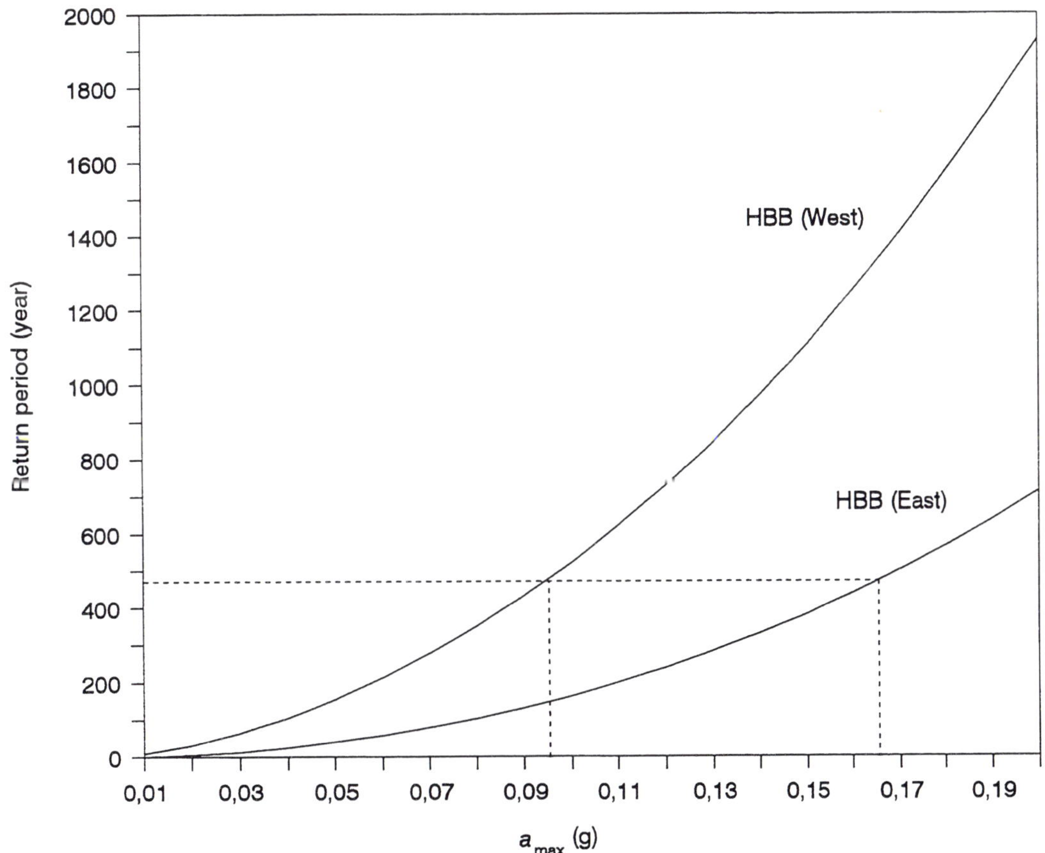

Figure 3.5   Return period for a_{max} (g) at a site located at the centre of a very large seismogenic source zone.

Table 3.4   Numerical results obtained from equations 3.23 to 3.26 using the HBB attenuation relations for eastern and western Canada.

Parameter	HBB relation for eastern Canada	HBB relation for western Canada
b_1	-5,66	-4,59
b_2	1,30	1,30
b_3	1,10	1,50
N_o	0,00848	0,00848
c_1	0,00348	0,01019
K	$0,00000538(a_{max})^{-1,42}$	$0,00002483(a_{max})^{-1,42}$
F	-0,43462	0,13462
d_o(km)	$22,80(a_{max})^{-0,90909}$	$20,27(a_{max})^{-0,66667}$
$a_{max\ possible}$	1,15 g	1,02 g
N	$-0,00003889(a_{max})^{-1,42}$ $(3,68 - 3,89(a_{max})^{-0,39511})$	$-0,00057945(a_{max})^{-1,42}$ $(0,67 - 0,67(a_{max})^{0,08975})$

3.4.3 Probability of Excedance of a Given Seismicity Level

The recurrence rate, N, and the return period, T_r, are obtained from the historical seismicity of a region. Predictions can be obtained by assuming that the sequence of earthquake occurrence follows a theoretical probability distribution model. In probabilistic seismic hazard analysis, Poisson's probability model, which assumes that earthquakes are space and time independent (seismogenic model), is usually considered. According to this model, and referring to the historical seismicity of the region, the probability, $P_n(a_{max})_t$, of having n earthquakes producing a peak ground acceleration greater or equal to a_{max} at a given site in a certain time t is given by:

$$P_n\left(a_{max}\right)_t = \frac{(Nt)^n \, e^{-Nt}}{n!} \tag{3.28}$$

where N = number of annual earthquakes (recurrence rate) producing a peak ground acceleration greater or equal to a_{max} at the site.

The probability, $P(a_{max})_t$, of having at least one earthquake causing an acceleration greater or equal to a_{max} at the site during a given time t is then equal to 1 - the probability, $P_o(a_{max})_t$, of having no earthquake causing an acceleration greater or equal to a_{max} at the site.

$$P\left(a_{max}\right)_t = 1 - e^{-Nt} \tag{3.29}$$

If the annual probability is considered ($t = 1$ year):

$$P\left(a_{max}\right) = 1 - e^{-N} \tag{3.30}$$

For small values of N (a strong earthquake is nevertheless a rare phenomenon):

$$P\left(a_{max}\right)_1 \approx N \tag{3.31}$$

3.4.4 General Application of Probabilistic Seismic Hazard Analysis in Canada

The probabilistic seismic hazard analysis procedure can be used with any seismic parameter if the attenuation relation for this parameter is known. Generally, a site is influenced by many seismic sources of different geometries and surface areas. Here, the integral to evaluate the recurrence rate, N, cannot be solved closed-form as in the previous example; instead, a numerical integration must be carried out. The computer program developed by McGuire (1976) is used in the general application of the probabilistic seismic hazard analysis in Canada.

Figure 3.6 shows a flowchart describing the general application of the probabilistic seismic hazard analysis procedure for the peak ground horizontal acceleration, a_{max}. The steps are the same as described in section 3.4.1. For a value of a_{max} and a given site, all contributions to the total recurrence rate, N, are considered by dividing each seismogenic source in sectors of smaller surface areas. The contributions of all sectors from all sources, ΔN, are then added to obtain the final recurrence rate, N.

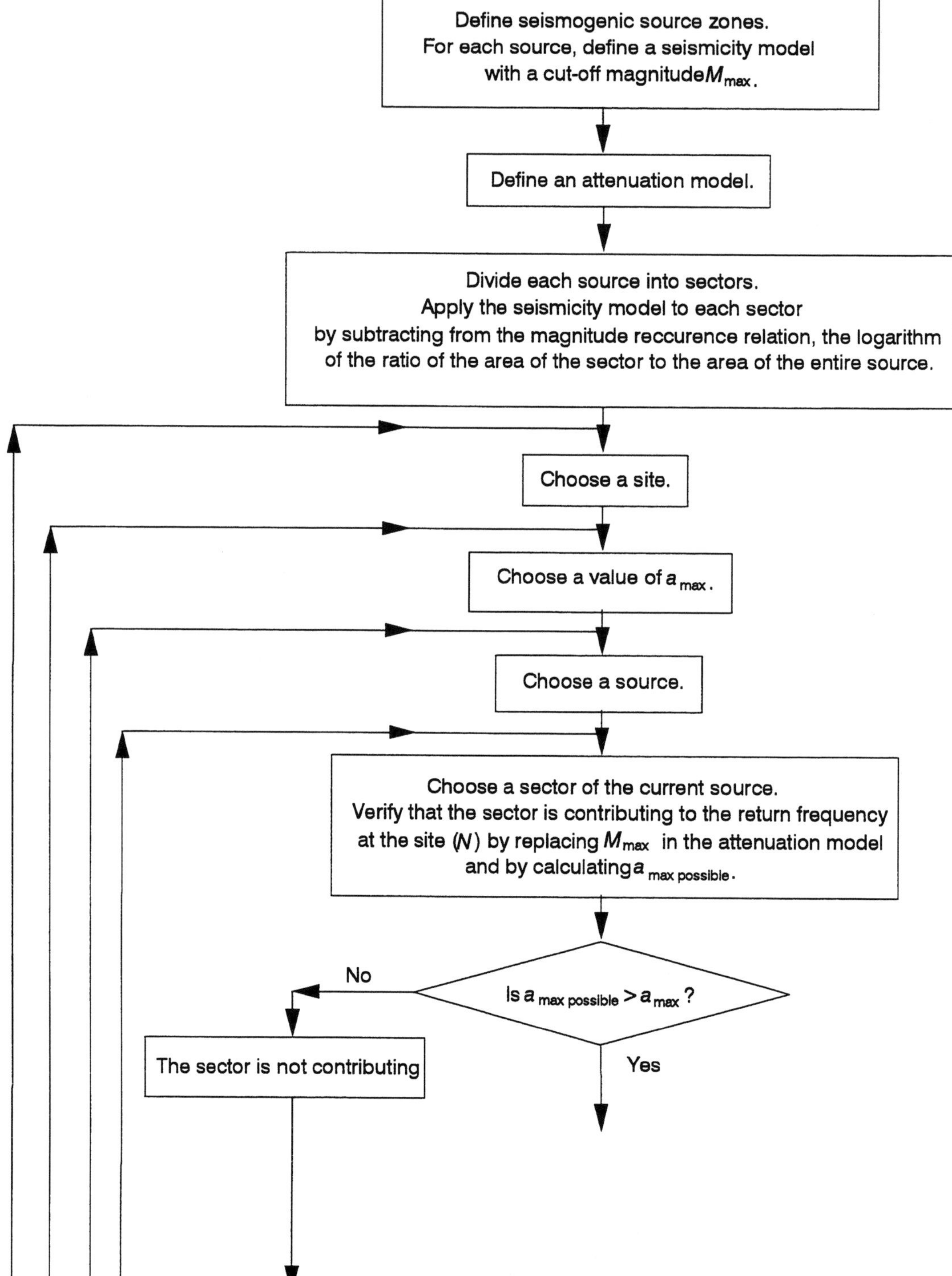

Figure 3.6   Flowchart for the general application of the probabilistic seismic hazard analysis procedure.

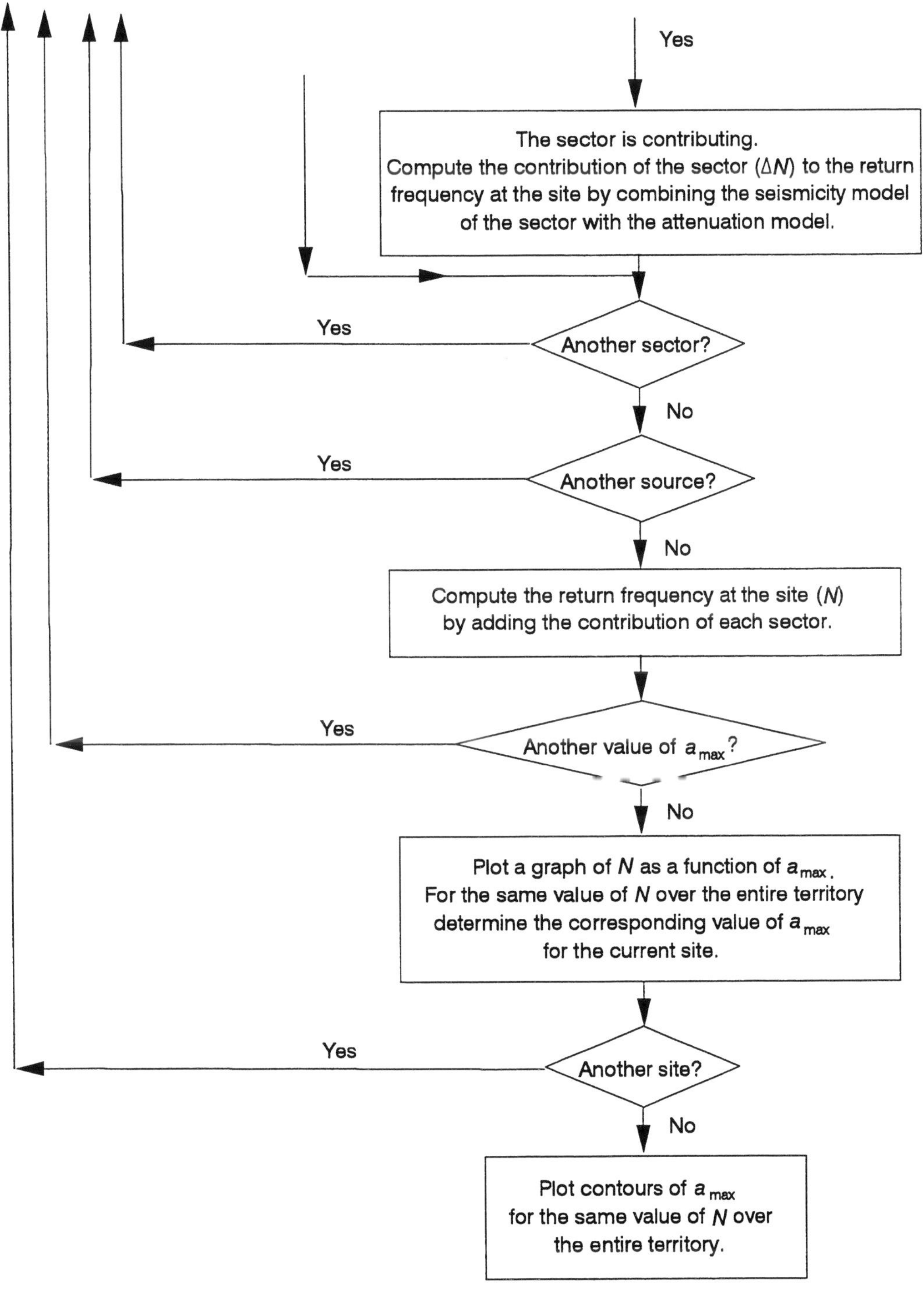

Figure 3.6 (continued)   Flowchart for the general application of the probabilistic seismic hazard analysis procedure.

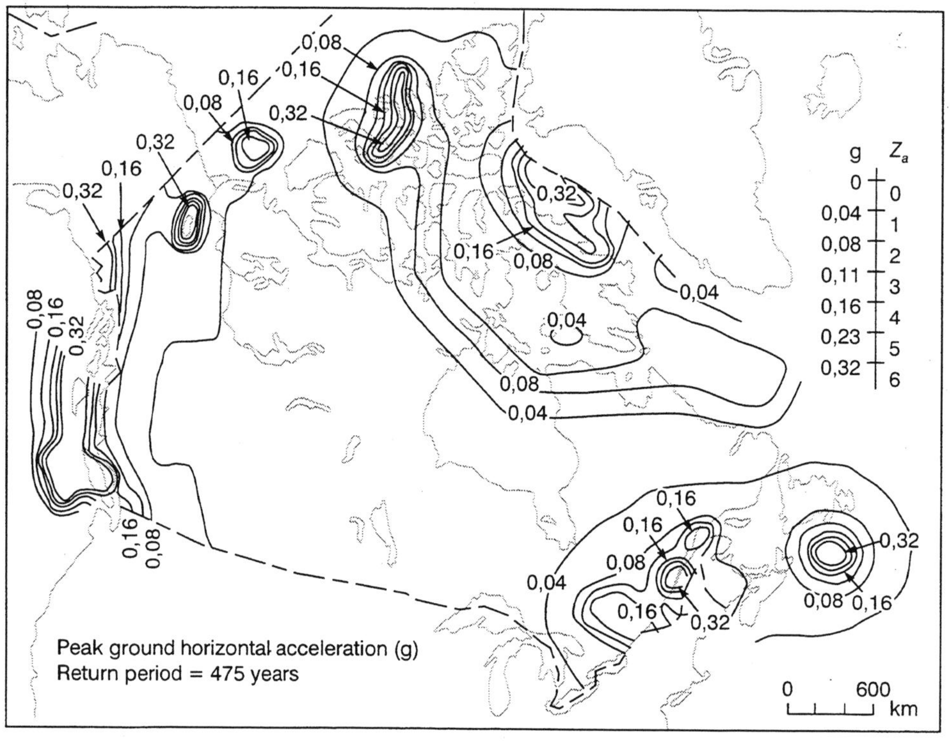

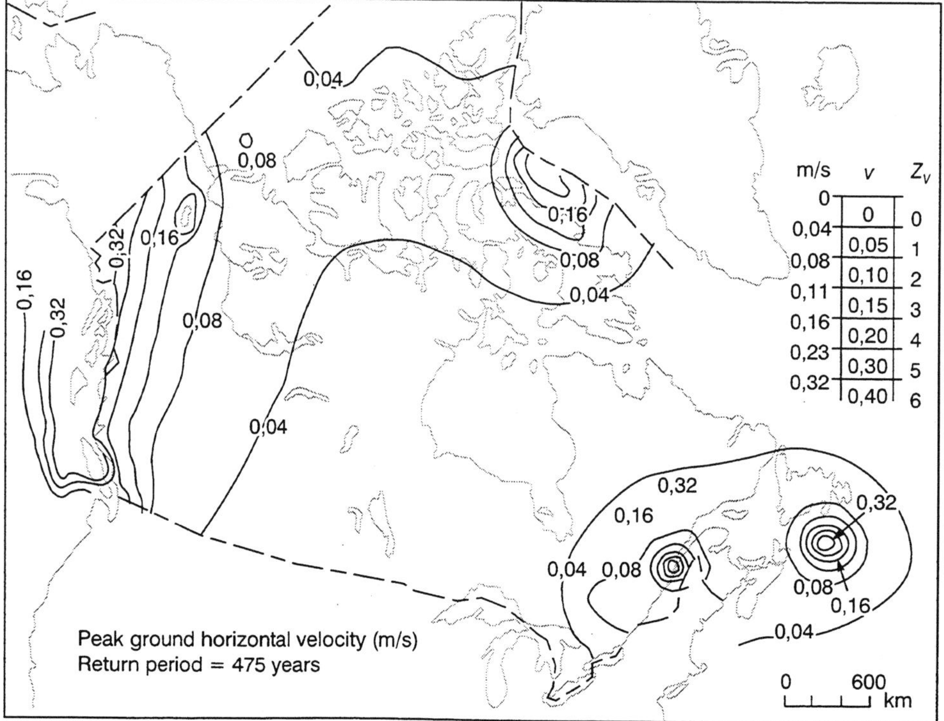

Figure 3.7   Seismic zoning maps of Canada (from the commentary of NBCC, 1995).

To develop a seismic zoning map for a specific seismic parameter (acceleration, velocity, etc.), the value of that parameter is calculated for a constant recurrence rate, N, at many sites. Then, contour lines are plotted to define the design values of the seismic design parameter for the entire region. This method was used to construct the seismic zoning maps included in the 1985, 1990, and 1995 editions of the NBCC. These maps have contours of peak ground horizontal accelerations and velocities based on a constant probability of excedance for the entire Canadian territory. The value of the recurrence rate used in these calculations is $N = 0{,}0021$ (probability of excedance of 10% in 50 years). This recurrence rate corresponds to a return period of 475 years. The seismic zoning maps of Canada do not take into account the local geotechnical conditions as the HBB attenuation relations used are applicable only to rock sites. If those local factors are to be considered, micro zoning maps have to be elaborated for specific sites.

Figure 3.7 shows the seismic zoning maps included in the commentary of the 1995 edition of the code. Regions showing the highest seismic hazard are found on the west coast of British Columbia, the St. Lawrence Valley and the Charlevoix region in Quebec, the Arctic, and the Grand Banks region of Newfoundland. Use of these maps is discussed in chapter 6.

3.5 PROBLEMS

3.1 The historical seismic data for a region are: Richter's local magnitude 6,0, 6,5, 5,5, 5,2, 7,0. These data were recorded between 1960 and 1979 (20 years) taking into account only earthquakes of magnitude greater than or equal to five.

a) Using only these data, compute the number of earthquakes per year with a magnitude greater than or equal to five. What is the corresponding return period?

b) On graph paper, construct a recurrence relation of the type:
$$\log_{10} N = A - bM$$
where N = number of earthquakes per year with a magnitude greater or equal to M
M = Richter's local magnitude
A, b = seismic constants to be determined by regression

c) Using the recurrence relation constructed in b), compute the number of earthquakes per year with a magnitude greater than or equal to five. Compare your result with the one obtained in a).

d) Using the recurrence relation constructed in b), compute the number of earthquakes per year with a magnitude greater than or equal to 7,5. Why is it hazardous, but also necessary, to extrapolate beyond the historical seismic data?

3.2 The Gutenberg-Richter model best representing the seismicity of the entire world has the following parameters:
$$A = 7{,}7$$
$$b = 0{,}9$$
$$\log = \log_e = \ln$$

a) Compute the return period of an earthquake with a magnitude $M \geq 6$.
b) Based on Poisson's probabilistic model, what is the probability of having three earthquakes with $M \geq 8$ next year?
c) What is the return period of an earthquake having a magnitude of between 7 and 8?

3.3 Figure 3.8 shows a site located symmetrically from a known fault. This fault can be considered a linear seismogenic fault on which shallow earthquakes can occur at any location with equal probability. The following equation defines the magnitude-recurrence relation for this fault.

$$\ln N = A - bM_L$$

where A = 2,00
      b = 1,74
      M_L = Richter's local magnitude
      N = number of earthquakes, per year and per unit length, with a magnitude greater or equal to M.

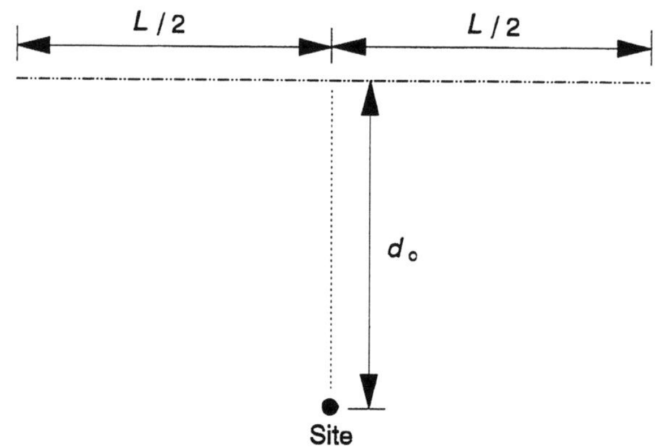

Figure 3.8 Site located symmetrically from a known fault.

The attenuation relation for the peak ground horizontal acceleration in the region is given by:

$$a_{max} = ce^{dM_L} R^{-f}$$

where a_{max} = peak ground horizontal acceleration (g)
      M_L = Richter's local magnitude
      R = epicentral distance (km)
      c = 0,01
      d = 1,305
      f = 1,5

a) Derive a general expression for the number of earthquakes, N, per year producing a peak ground horizontal acceleration at the site greater than or equal to a_{max}.
b) If the length of the fault, L, is 10 km, plot a graph showing the variation of a_{max}, having a 475-year return period, with the distance d_o. Use values of d_o between 10 and 100 km.

3.4   Figure 3.9 shows the seismicity of the San Francisco area for the last 163 years. It is assumed that the seismicity of the region can be modelled by two rectangular seismogenic source zones. The first source zone is associated with the San Andreas fault. The second source zone combines the Hayward and Calaveras faults.

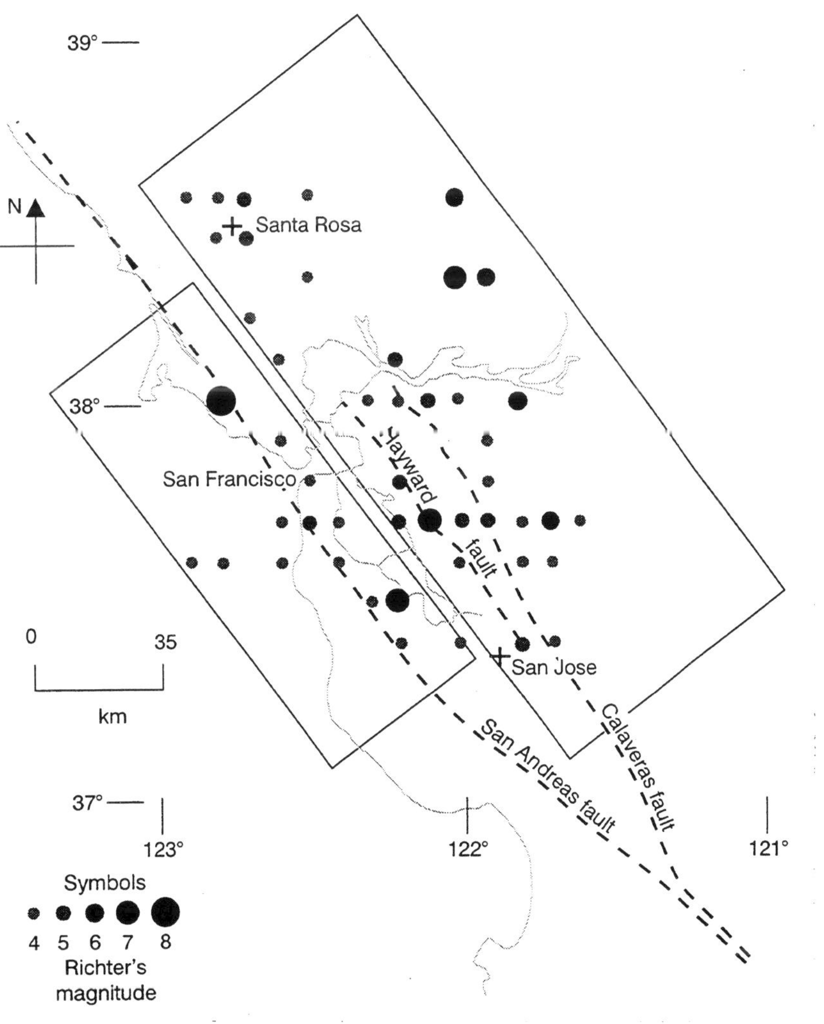

Figure 3.9   Seismicity of the San Francisco region.

a) Construct a magnitude-recurrence relation, $\ln N$ vs M, for each seismogenic source zone.

b) Using the attenuation relation constructed in problem 2.5, plot a graph showing the variation of the peak ground horizontal acceleration with the return period for the cities of San Jose, San Francisco, and Santa Rosa. Use only one sector per source zone and a constant focal depth of 15 km for the region.

c) For each city, estimate the peak ground horizontal acceleration having a 100- year return period.

3.5   A nuclear power plant must be located in a region influenced by three distinct seismotectonic source zones capable of producing shallow earthquakes (figure 3.10). These source zones are small and can be considered point sources. Figure 3.10 also shows two possible sites for the power plant. From a seismic hazard point of view, which site should be selected? Consider a design earthquake with a return period of 10 000 years. Justify your answer.

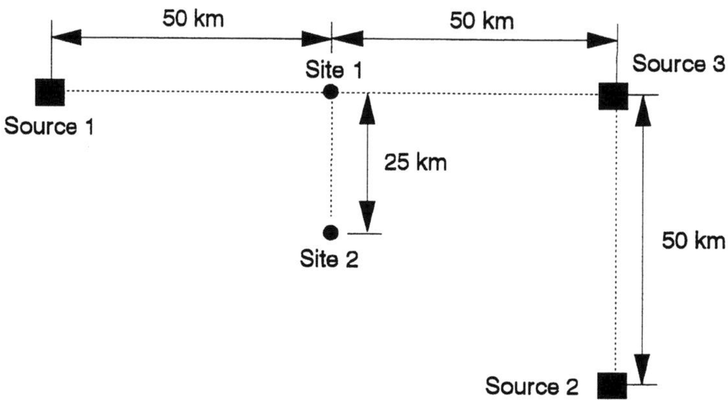

Figure 3.10   Possible locations for a nuclear power plant.

The seismicity model for the three sources can be represented by the following equation:

$$\log_{10} N_i = A_i - b_i M_L$$

$$i = 1, 2, 3$$

where N_i = number of earthquakes per year from source $\# i$ having a magnitude greater or equal to M

M_L = Richter's local magnitude

A_1 = 6,30

b_1 = 1,30

A_2 = 5,94

b_2 = 1,14

A_3 = 5,05

b_3 = 1,09

The attenuation relation for the region is given by:

$$a_{max} = 0{,}01e^{1{,}3M_L}\,d^{-1{,}5}$$

where a_{max} = peak ground horizontal acceleration (g)

M_L = Richter's local magnitude

d = hypocentral distance (km)

3.6 REFERENCES

Adams, J., et al. (1995). *Trial Seismic Hazard Maps of Canada-1995: Preliminary Values for Selected Canadian Cities*. Geological Survey of Canada, Open File 3029.

Anderson, J.G., and Luco, J.E. (1983). "Consequences of Slip Rate Constraints on Earthquake Occurrence Relations." *Bulletin of the Seismological Society of America*, 73(2): 471–96.

Basham, P.W., et al. (1982). *New Probabilistic Strong Seismic Ground Motion Map of Canada: A Compilation of Earthquake Source Zones, Methods and Results*. Earthphysics Branch Open File, Report No. 82-83. Ottawa: Energy, Mines and Resources.

— (1985). "New Probabilistic Strong Seismic Ground Motion Map of Canada." *Bulletin of the Seismological Society of America*, 75(2): 563–95.

Cornell, C.A. (1968). "Engineering Seismic Risk Analysis." *Bulletin of the Seismological Society of America*, 58(5): 1583–1606.

EERI Committee on Seismic Risk. (1984). "Glossary of Terms for Probabilistic Seismic Risk and Hazard Analysis." *Earthquake Spectra*, 1(1): 33–40.

Gutenberg, B., and Richter, C.F. (1965). *Seismicity of the Earth*. New York: Hafner.

Hasegawa, H.S., Basham, P.W., and Berry, M.J. (1981). "Attenuation Relations for Strong Seismic Ground Motion in Canada." *Bulletin of the Seismological Society of America*, 71(6): 1943–62.

Kaila, K.L., and Narain, H. (1971). "A New Approach for the Preparation of Quantitative Seismicity Maps." *Bulletin of the Seismological Society of America*, 61(5): 1275–91.

McGuire, R.K. (1976). *FORTRAN Computer Program for Seismic Risk Analysis*. Open File Report 76–67. Washington, DC: U.S. Geological Survey.

National Building Code of Canada and Commentary. 8th ed. (1980), 9th ed. (1985), 10th ed. (1990), 11th ed. (1995). Ottawa: National Research Council of Canada.

CHAPTER 4

Elements of
Structural Dynamics

4.1 INTRODUCTION

A dynamic analysis is required to evaluate realistically the behaviour of a structure during an earthquake. Structural engineers are familiar with the static analysis of structures, for which a single load pattern is applied and a unique, time-independent solution is obtained. A dynamic analysis, on the other hand, generates a sequence of time-varying solutions. The main purpose of this chapter is to present the fundamental notions of structural dynamics that are relevant to structural engineers.

4.1.1 Reason for Structural Dynamics

Most loadings applied to civil-engineering structures, including seismic loads, are usually considered equivalent static loads. The amplitude, direction, and location of a static load do not vary with time. Therefore, the inertia effects (the mass times its acceleration) are not considered significant. The amplitude, direction, and location of a dynamic load vary with time. Inertia effects can therefore be generated by dynamic loads. The main purpose of structural dynamics is to evaluate the time variations of stresses and deformations in structures caused by arbitrary dynamic loads. From this perspective, the static analysis of structures is a particular case of structural dynamics.

4.1.2 Examples of Dynamic Loads

Almost all structures are submitted to dynamic loads during their life span. Two types of dynamic loadings can be differentiated: random loads and deterministic loads.

- Random loads

Random loads are described by statistical parameters (mean, standard deviation, frequency contents, etc.). Examples of random loads are:
- vibrations on a dance floor;
- future earthquakes;
- wind pressures on a building.

- Deterministic loads

The amplitude, direction, and location of a deterministic load are known at all times. Two classes of deterministic loads can be differentiated: periodic loads and nonperiodic loads.

- Periodic loads

A periodic (or cyclic) load is repetitive and exhibits the same time variation for many cycles.

- Nonperiodic loads

Nonperiodic loads can be short (impulse) or can last a long time, but without repetition.

This chapter deals with the dynamic analysis of structures subjected to deterministic (periodic or nonperiodic) loads. Examples of deterministic loads are presented in figure 4.1.

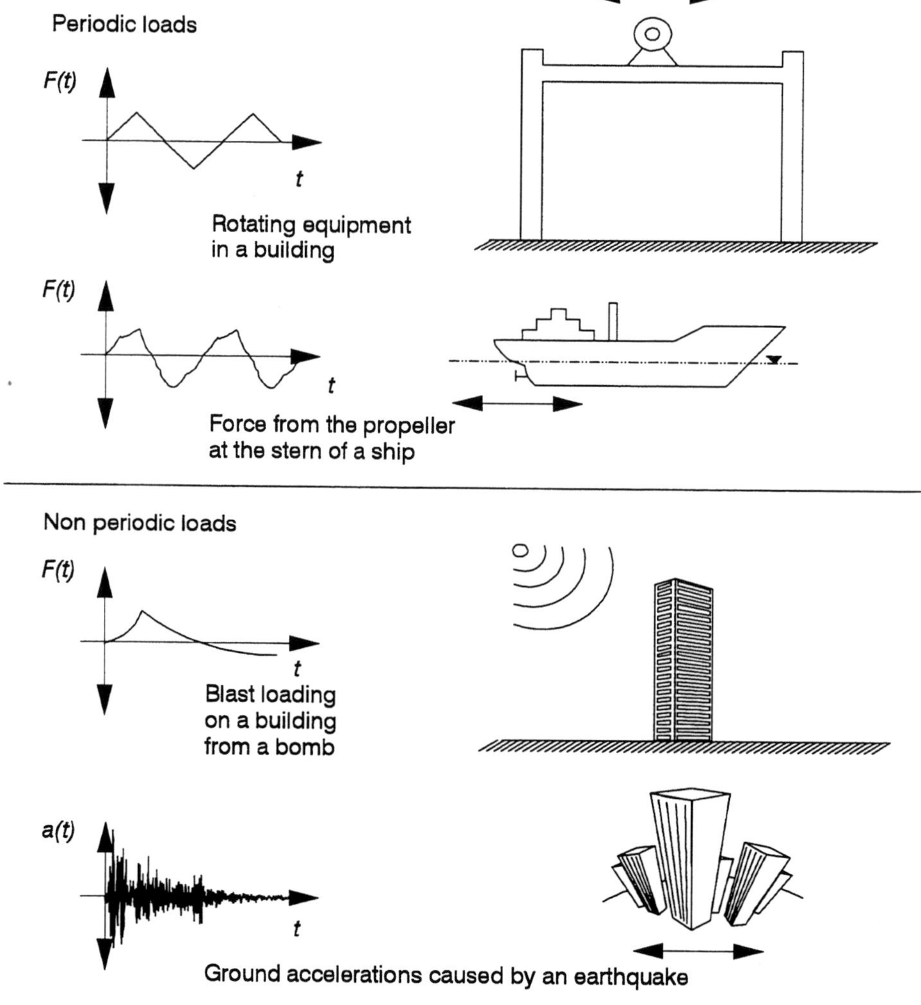

Figure 4.1   Examples of dynamic and deterministic loads.

4.1.3 Dynamic Degrees of Freedom (DDOF)

The number of dynamic degrees of freedom is the smallest number of coordinates required to define the position of all mass particles in a system. Usually, several components of motion are negligible with respect to others. Therefore, a system that may initially appear complicated can be modelled with only a few degrees of freedom. The choice of the relevant degrees of freedom is a crucial step in a dynamic analysis. For example, let us consider the pendulum in figure 4.2, made of a rock suspended by a hinged bar. If all possible deformations of the bar and the rock were considered, this system would have an infinite number of DDOF. However, the following observations can be made:

- The bar is much lighter than the rock (its mass is negligible).
- The in-plane deformations of the rock are much smaller than its rigid body motion.
- The rock's dimensions are small compared with the length of the bar (lumped mass).
- The elongation of the bar is small compared with its rigid body motion.

Thus, the system can be reduced to one DDOF: the bar's rotation around the hinge.

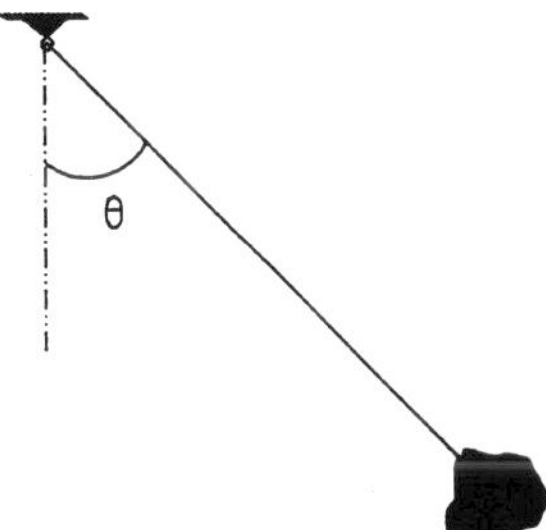

Figure 4.2 A dynamic system: the pendulum.

The laterally loaded portal frame shown in figure 4.3 represents a more complex dynamic system. Again, if the spatial distribution of the mass is considered, the system has an infinite number of DDOF. Nevertheless, the system can be reduced to one DDOF by making the following assumptions:

- The mass of the frame can be concentrated at the roof level.
- The vertical motion of the frame is very small compared with its lateral motion.
- The members of the frame exhibit some lateral stiffness but are massless.

Usually, the dynamic analysis of a structure can be performed with only a few dynamic degrees of freedom.

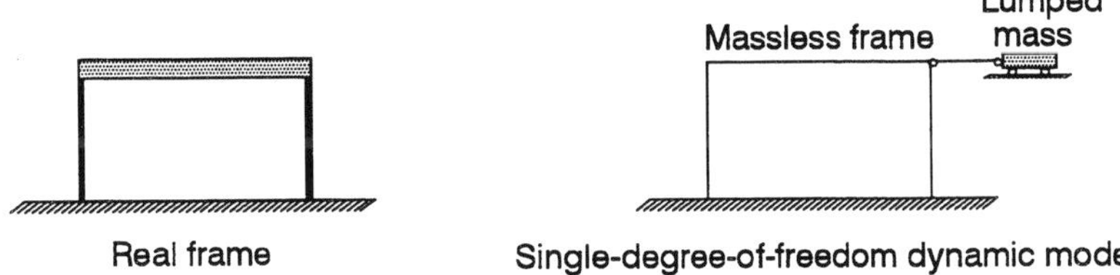

Figure 4.3 Simple portal frame modelled as a single-degree-of-freedom system.

4.1.4 Distinction between Mass and Weight

In structural dynamics, the distinction between mass and weight must be clearly established. Mass measures the quantity of matter in a body. Weight measures the force needed to produce a given acceleration on a body. A well-known equation relating mass and weight is:

$$\text{weight} = (\text{mass})\ (\text{acceleration of gravity (g)}) \tag{4.1}$$

The English system of measurement is based on the fundamental units of force, length, and time. All other units are derived from these basic units. In this system, the pound-mass (lbm) is the unit of mass: the mass for which a pound-force (lbf) produces an acceleration equal to the acceleration of gravity ($32{,}174$ ft/s^2 or $386{,}4$ in/s^2).

$$1\ \text{lbm} = \frac{1\ \text{lbf}}{32{,}174\ \text{ft/s}^2} \tag{4.2}$$

A unit of mass that is also used in the English system of measurement (particularly in hydraulics) is the **slug**: the mass for which 1 lbf produces an acceleration of 1 ft/s^2.

$$1\ \text{slug} = \frac{1\ \text{lbf}}{1\ \text{ft/s}^2} = 32{,}174\ \text{lbm} \tag{4.3}$$

The International Metric System (SI) is much more consistent than the English system of measurement, and it may someday be used worldwide. The fundamental units of the SI system are the kilogram (kg) for mass, the metre (m) for length, and the second (s) for time. The unit of force is the Newton (N), representing the force required to produce an acceleration of 1 m/s^2 to a 1 kg mass.

$$1\ \text{N} = (1\ \text{kg})\ (1\ \text{m/s}^2) \tag{4.4}$$

In structural dynamics, keeping track of the units in mathematical operations is important. One must remember that:

$$\begin{aligned} \text{force} &= (\text{mass})\ (\text{length})\ (\text{time}^{-2}) \\ \text{mass} &= (\text{force})\ (\text{length}^{-1})\ (\text{time}^{2}) \end{aligned} \tag{4.5}$$

To avoid confusion, units of mass can be defined through the units of force, length, and time:

$$\text{unit of mass (English system)} = \frac{\text{lbf} \cdot \text{s}^2}{\text{ft}} \tag{4.6}$$

$$\text{unit of mass (SI)} = \frac{\text{N} \cdot \text{s}^2}{\text{m}}$$

4.1.5 Newton's Second Law

Sir Isaac Newton (1642–1727) postulated the laws of motion that formed the cornerstone of structural dynamics. He introduced the notion of a body's momentum: the product of its mass, m, and its velocity, v. He stated the principle that when a force, F, acts on a body, the rate of change of its momentum must be equal to the applied force.

$$F = \frac{d(mv)}{dt} \tag{4.7}$$

Note that both the momentum, mv, and the applied force, F, vary with time.

Equation 4.7 expresses Newton's second law and is the fundamental tool of structural dynamics. Only systems with a constant mass are considered herein. For this case, equation 4.7 can be written as:

$$F = m\frac{dv}{dt} = ma \tag{4.8}$$

This expression represents the common equation "force equals mass times acceleration." Note that equation 4.8 is valid for a system with a constant mass.

The Newtonian notation is used throughout this chapter. A time derivative is denoted by a dot over the variable considered.

$$
\begin{aligned}
x &= \text{displacement} \\
\dot{x} &= \frac{dx}{dt} = \text{velocity} \\
\ddot{x} &= \frac{d\dot{x}}{dt} = \frac{d^2x}{dt^2} = \text{acceleration}
\end{aligned}
\tag{4.9}
$$

Another way of visualizing a dynamic problem is illustrated in figure 4.4. When a dynamic force, $F(t)$, is applied to a beam, displacements, $y(t)$, are produced. Deriving these displacements with respect to time gives rise to velocities, $\dot{y}(t)$. Furthermore, the time derivatives of these velocities produce accelerations, $\ddot{y}(t)$. Since the mass of the beam, m (mass per unit length), is uniformly distributed along the span, the product $m\ddot{y}$ represents inertia forces that could be added to or subtracted from the applied force $F(t)$.

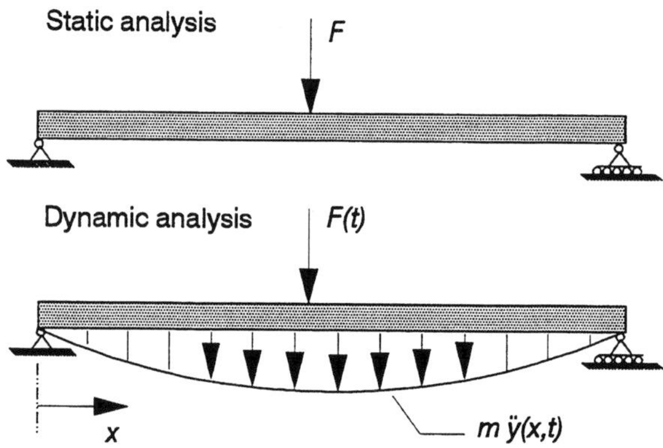

Figure 4.4 Illustration of structural dynamics for a simple beam.

4.1.6 Steps in a Dynamic Analysis

The main steps of a dynamic analysis of a structure are:

1. definition of the dynamic loading;

2. modelling of the structure:
   - identification of the important DDOF,
   - evaluation of the structural properties (mass, stiffness, damping, yield strength, etc.),
   - characterization of the soil properties;

3. dynamic analysis;

4. appraisal of results;

5. combination of the dynamic response with the responses arising from the other (static) loads to obtain the maximum response values;

6. design (or verification) of the structure.
   Dynamic analysis (step 3) is the main subject of this chapter. The method used to carry out the dynamic analysis depends on the structural characteristics and the dynamic loading (fig. 4.5).

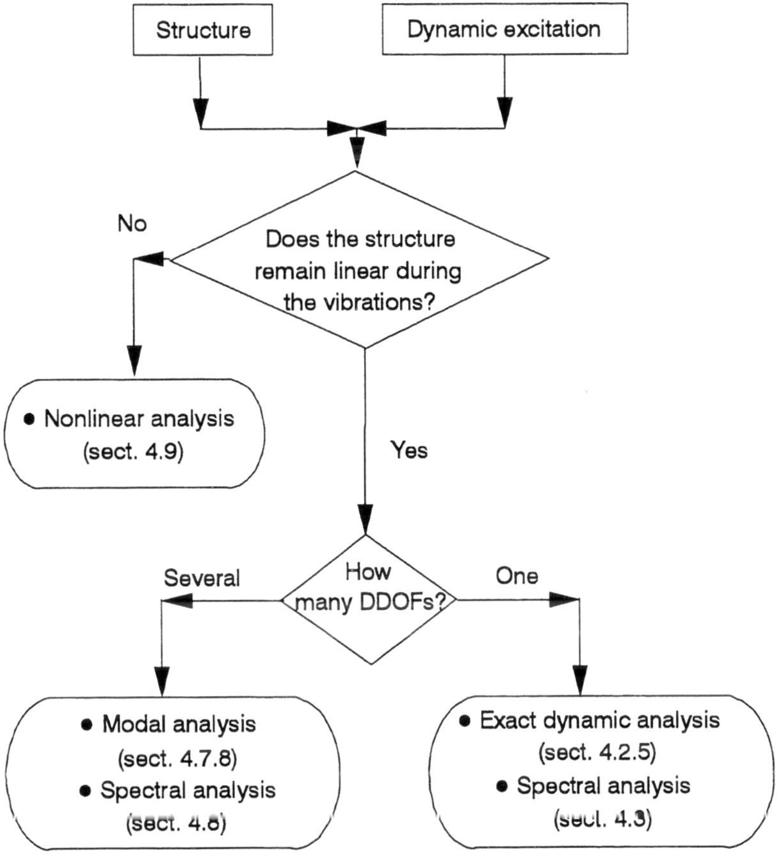

Figure 4.5 Dynamic analysis procedures.

4.1.7 Structural Dynamics and Applied Mechanics

Dynamic analysis provides the solution to a problem involving applied mechanics, which is part of physics. Applied mechanics builds a model of the physical world and predicts the behaviour of the model and not the real world. To make this prediction, the model is based on premises and axioms. Dynamic analysis, within the framework of applied mechanics, is illustrated in figure 4.6.

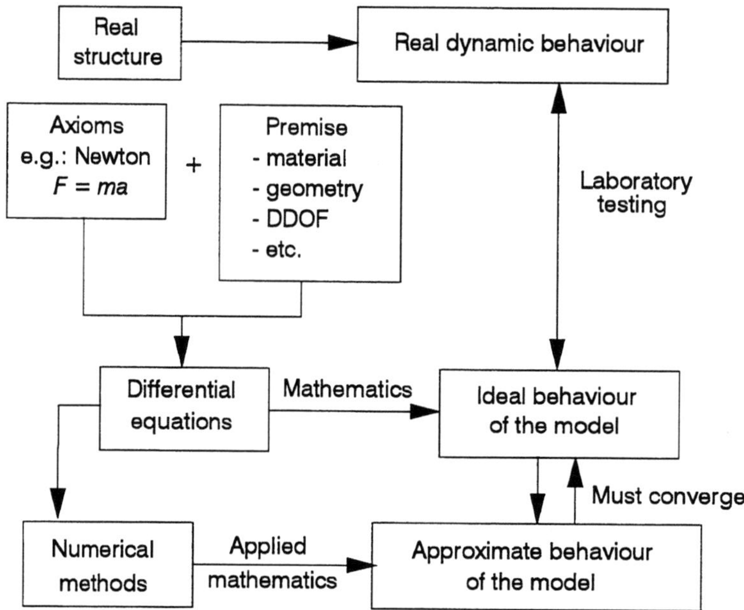

Figure 4.6 Structural dynamics within the framework of applied mechanics.

4.2 DYNAMIC ANALYSIS OF SINGLE-DEGREE-OF-FREEDOM STRUCTURES

This section presents the dynamic analysis of linear systems with only one DDOF. This system is subjected to arbitrary dynamic loads, including earthquakes. The notion of response spectrum for rapid seismic analysis of a single-degree-of-freedom (SDOF) system is discussed in section 4.3.

The dynamic analysis of an SDOF system is extremely important for the following reasons:

- Certain types of dynamic response can be identified and basic notions can be identified.
- Certain civil-engineering structures can be modelled as SDOF systems.
- The dynamic response of SDOF systems is vital for determination of the dynamic response of more complex multi-degrees-of-freedom (MDOF) structures (section 4.7).
- The spectral analysis is based on response spectra obtained from SDOF systems subjected to earthquake accelerograms at their base.

4.2.1 Mathematical Model

Representing a structure by an equivalent SDOF system is a delicate task. The accuracy of the solution will depend on the properties of the mathematical model. In this section, we consider a SDOF system as a system for which all inertia forces are concentrated in a single mass. The mathematical model of a SDOF system is illustrated in figure 4.7.

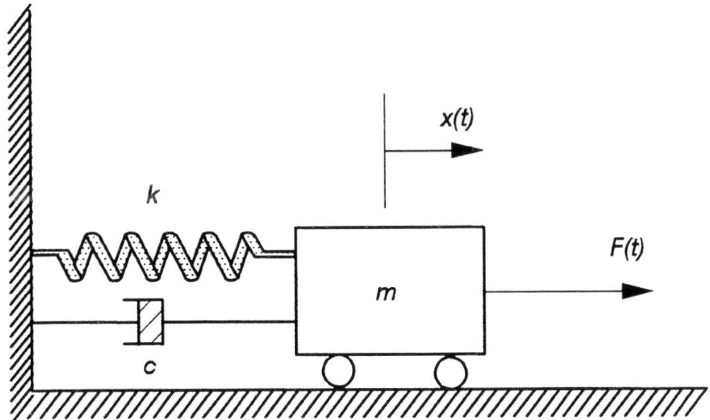

Figure 4.7 Mathematical model of a SDOF system.

where   *F(t)* = a time-varying dynamic load

   x(t) = time-varying relative displacement of the mass with respect to the base corresponding to F(t)

   m   = mass of the system (weight/g)

   k   = stiffness of the system (static force required to induce a unit displacement). For a building subjected to lateral loads, *k* represents the lateral stiffness of the structure

   c   = equivalent viscous damping coefficient of the system (dynamic force required to produce a unit velocity). This viscous damping is introduced to represent the energy dissipated during vibration of the system. This energy dissipation arises from many sources which are not easy to separate. The viscous damping model is retained because it is mathematically convenient, as we will see below.

4.2.2 Examples of Lateral Stiffness Coefficients

For a building modelled as a SDOF system and subjected to lateral loads (from wind or earthquakes), the stiffness coefficient *k* represents the total lateral stiffness of the structure. Several structural elements contribute to this stiffness. To be able to solve practical problems, it is worthwhile to review the force-displacement relation (stiffness) of common structural elements.

a) Column fixed at both ends (fig. 4.8)

$$k = \frac{V}{\Delta} = \frac{12\,EI}{L^3} \qquad\qquad (4.10)$$

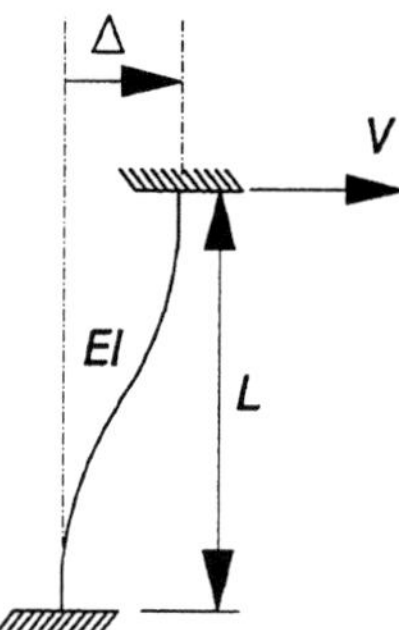

Figure 4.8 Column fixed at both ends.

b) Column supported at the base and fixed at floor level (fig. 4.9)

$$k = \frac{V}{\Delta} = \frac{3EI}{L^3}$$

(4.11)

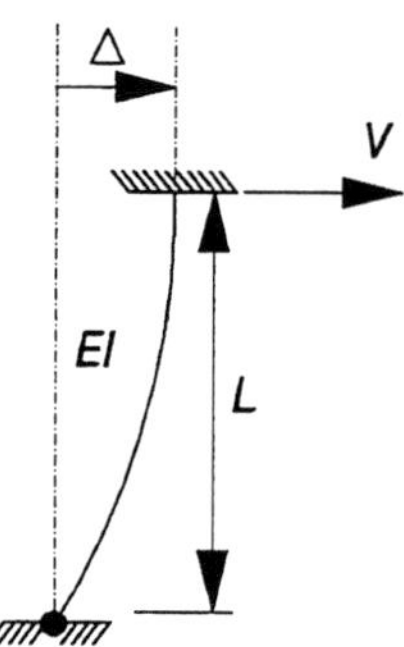

Figure 4.9 Column supported at the base and fixed at floor level.

c) Diagonal bracing member supported at both ends (fig. 4.10)

$$k = \frac{V}{\Delta} = \frac{EA}{L}\cos^2\theta \, \sin\theta = \frac{EA}{L_D}\cos^2\theta$$

(4.12)

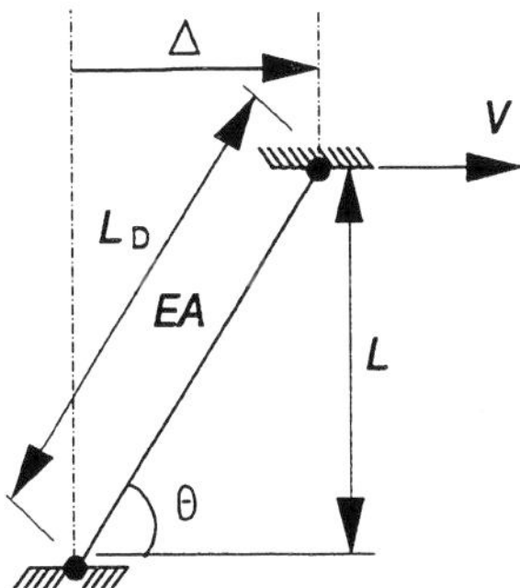

Figure 4.10   Diagonal bracing member supported at both ends.

d) Cantilever wall (fig. 4.11)

$$k = \frac{V}{\Delta} = \frac{3\,EI}{h^3\left[1 + 0,6(1+v)\dfrac{d^2}{h^2}\right]} \tag{4.13}$$

where v is Poisson's ratio.

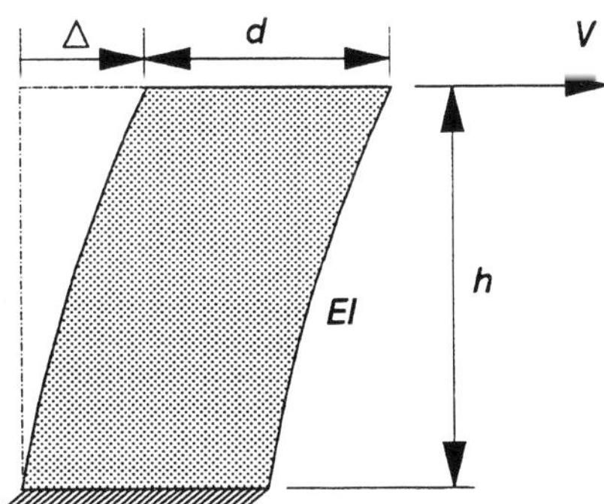

Figure 4.11   Cantilever wall.

4.2.3 Equations of Motion

Arbitrary dynamic loading. Let us consider the free-body diagram of a SDOF system, as shown in figure 4.12. By applying Newton's second law, we get:

$$F(t) - kx(t) - c\dot{x}(t) = m\ddot{x}(t) \tag{4.14}$$

or

$$m\ddot{x}(t) + c\dot{x}(t) + kx(t) = F(t) \tag{4.15}$$

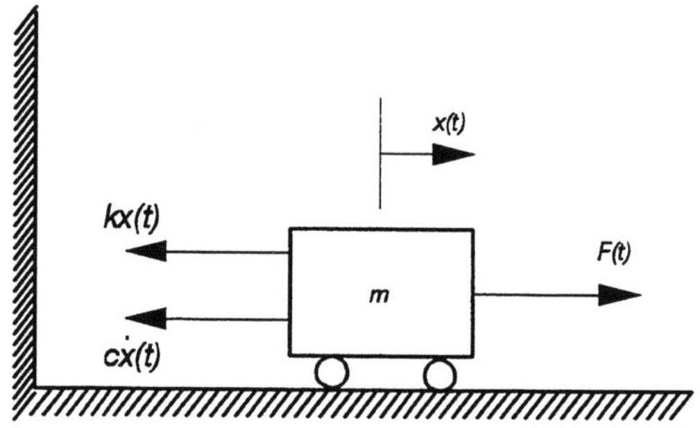

Figure 4.12   Free-body diagram of a SDOF system.

The relation 4.15 is the equation of motion (dynamic equilibrium equation) of a SDOF system. This equation is the fundamental expression we wish to solve for an arbitrary dynamic load, *F(t)*. Equation 4.15 assumes that the properties of the system *m, k, c* remain constant during the response (vibrations) of the system; in other words, the system is linear. This means, for example, that the material remains within its elastic limit, and therefore no yielding is possible.

Base excitation (earthquake problem). For a structure subjected to an earthquake base motion, the dynamic input is not the result of an explicit force applied to the system, but is caused by an implicit inertia force resulting from the movement of the base. To derive the equation of motion for the earthquake problem, we consider a SDOF system subjected to a displacement of its base, $x_s(t)$ (fig. 4.13). The DDOF, *x(t)*, represents the relative displacement of the system with respect to the base. Therefore, the elastic force generated by the stiffness of the system is proportional to *x(t)*, the damping force is proportional to $\dot{x}(t)$, but the inertia force generated by the mass of the system is proportional to the total acceleration of the system $\ddot{x}(t) + \ddot{x}_s(t)$.

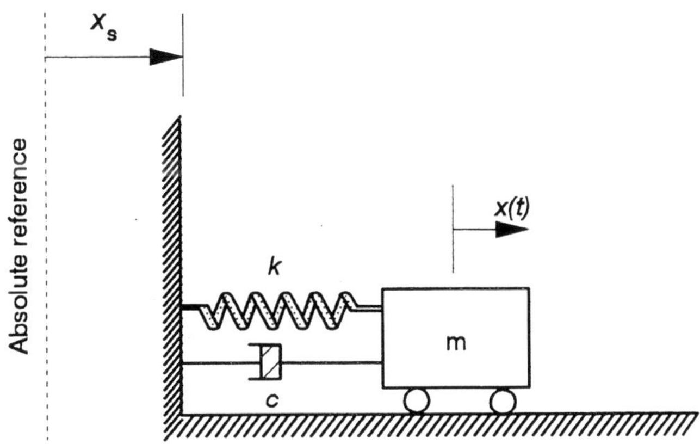

Figure 4.13   SDOF system subjected to a base motion.

Applying Newton's second law to this system yields:

$$- kx(t) - c\dot{x}(t) = m\,[\ddot{x}_s(t) + \ddot{x}(t)] \tag{4.16}$$

or

$$m\,\ddot{x}(t) + c\dot{x}(t) + kx(t) = -m\,\ddot{x}_s(t) \tag{4.17}$$

Equation 4.17 is identical to equation 4.15 except for the dynamic load, $F(t)$, replaced by a fictitious dynamic load, $-m\ddot{x}_s(t)$. This result is very important as it shows that the seismic problem (translatory motion at the base) is identical to the dynamic problem with a fixed base. The dynamic load is simply replaced by $-m\ddot{x}_s(t)$ (fig. 4.14).

The time variation of the ground acceleration (not the displacement), called the accelerogram, is needed to determine the seismic response of a SDOF system.

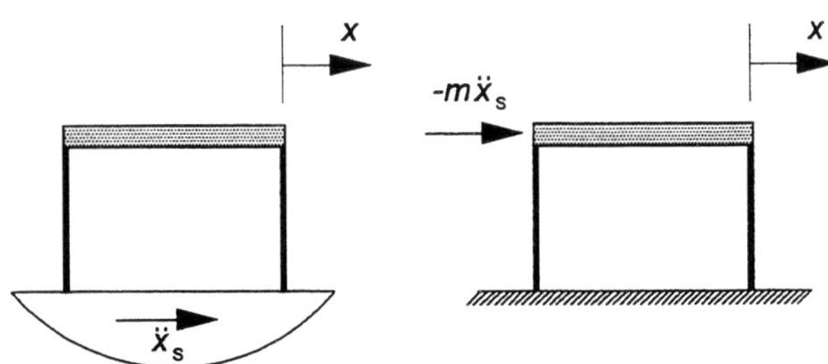

Figure 4.14   Equivalency of the seismic problem.

4.2.4 Free Vibrations

The simplest dynamic response of a SDOF system occurs when the system is in free vibration. Free vibration is the result solely of initial conditions (displacement or velocity) without external dynamic excitation. The free-vibration response is very important for finding the fundamental characteristics of a system: the natural vibration period.

Undamped systems. The equation of motion for an undamped SDOF system in free vibration is written as follows:

$$m\ddot{x}(t) + kx(t) = 0 \tag{4.18}$$

It can also be written as:

$$\ddot{x}(t) + \omega^2 x(t) = 0 \tag{4.19}$$

where

$$\omega^2 = \frac{k}{m} \tag{4.20}$$

The general (homogeneous) solution for equation 4.20 is the following:

$$x(t) = A \sin \omega t + B \cos \omega t \tag{4.21}$$

in which the integration constants A and B depend on the initial conditions of the system. Let us suppose that the initial displacement is x_0 and that the initial velocity is $\dot{x}_0$. This yields:

$$A = \frac{\dot{x}_0}{\omega} \tag{4.22}$$

$$B = x_0$$

The general solution is then written as:

$$x(t) = \frac{\dot{x}_0}{\omega} \sin \omega t + x_0 \cos \omega t \tag{4.23}$$

Figure 4.15 represents this general solution in the form of a time-varying curve.

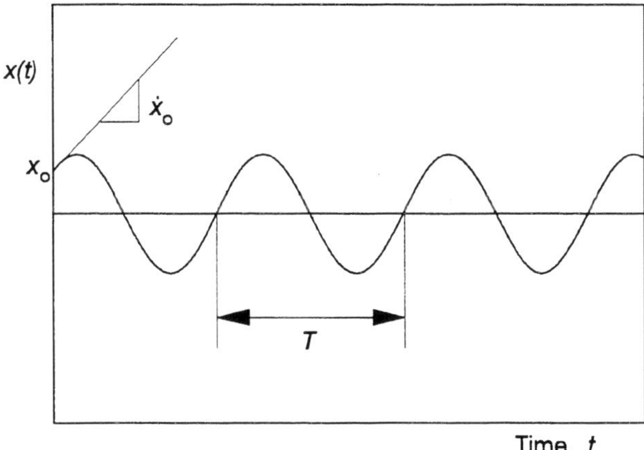

Figure 4.15   Free vibration of an undamped SDOF system.

This result illustrates many fundamental concepts:

- The amplitude is constant, meaning that the vibrations would continue indefinitely. Because this is physically impossible, the concept of damping becomes necessary.
- The required time T (in seconds) to go through a complete free-vibration cycle is called the natural period of the system.
- The quantity ω is called the angular frequency of vibration and is measured in radians per second.
- The natural frequency f of the system is defined as the reciprocal of the period and is measured in cycles per second or Hertz.

These fundamental properties $T,\ \omega, f$ only depend on the mass and the stiffness of the system and are related by:

$$T = \frac{2\pi}{\omega} = \frac{1}{f}$$

$$T = 2\pi\sqrt{\frac{m}{k}};\ \omega = \sqrt{\frac{k}{m}};\ f = \frac{1}{2\pi}\sqrt{\frac{k}{m}}$$

(4.24)

Therefore, if two structures have the same stiffness but different masses, the structure with the greater mass will have a longer natural period of vibration and a lower natural frequency. However, if the two structures have the same mass but different stiffness, the structure with the greater stiffness will have a shorter natural period of vibration and a higher natural frequency.

Damped systems. In reality, the damping of a structure in free vibration tends to shorten the amplitude of the vibrations over time. There are different types of damping:

- viscous damping, in which the damping force is proportional to the velocity;
- friction damping, in which the damping force is constant;
- internal damping, in which the force is proportional to the amplitude of the displacement.

In civil-engineering structural dynamics, viscous damping is almost always used because of its mathematical convenience. In this case, the equation of motion for a linear, damped SDOF system in free vibration is:

$$m\ddot{x}(t) + c\dot{x}(t) + kx(t) = 0 \tag{4.25}$$

or

$$\ddot{x}(t) + \frac{c}{m}\dot{x}(t) + \frac{k}{m}x(t) = 0 \tag{4.26}$$

If we define:

$$\omega = \sqrt{\frac{k}{m}} = \text{circular frequency}$$

$$\zeta = \frac{c}{2\omega m} = \text{critical damping ratio} \geq 0 \tag{4.27}$$

it yields:

$$\ddot{x}(t) + 2\zeta\omega\dot{x}(t) + \omega^2 x(t) = 0 \tag{4.28}$$

The general (homogeneous) solution to equation 4.28 is:

$$x(t) = Ce^{\lambda t} \tag{4.29}$$

Substituting equation 4.29 into equation 4.28 yields:

$$Ce^{\lambda t}(\lambda^2 + 2\zeta\omega\lambda + \omega^2) = 0 \tag{4.30}$$

which is a quadratic equation in λ with the following roots:

$$\lambda = -\zeta\omega \pm \omega\sqrt{\zeta^2 - 1} \tag{4.31}$$

According to equation 4.31, the solution depends on the sign of the square root. There are three possible solutions:

$$\zeta > 1 \ = \ \text{overdamped system}$$
$$\zeta = 1 \ = \ \text{critically damped system}$$
$$\zeta < 1 \ = \ \text{underdamped system}$$
(4.32)

a) Overdamped system ($\zeta > 1$)

In this case, the roots of equation 4.31 are negative and real.

$$\lambda = -\zeta\omega \pm \omega_s$$
(4.33)

where

$$\omega_s = \omega\sqrt{\zeta^2 - 1}$$
(4.34)

which yields as a general solution:

$$x(t) = e^{-\zeta\omega t} \left[C_1 e^{\omega_s t} + C_2 e^{-\omega_s t} \right]$$
(4.35)

This solution is illustrated graphically in figure 4.16. It is important to note that vibrations are impossible. In fact, this case has no practical significance in the dynamic analysis of civil-engineering structures.

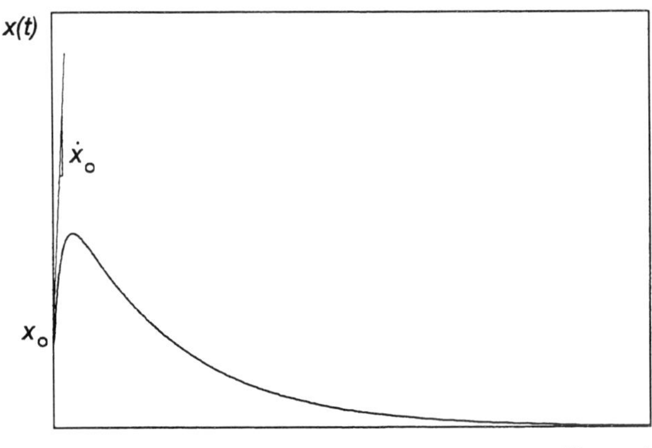

Figure 4.16 Typical response of an overdamped system.

b) Critically damped system ($\zeta = 1$)

In this case, the two roots are equal and the general solution becomes:

$$x(t) = [\ C_1 + C_2 t\ ]\ e^{-\omega t} \tag{4.36}$$

With the initial conditions x_0 and $\dot{x}_0$ at $t = 0$, it yields:

$$x(t) = x_0 \left[1 + (\omega + \frac{\dot{x}_0}{x_0})t \right] e^{-\omega t} \tag{4.37}$$

The graph in figure 4.17 illustrates this solution. Again, no vibration occurs. This case, however, represents the boundary between an oscillatory response and a monotonic response (without vibration).

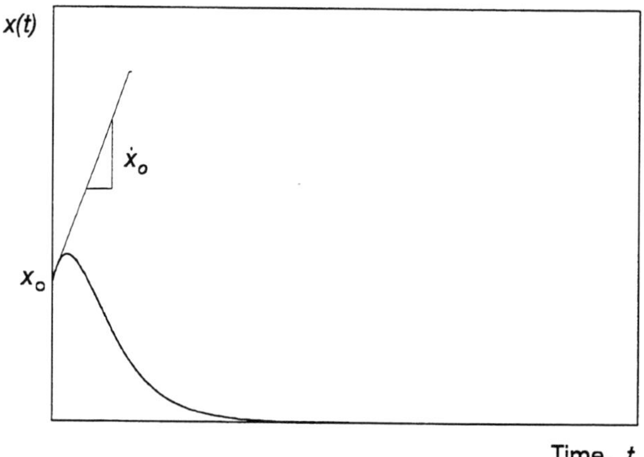

Figure 4.17   Typical response of a critically damped system.

In this case, we define:

$$\zeta_{cr} = \frac{C}{C_{cr}} = 1 = \textbf{critical damping}$$
$$C_{cr} = 2\,\omega m = \textbf{critical damping coefficient} \tag{4.38}$$

This case involving critical damping has no practical value. Civil-engineering structures are always considered to be underdamped systems with a viscous damping ratio less than 20% critical ($\zeta < 0{,}2$).

c) Underdamped system ($\zeta < 1$)

In this case, the roots of equation 4.31 are complex values.

$$\lambda = -\zeta\omega \pm i\omega_d \tag{4.39}$$

where

$$i = \sqrt{-1}$$

$$\omega_d = \omega\sqrt{1-\zeta^2} \tag{4.40}$$

The solution of equation 4.26 is thus:

$$x(t) = e^{-\zeta\omega t}(C_1 e^{i\omega_d t} + C_2 e^{-i\omega_d t}) \tag{4.41}$$

Using Euler's relations, the solution is simplified to:

$$x(t) = e^{-\zeta\omega t}(A\,\sin\omega_d t + B\,\cos\omega_d t) \tag{4.42}$$

To evaluate constants A and B, the initial conditions are set (at $t - 0$):

$$x(0) = x_0; \quad \dot{x}(0) = \dot{x}_0 \tag{4.43}$$

It then yields:

$$x(t) = e^{-\zeta\omega t}\left[\frac{\dot{x}_0 + \zeta\omega x_0}{\omega_d}\sin\omega_d t + x_0\cos\omega_d t\right] \tag{4.44}$$

This equation can also be written as:

$$x(t) = X e^{-\zeta\omega t}\sin(\omega_d t + \Phi) \tag{4.45}$$

where X represents the amplitude of the response and Φ the phase angle:

$$X = \frac{1}{\omega_d}\sqrt{(x_0\,\omega_d)^2 + (\dot{x}_0 + \zeta\omega x_0)^2}$$

$$\Phi = \tan^{-1}\left(\frac{\omega_d x_0}{\dot{x}_0 + \zeta\omega x_0}\right) \tag{4.46}$$

The graph in figure 4.18 shows the solution given by equation 4.46. The motion is harmonic and has an angular frequency ω_d. The amplitude decreases because of the term $e^{-\zeta\omega t}$.

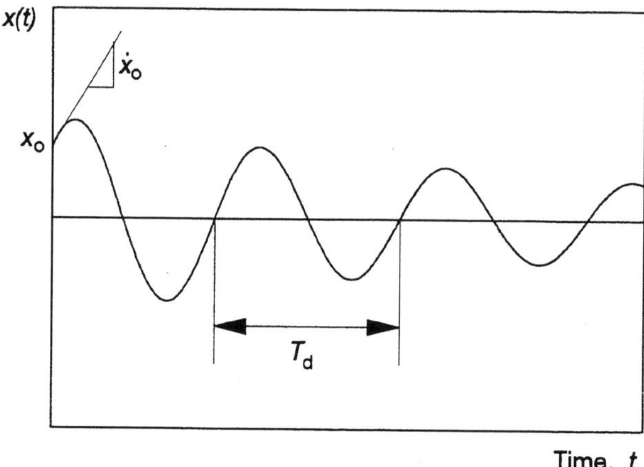

Figure 4.18 Typical response of an underdamped system.

We define:

$$\omega_d = \omega\sqrt{1 - \zeta^2} = \text{damped circular frequency}$$

$$f_d = \frac{\omega_d}{2\pi} = \text{damped natural frequency}$$

$$T_d = \frac{1}{f_d} = \text{damped natural period} \tag{4.47}$$

In practice, however, it is not necessary to make a distinction between undamped and damped natural frequencies. For example, consider the following system with a damping ratio of 20% critical:

$$\omega_d = \omega\sqrt{1 - 0{,}2^2} = \omega\sqrt{0{,}96} = 0{,}980\,\omega \approx \omega \tag{4.48}$$

For a damping ratio < 20% critical, which represents an upper boundary for most civil-engineering structures, the natural frequency is basically not altered by the presence of damping.

Logarithmic decrement. Testing a structure in free vibration is an excellent way to determine its critical damping ratio. For this purpose, let us examine the response shown in figure 4.18. The interval between two response peaks is constant and equal to $2\pi/\omega_d$. The ratio of the amplitude of two consecutive peaks x_n/x_{n+1} is the same for any value of n. Using equation 4.45, it yields:

$$\frac{x_n}{x_{n+1}} = e^{-\zeta\omega(t_n - t_{n+1})} = e^{\zeta\omega T_d} \tag{4.49}$$

The natural logarithm of this ratio is called the logarithmic decrement δ:

$$\delta = \ln\left(\frac{x_n}{x_{n+1}}\right) = \zeta\omega T_d = \frac{2\pi\zeta}{\sqrt{1-\zeta^2}} \tag{4.50}$$

for which the critical damping ratio is defined by:

$$\zeta = \frac{\delta}{2\pi\sqrt{\left[1 + \left(\frac{\delta}{2\pi}\right)^2\right]}} \approx \frac{\delta}{2\pi} \tag{4.51}$$

If the decrement is slow, the result is more accurate if peaks separated by many cycles p are compared rather than two consecutive peaks. In this case, it yields:

$$\delta = \frac{1}{p}\ln\frac{x_n}{x_{n+p}} \tag{4.52}$$

Examples

Example 1: Natural frequency of an industrial building (after Anderson, 1989)
A SDOF model of a symmetrical steel industrial building is constructed, as shown in figure 4.19. Its natural period of vibration is estimated for each principal direction. The vertical bracing members are made of circular rods with a 25 mm diameter. These rods are very slender and cannot resist any compression load. All columns are made of W200 × 36 steel sections. Table 4.1 shows the dead load on the building.

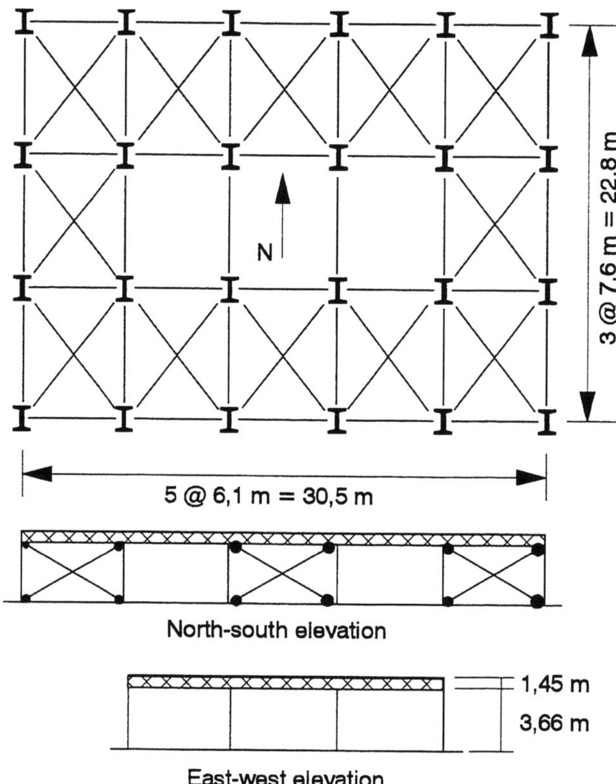

Figure 4.19   Single-storey industrial building (after Anderson, 1989).

Table 4.1   Dead load on an industrial building (adapted from Anderson, 1989).

Load on the roof (kPa)	
Roof material	0,43
Lighting, ceiling, mechanical equipment	0,29
Roof trusses	0,12
Roof joists	0,10
Diagonal bracing	0,10
Half-height of columns (1,83 m)	0,02
Total	1,06
Load on the walls (kPa)	
Windows, frames	0,19
Siding	0,29
Total	0,48

The first step in the analysis is to calculate the mass of the equivalent SDOF system. The total weight at roof level is:

$$W = (1,06 \text{ kPa})(30,5 \text{ m})(22,8 \text{ m}) + (0,48 \text{ kPa})2(30,5 \text{ m} + 22,8 \text{ m})(1,83 \text{ m})$$

$$W = 830 \text{ kN}$$

The corresponding mass is:

$$m = \frac{W}{g} = \frac{830 \text{ kN}}{9,81 \text{ m/s}^2} = 85 \frac{\text{kN}-\text{s}^2}{\text{m}} \tag{4.54}$$

The next step is to determine the total lateral stiffness of the building in each direction. In the north-south direction, moment resisting frames act as the lateral load resisting system. As the stiffness of the beams is much greater than the stiffness of the columns, we can write:

$$k_i = \frac{12\ EI}{L^3} = \frac{(12)(200 \times 10^6 \text{ kN/m}^2)(34,4 \times 10^{-6}\text{m}^4)}{(3,66\,\text{m})^3}$$

$$k_i = 1684 \text{ kN/m} \tag{4.55}$$

$$k = \sum_{i=1}^{24} k_i = 24(1684 \text{ kN/m}) = 40\,416 \text{ kN/m}$$

In the east-west direction, the lateral loads are carried by diagonal bracing.

$$k_i = \frac{AE}{L_d}\cos^2\theta$$

$$A = \frac{\pi d^2}{4} = 490 \times 10^{-6}\text{m}^2$$

$$L_d = \sqrt{3,66^2 + 6,10^2} = 7,11\,\text{m} \tag{4.56}$$

$$\theta = \tan^{-1}\left(\frac{3,66}{6,10}\right) = 31°; \ \cos(31°) = 0,858$$

$$k_i = \frac{(490 \times 10^{-6}\text{m}^2)(200 \times 10^6\text{kN/m}^2)(0,858)^2}{(7,11 \text{ m})} = 10\,147 \text{ kN/m}$$

$$k = (6)(10\,147 \text{ kN/m}) = 60\,881 \text{ kN/m}$$

Finally, the natural frequencies are calculated:

- North-south

$$\omega = \sqrt{\frac{k}{m}} = \sqrt{\frac{40\,416 \ \text{kN/m}}{85 \ \text{kN} \bullet \text{s}^2/\text{m}}} = 21,80 \ \text{rad/s}$$

$$T = \frac{2\pi}{\omega} = \frac{2\pi}{21,8} = 0,288 \ \text{s} \tag{4.57}$$

$$f = \frac{1}{T} = 3,47 \ \text{Hz}$$

- East-west

$$\omega = \sqrt{\frac{k}{m}} = \sqrt{\frac{60\,881 \ \text{kN/m}}{85 \ \text{kN} \bullet \text{s}^2/\text{m}}} = 26,76 \ \text{rad/s}$$

$$T = \frac{2\pi}{\omega} = \frac{2\pi}{26,76} = 0,235 \ \text{s} \tag{4.58}$$

$$f = \frac{1}{T} = 4,26 \ \text{Hz}$$

It has been assumed that the free vibrations are uncoupled in each principal direction. In other words, torsional vibrations are not considered.

Example 2: Free-vibration test of a simple building
A one-storey building is modelled with a rigid floor supported by massless columns (fig. 4.3). To evaluate the dynamic properties of the building (total weight, natural frequency, and damping ratio), the building is tested in free vibration. A hydraulic jack moves the floor laterally and releases it suddenly. During the test, a force of 89 kN is required to move the floor by 5 mm. One cycle after initial release of the floor, the amplitude of vibrations is only 4 mm. The time required for this cycle is 1,40 s.
The stiffness of the system is:

$$k = \frac{89 \ \text{kN}}{5 \ \text{mm}} = 17 \ \frac{\text{kN}}{\text{mm}} \tag{4.59}$$

The weight of the floor is:

$$W = \frac{g\,k\,T^2}{(2\,\pi)^2} = \frac{(9810\,\text{mm/s}^2)(17\text{kN/mm})(1,4\text{s})^2}{(2\,\pi)^2} \tag{4.60}$$

$$W = 8\,280\ \text{kN}$$

The natural frequency is simply:

$$f = \frac{1}{T} = 0,714\,\text{Hz} \tag{4.61}$$

The logarithmic decrement is:

$$\delta = \ln\frac{5\ \text{mm}}{4\ \text{mm}} = 0,223 \tag{4.62}$$

The damping ratio becomes:

$$\zeta = \frac{\delta}{2\,\pi} = 0,0355 = 3,55\%\ \text{critical} \tag{4.63}$$

4.2.5 Forced Vibrations

In this section, we solve the equation of motion for a linear SDOF system subjected to an arbitrary dynamic force $F(t)$.

$$m\,\ddot{x}(t) + c\,\dot{x}(t) + k\,x(t) = F(t) \tag{4.64}$$

Vibration under a short-duration impulse. To develop a general method to solve equation 4.64 under an arbitrary dynamic load, we first consider the response of a system under a short-duration impulse (impact) (fig. 4.20).

If the contact duration, t_1, is very short compared to the natural period of the system, T, then the impulse can be treated as a sudden change of velocity. In other words, the system does not have time to react during the time interval t_1. When the system is at rest at $t = 0$, the fundamental equation relating the impulse to the momentum is:

$$\int_0^t F(t)\,dt = m\,\dot{x}(t) \qquad 0 \leq t \leq t_1 \tag{4.65}$$

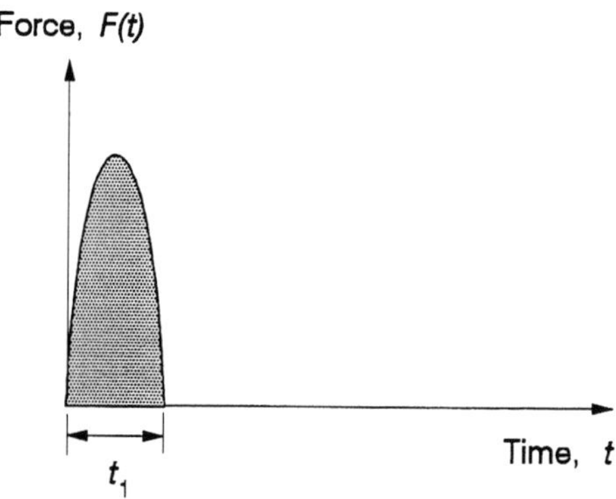

Figure 4.20   Short-duration impulse.

After the end of the impulse $(t > t_1)$, the system is in free vibration with the following initial conditions:

$$\dot{x}(t_1) = \frac{1}{m} \int_0^{t_1} F(t)\,dt$$

$$x(t_1) = 0 \text{ (negligible)}$$

(4.66)

The response of the system is given by equation 4.44:

$$x(t-t_1) = \frac{1}{m\,\omega_d} \int_0^{t_1} F(t)\,dt\, e^{-\zeta\omega(t-t_1)} \sin \omega_d(t-t_1)$$

(4.67)

Vibrations under an arbitrary dynamic load; Duhamel's integral. The response of a SDOF system under a short-duration impulse can be used to deduce the response of the same system under an arbitrary load, $F(t)$. All dynamic arbitrary loads can be represented by a succession of impulses (fig. 4.21). Now, let us consider a specific impulse, ending at time τ after the beginning of the load application. This impulse has an infinitesimal duration $d\tau$. The area under the impulse corresponds to $F(\tau)d\tau$. This specific impulse produces a unit response in free vibration at time t given by equation 4.67:

$$dx(t) = \frac{1}{m\,\omega_d} F(\tau)\,d\tau e^{-\zeta\omega(t-\tau)} \sin \omega_d(t-\tau) \qquad t \geq \tau$$

(4.68)

The total response of the system at time t is obtained by superimposing (integrating) the unit responses until time t:

$$x(t) = \frac{1}{m\,\omega_d} \int_0^t F(\tau)\, e^{-\zeta\omega(t-\tau)} \sin \omega_d(t-\tau)\, d\tau \qquad (4.69)$$

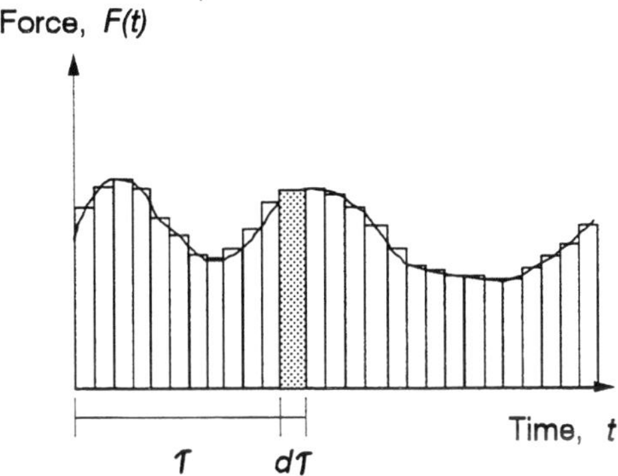

Figure 4.21 Dynamic arbitrary load represented by a succession of rectangular impulses.

Equation 4.69 is called Duhamel's integral. It is important to note that the principle of superposition is used to deduce it. Therefore, equation 4.69 is valid only for a linear system. If a dynamic load, $F(t)$, is represented by a mathematical function, equation 4.69 is then integrated directly. In practice, and especially in the case of a seismic excitation at the base, it is often necessary to use a numerical integration. It should be noted that if the system is not at rest at $t = 0$, the solution in free vibration of equation 4.44 must be superimposed.

$$x(t) = e^{-\zeta\omega t}\left[\frac{\dot{x}_0 + \zeta\omega x_0}{\omega_d}\sin\omega_d t + x_0\cos\omega_d t\right] +$$

$$\frac{1}{m\,\omega_d}\int_0^t F(\tau)\, e^{-\zeta\omega(t-\tau)}\sin\omega_d(t-\tau)\, d\tau \qquad (4.70)$$

Examples

Example 1: Undamped system subjected to a constant load (fig. 4.22)

$$F(t) = F_o$$
$$x_0 = 0$$
$$\dot{x}_0 = 0 \tag{4.71}$$

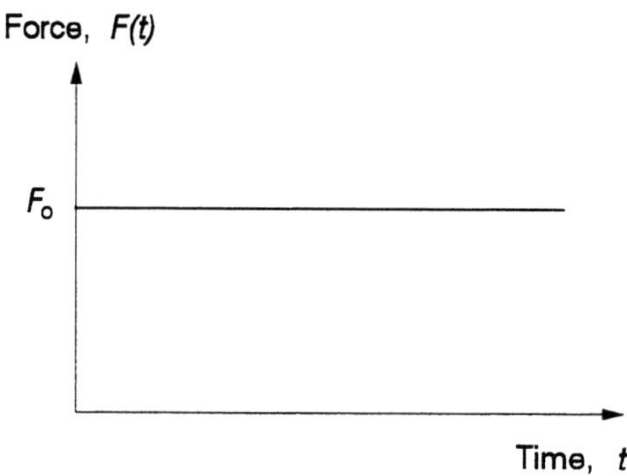

Figure 4.22   Constant dynamic force F_o.

Using equation 4.70 yields:

$$x(t) = \frac{F_o}{m\,\omega} \int_0^t \sin \omega(t - \tau)\, d\tau$$

$$x(t) = \frac{F_o}{m\,\omega^2}(1 - \cos \omega t) \tag{4.72}$$

We observe that:

$$\frac{F_o}{m\,\omega^2} = \frac{F_o}{k} = x_{st} = \text{static displacement} \tag{4.73}$$

Then, the response of the system is:

$$x(t) = x_{st}(1 - \cos \omega t)$$

The graphical representation of this response in figure 4.23 shows that the maximum displacement is twice the displacement created by a static load F_o. As a result, all internal forces, strains and others, double for this kind of dynamic load. Figures 4.23 and 4.18 show some similarities, except for the vertical axis, which is displaced by a value of F_o/k.

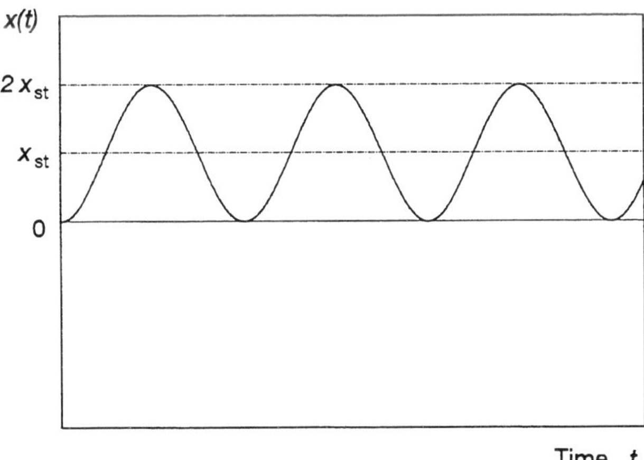

Figure 4.23   Response of an undamped system to a constant load F_o.

Example 2: Undamped system subjected to a sinusoidal load

$$F(t) = F_o \sin \bar{\omega}t$$
$$x_0 = 0$$
$$\dot{x}_0 = 0$$

(4.75)

Replacing the variables of equation 4.75 in equation 4.70, the response of the system is as follows:

$$x(t) = \frac{F_o}{m\,\omega} \int_0^t \sin \bar{\omega}\tau \sin \omega(t-\tau)\,d\tau$$

$$x(t) = \frac{F_o}{k} \left[\int_0^t \sin \bar{\omega}\tau (\sin \omega t \cos \omega\tau - \cos \omega t \sin \omega\tau)\,d\tau \right]$$

(4.76)

$$x(t) = \frac{F_o}{k} \left[\sin \omega t \int_0^t \sin \bar{\omega}\tau \cos \omega\tau\,d\tau - \cos \omega t \int_0^t \sin \bar{\omega}\tau \sin \omega\tau\,d\tau \right]$$

Evaluating the integrals yields:

$$x(t) = \frac{F_o}{k}\frac{1}{1-\beta^2}(\sin \bar{\omega}t - \beta \sin \omega t)$$

$$\beta = \frac{\bar{\omega}}{\omega} = \text{frequency ratio} \tag{4.77}$$

It is important to know the different terms of the response:

$$\frac{F_o}{k} = x_{st} = \text{static displacement}$$

$$\frac{1}{1-\beta^2} = \text{dynamic amplification factor} \tag{4.78}$$

$$\sin \bar{\omega}t = \text{steady-state response}$$

$$\sin \omega t = \text{transient response}$$

In practice, the transient response disappears quickly when damping is present. However, for the hypothetical case of an undamped system, this response continues indefinitely. Considering only the steady-state response, there is a relation between the absolute value of the amplification factor and the frequency ratio (fig. 4.24).

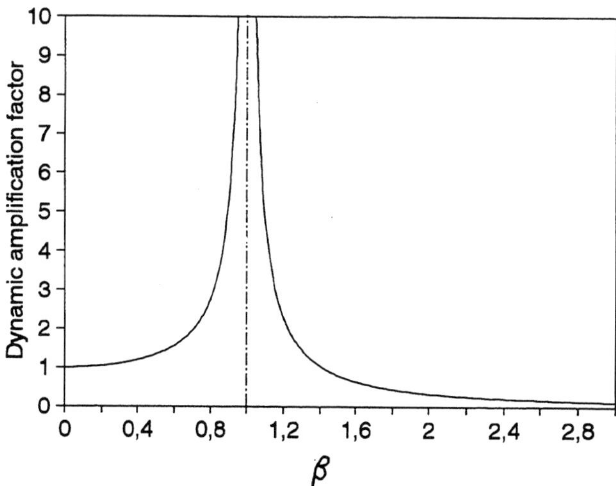

Figure 4.24 Frequency response of an undamped SDOF system subjected to a sinusoidal load.

The response of the system becomes undefined when the excitation frequency equals the natural frequency of the system $(\beta=1)$. This phenomenon is called resonance. In practice, the amplitude at resonance has a finite value because damping is present.

Example 3: Damped system subjected to a sinusoidal load

$$F(t) = F_o \sin \bar{\omega} t$$
$$x_0 = 0$$
$$\dot{x}_0 = 0 \tag{4.79}$$

Replacing the variables of equation 4.79 in equation 4.70, the response of the system is as follows:

$$x(t) = \frac{F_o}{m\,\omega_d} \int_0^t e^{-\zeta\omega(t-\tau)} \sin\bar{\omega}\tau \sin\omega_d(t-\tau)\,d\tau$$

$$x(t) = \frac{F_o}{m\,\omega_d} e^{-\zeta\omega t} \int_0^t e^{\zeta\omega\tau}\sin\bar{\omega}\tau(\sin\omega_d t\cos\omega_d\tau - \cos\omega_d t\sin\omega_d\tau)\,d\tau \tag{4.80}$$

$$x(t) = \frac{F_o}{m\,\omega_d} e^{-\zeta\omega t}\left(\sin\omega_d\,t\int_0^t e^{\zeta\omega\tau}\sin\bar{\omega}\tau\cos\omega_d\tau\,d\tau - \cos\omega_d\,t\int_0^t e^{\zeta\omega\tau}\sin\bar{\omega}\tau\sin\omega_d\tau\,d\tau\right)$$

Evaluating the integrals, the general response of the system is:

$$x(t) = x_{st}\,\alpha\beta e^{-\zeta\omega t}[(2\zeta^2-(1-\beta^2))\sin\omega_d t + 2\zeta\cos\omega_d t]$$
$$+x_{st}\,\alpha[(1-\beta^2)\sin\bar{\omega} t - 2\zeta\beta\cos\bar{\omega} t] \tag{4.81}$$

where

$$\beta = \frac{\bar{\omega}}{\omega} = \text{frequency ratio}$$

$$x_{st} = \frac{F_o}{k} = \text{static displacement} \tag{4.82}$$

$$\alpha = \frac{1}{(1-\beta^2)^2 + (2\zeta\beta)^2}$$

The first term of equation 4.81 represents the transient response of the system. This term is damped out quite quickly and does not contribute to the solution after a few cycles.

The second term of equation 4.81 represents the forced-vibration (steady-state) response of the system to the excitation frequency. This steady-state response is written as:

$$x(t) = x_{st}\sqrt{\alpha}\sin(\bar{\omega}t + \Phi)\tag{4.83}$$

where

$$x_{st}\sqrt{\alpha} = \text{forced response amplitude}$$

$$\Phi = \tan^{-1}\frac{2\zeta\beta}{1-\beta^2} = \text{phase}\quad 0 < \Phi < 180°\tag{4.84}$$

The ratio of the amplitude of the forced-vibration response to the static displacement is called the Dynamic Amplification Factor (DAF).

$$DAF = \frac{x_{st}\sqrt{\alpha}}{x_{st}} = \sqrt{\alpha} = \frac{1}{\sqrt{(1-\beta^2)^2 + (2\zeta\beta)^2}}\tag{4.85}$$

The DAF is a function of the frequency ratio and the damping (fig. 4.25). As we can see, if there is no damping and the frequency ratio is equal to one (resonance phenomenon), theoretically, the dynamic amplification becomes equal to infinity. This result corresponds to the result of example 2 and is strictly theoretical, since all systems have internal damping, even if minimal. The dynamic amplification at resonance is:

$$DAF_{max}\big|_{\beta=1} = \frac{1}{2\zeta}\tag{4.86}$$

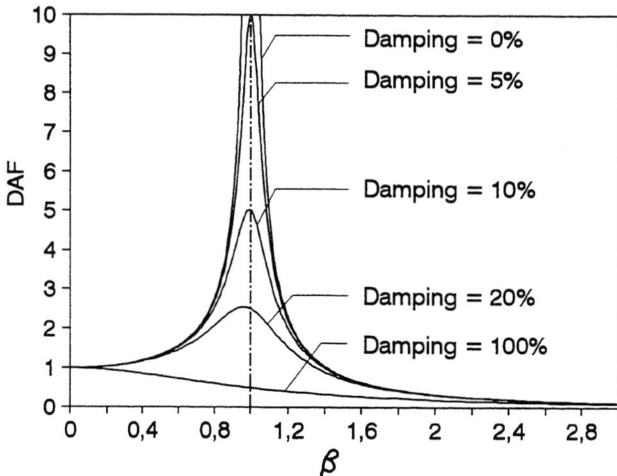

Figure 4.25 Frequency response of a damped SDOF system subjected to a sinusoidal load.

We also note in figure 4.25 that

$$DAF = 1 \quad \text{when } \beta = 0$$
$$DAF = 0 \quad \text{when } \beta = \infty$$

(4.87)

Physically, the system moves very slowly and strictly follows the applied load when the excitation frequency is very low compared to the natural frequency of the system. However, when the excitation frequency is very high compared to the natural frequency of the system, the mass of the system cannot "follow" the rapid fluctuations of the load; therefore, the system remains stationary.

Similarly, the phase angle, ϕ, is a function of β and the damping (fig. 4.26). At resonance, the phase angle is:

$$\Phi_{\beta=1} = \frac{\pi}{2}$$

(4.88)

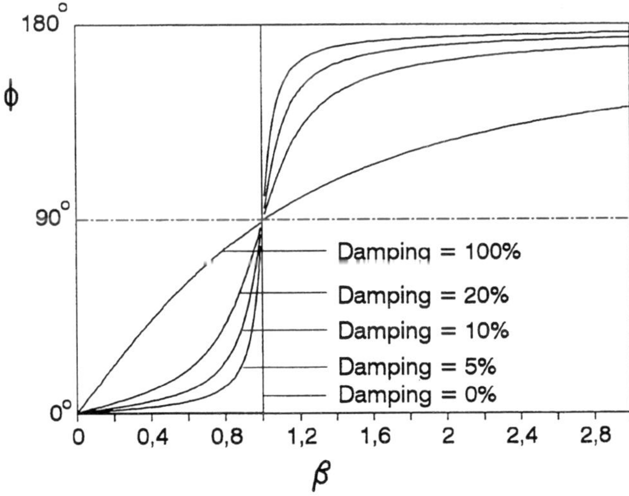

Figure 4.26   Variation of the phase angle with damping and the frequency ratio.

Estimation of damping by the half-bandwidth method. As seen in example 3 (equation 4.81), the frequency response of a SDOF system subjected to a sinusoidal dynamic load (fig. 4.25) is controlled mostly by the amount of damping in the system. This amount of damping can be estimated by building an experimental frequency response curve and by using the half-bandwidth method in the region of the resonance peak. To use this method, the frequency ratios are determined on each side of the resonance peak *(β = 1)*. The amplitude at these frequency ratios is equal to the amplitude at resonance divided by $\sqrt{2}$.

$$\frac{1}{\sqrt{2}}\frac{x_{st}}{2\,\zeta} = x_{st}\sqrt{\alpha} = x_{st}\sqrt{\frac{1}{\left(1-\beta^2\right)^2+\left(2\,\zeta\beta\right)^2}} \tag{4.89}$$

This equation is squared to yield:

$$\frac{1}{8\,\zeta^2} = \frac{1}{\left(1-\beta^2\right)^2+\left(2\,\zeta\beta\right)^2} \tag{4.90}$$

The square of the frequency ratio, β^2, can now be isolated.

$$\beta^2 = 1 - 2\,\zeta^2 \pm 2\,\zeta\sqrt{1+\zeta^2} \tag{4.91}$$

When the damping ratio is less than 20% critical, the term ζ^2 under the square root can be omitted, which yields two roots for the frequency ratio.

$$\begin{aligned}\beta_1 &\approx 1 - \zeta - \zeta^2\\ \beta_2 &\approx 1 + \zeta - \zeta^2\end{aligned} \tag{4.92}$$

By subtracting these two roots, the damping ratio can be estimated.

$$\zeta \approx \frac{\beta_2-\beta_1}{2} = \frac{1}{2}\left(\frac{\omega_2-\omega_1}{\omega_r}\right) \tag{4.93}$$

where ω_r is the resonant frequency and ω_1 and ω_2 are frequencies for which the amplitude of the forced vibrations response is equal to the amplitude at resonance divided by $\sqrt{2}$. Once the experimental frequency response curve is established, equation 4.93 can be applied directly, as shown in figure 4.27. However, the accuracy of the results greatly depends on the resolution of the frequency response curve close to the resonance (Wilson, 1991).

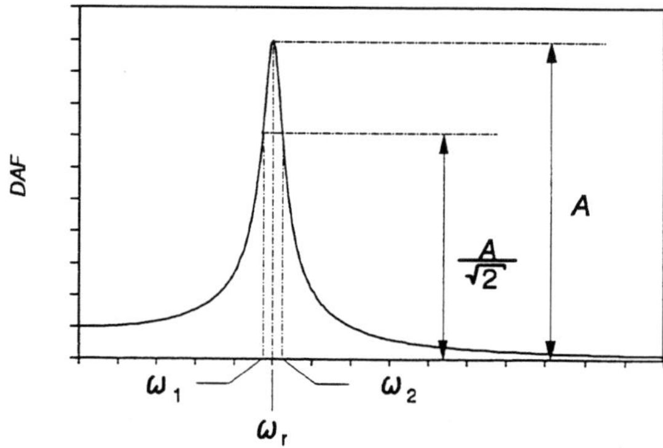

Figure 4.27 Use of the half-bandwidth method to estimate damping.

4.3 ELASTIC EARTHQUAKE RESPONSE SPECTRA

4.3.1 Definition

The response of a SDOF structure, caused by an earthquake accelerogram, $\ddot{x}_s(t)$, is obtained simply by replacing the dynamic load, $F(t)$, in equation 4.69 with a fictitious dynamic load equal to $-m\ddot{x}_s(t)$.

$$x(t) = -\frac{1}{\omega_d}\int_0^t \ddot{x}_s(\tau)\,e^{-\zeta\omega(t-\tau)}\sin\omega_d(t-\tau)\,d\tau \tag{4.94}$$

Usually, the accelerogram cannot be defined with a simple mathematical expression. Therefore, equation 4.94 requires a numerical integration. The response of a SDOF structure to a given accelerogram, $\ddot{x}_s(t)$, is a function of the damping, ζ, and the natural frequency, ω. For a given structure (ω and ζ), the absolute value of the maximum response, $|x|_{max}$, is evaluated using equation 4.94 for a specific accelerogram (earthquake) and plotted on a graph. Many coordinates are obtained on the graph by changing ω and ζ. The resulting graph is called a relative displacement response spectrum. In general, the response spectrum is a fundamental characteristic of the accelerogram in terms of the maximum response it produces on a linear SDOF system. There are different types of response spectra:

$$
\begin{aligned}
S_D &= |x|_{max} &&= \text{relative displacement response spectrum (spectral displacement)}\\
S_V &= |\dot{x}|_{max} &&= \text{relative velocity response spectrum (spectral velocity)}\\
S_{Ar} &= |\ddot{x}|_{max} &&= \text{relative acceleration response spectrum}\\
S_{Da} &= |x + x_s|_{max} &&= \text{absolute displacement response spectrum}\\
S_{Va} &= |\dot{x} + \dot{x}_s|_{max} &&= \text{absolute velocity response spectrum}\\
S_A &= |\ddot{x} + \ddot{x}_s|_{max} &&= \text{absolute acceleration response spectrum (spectral acceleration)}
\end{aligned}
$$

The earthquake spectra that are most useful in earthquake engineering are S_D, S_V and S_A. Figure 4.28 shows an example of these response spectra for one of the accelerograms recorded in Chicoutimi North during the 1988 Saguenay earthquake.

4.3.2 Properties of Response Spectra

Response spectra have the following properties:

1. They give the maximum response values of a SDOF system subjected to a given earthquake accelerogram.
2. They give the maximum response values in each mode of a MDOF system subjected to a given earthquake accelerogram (this will be discussed below).
3. They indicate the frequency distribution of the seismic energy of a given earthquake accelerogram, meaning that the response of a SDOF system is amplified when the seismic energy is close to its natural frequency.

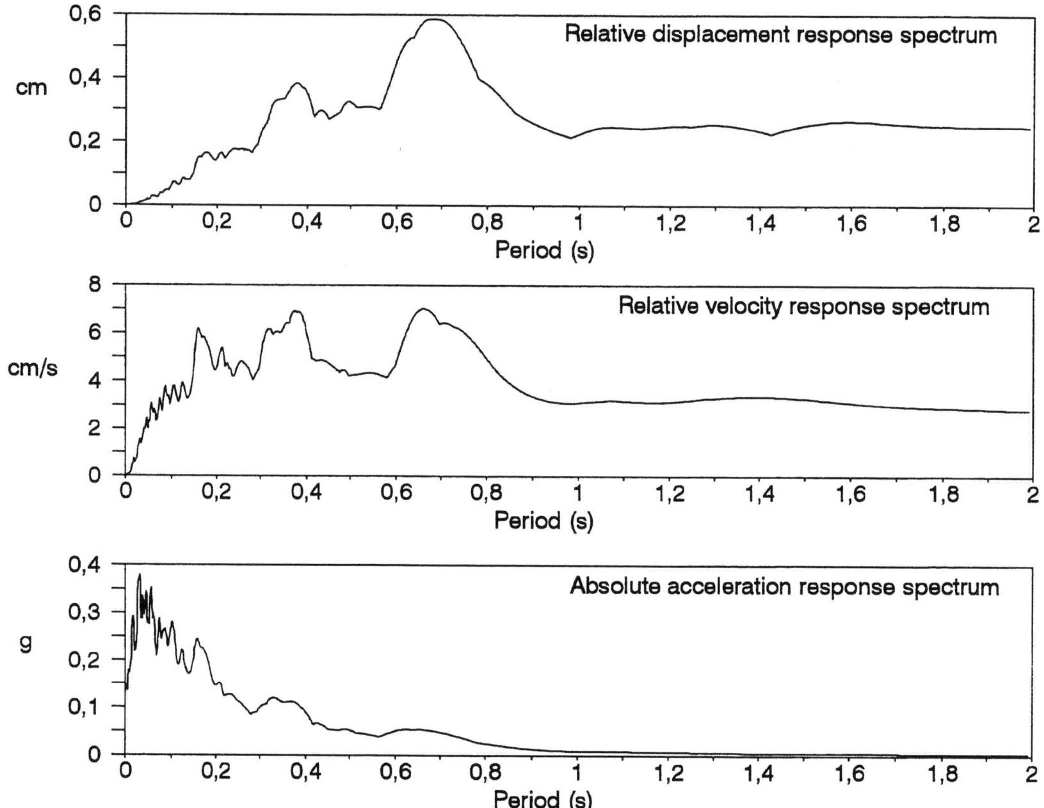

Figure 4.28   Response spectra at 5% damping for the N24E component of the accelerogram recorded at Chicoutimi North during the 1988 Saguenay earthquake.

4.3.3 Exact Response Spectra

The relative displacement response spectrum is obtained directly by Duhamel's integral given by equation 4.69:

$$S_D = |x|_{max} = \left| -\frac{1}{\omega_d} \int_0^t \ddot{x}_s(\tau) e^{-\zeta\omega(t-\tau)} \sin\omega_d(t-\tau)\,d\tau \right|_{max} \tag{4.95}$$

The acceleration and relative velocity are as follows:

$$\dot{x}(t) = \frac{dx(t)}{dt}$$

$$\ddot{x}(t) = \frac{d^2x(t)}{dt^2} \tag{4.96}$$

By convolution, if a time function, $F(t)$, is given by:

$$F(t) = \int_{u_0(t)}^{u_1(t)} f(t,\tau)d\tau \tag{4.97}$$

Then its time derivative becomes:

$$\frac{dF(t)}{dt} = \int_{u_0(t)}^{u_1(t)} \frac{\partial f(t,\tau)}{\partial t}d\tau + \frac{du_1(t)}{dt}f(u_1(t),\ t) - \frac{du_0(t)}{dt}f(u_0(t),\ t) \tag{4.98}$$

Applying convolution to Duhamel's integral given in equation 4.69, we have:

$$f(t,\tau) = -\frac{1}{\omega_d}\ddot{x}_s(\tau)e^{-\zeta\omega(t-\tau)}\sin\omega_d(t-\tau)$$
$$u_0(t) = 0 \tag{4.99}$$
$$u_1(t) = t$$

The relative velocity is then:

$$\dot{x}(t) = -\int_0^t \ddot{x}_s(\tau)e^{-\zeta\omega(t-\tau)}\cos\omega_d(t-\tau)d\tau$$
$$+ \frac{\zeta}{\sqrt{1-\zeta^2}}\int_0^t \ddot{x}_s(\tau)e^{-\zeta\omega(t-\tau)}\sin\omega_d(t-\tau)d\tau \tag{4.100}$$

The relative velocity response spectrum is given by:

$$S_V = |\dot{x}(t)|_{\max} \tag{4.101}$$

Similarly, the relative acceleration is obtained by differentiating equation 4.100 with respect to time:

$$\ddot{x}(t) = 2\zeta\omega\int_0^t \ddot{x}(\tau)e^{-\zeta\omega(t-\tau)}\cos\omega_d(t-\tau)d\tau$$
$$+ \frac{\omega(1-2\zeta^2)}{\sqrt{1-\zeta^2}}\int_0^t \ddot{x}_s(\tau)e^{-\zeta\omega(t-\tau)}\sin\omega_d(t-\tau)d\tau - \ddot{x}_s(t) \tag{4.102}$$

The absolute acceleration response spectrum is then:

$$S_A = |\ddot{x}(t) + \ddot{x}_s(t)|_{max} \tag{4.103}$$

The RESAS (REsponse Spectra of AccelerogramS) software available on the web site http://www.polymtl.ca/pub computes these exact spectral responses. The user's manual for this software is found in appendix A.

4.3.4 Pseudo Response Spectra

Usually, a civil-engineering structure has low damping (lower than 20% critical). The following hypotheses can then be made:

$$\begin{aligned} &\cdot\ \zeta,\ \zeta^2 \approx 0 \\ &\cdot\ \omega_d \approx \omega \\ &\cdot\ \cos\omega_d(t-\tau)\ \text{can be replaced by}\ \sin\omega_d(t-\tau) \end{aligned} \tag{4.104}$$

With these assumptions, equation 4.100 becomes:

$$\dot{x}(t) \approx -\int_0^t \ddot{x}_s e^{-\zeta\omega(t-\tau)} \sin\omega_d(t-\tau)d\tau = \omega x(t) \tag{4.105}$$

The pseudo relative velocity response spectrum is then:

$$S_V = \omega S_D \tag{4.106}$$

With the same assumptions, equation 4.102 is written:

$$\ddot{x}+\ddot{x}_s \approx \omega\int_0^t \ddot{x}_s e^{-\zeta\omega(t-\tau)} \sin\omega_d(t-\tau)d\tau = -\omega^2 x(t) \tag{4.107}$$

The pseudo absolute acceleration response spectrum becomes:

$$S_A = \omega^2 S_D = \omega S_V \tag{4.108}$$

4.3.5 Comparison between Exact and Pseudo Response Spectra

Comparing the exact response spectra (equations 4.100, 4.102) with the pseudo response spectra (equations 4.106, 4.108) for different accelerograms yields the following tendencies:

- In a system with zero damping, the results are essentially identical for natural periods of less than one second ($T < 1$ s).
- When damping increases to 20% critical, the differences are within 20% but without any observable bias.

- The pseudo acceleration response spectrum is generally more precise than the pseudo velocity response spectrum.

These variations are within the acceptable range expected from a seismic analysis. However, pseudo response spectra must not be used for highly damped systems ($> 20\%$ critical) or for systems with long natural periods ($T >> 1$ s). Figure 4.29 compares exact response spectra and pseudo response spectra for one of the accelerograms recorded in Chicoutimi North during the 1988 Saguenay earthquake.

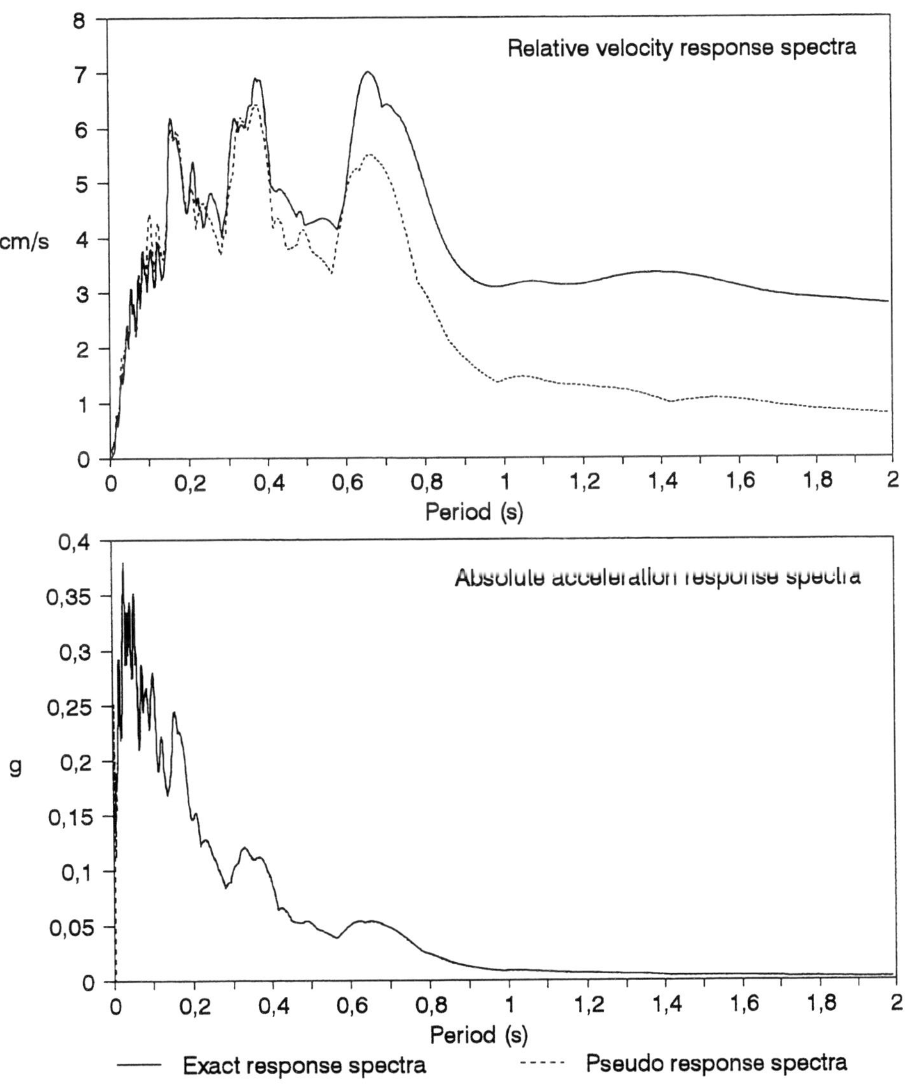

Figure 4.29 Exact response spectra and pseudo response spectra at 5% damping for the N24E component of the accelerogram recorded at Chicoutimi North during the 1988 Saguenay earthquake.

4.3.6 Tripartite Representation of Pseudo Response Spectra

In practice, the response spectra are represented by a graph with multiple logarithmic scales. This type of graph, as shown in figure 4.30, is called a tripartite graph. The advantage of the tripartite graph is that the same curve can display the following information:

- the exact relative displacement response spectrum;
- the pseudo relative velocity response spectrum;
- the pseudo absolute acceleration response spectrum.

To understand the tripartite graph, consider the variation of $\log_{10}S_v$ in function of $\log_{10}T$ for constant values of S_A or S_D.

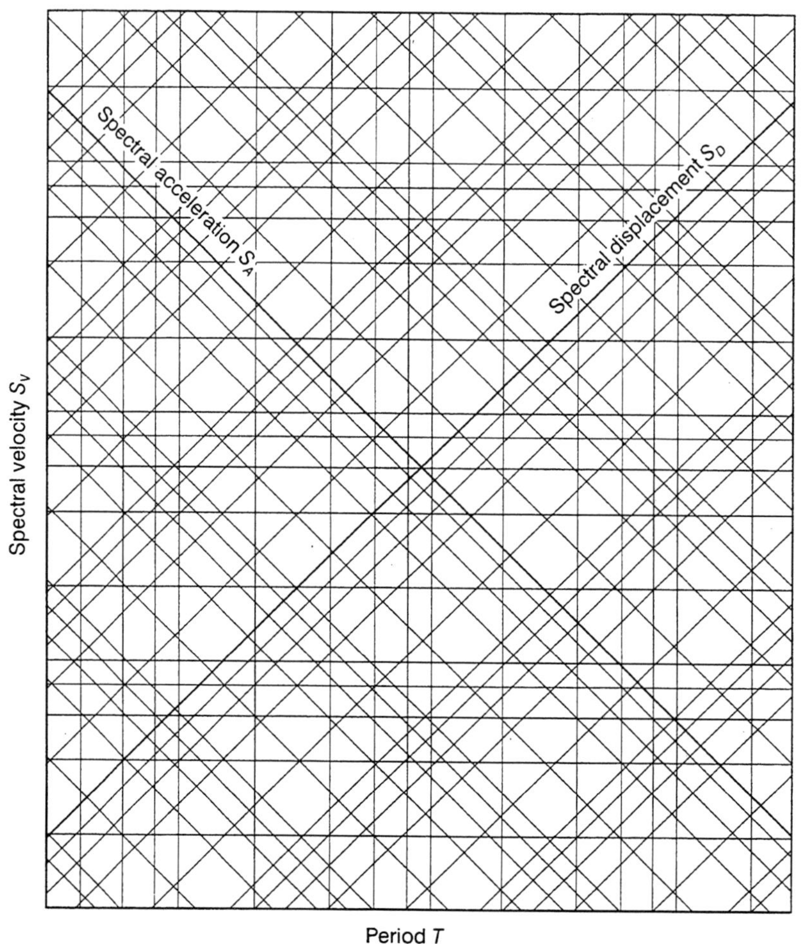

Figure 4.30   Tripartite representation of response spectra.

a) $S_A = \text{constant} = C_1$

If the pseudo acceleration response spectrum is equal to a constant, C_1, it can be written:

$$S_A = C_1 = \omega^2 S_D = \omega(\omega S_D) = \frac{2\pi}{T} S_V \tag{4.109}$$

Operating with the $\log_{10}$ on this equation yields:

$$\log_{10} C_1 = \log_{10} 2\pi - \log_{10} T + \log_{10} S_V \tag{4.110}$$

or

$$\log_{10} S_V = \log_{10} T + \left[\log_{10} C_1 - \log_{10} 2\pi\right] \tag{4.111}$$

then

$$\frac{d\left(\log_{10} S_V\right)}{d\left(\log_{10} T\right)} = +1 \tag{4.112}$$

This result indicates that a line at $+45°$ on the tripartite graph represents a constant spectral acceleration, S_A.

b) $S_D = \text{constant} = C_2$

If the relative displacement response spectrum is equal to a constant, C_2, it can be written:

$$S_D = C_2 = \frac{S_V}{\omega} = \frac{T}{2\pi} S_V \tag{4.113}$$

Operating with the $\log_{10}$ on this equation yields:

$$\log_{10} C_2 = \log_{10} T - \log_{10} 2\pi + \log_{10} S_V \tag{4.114}$$

or

$$\log_{10} S_V = -\log_{10} T + \left[\log_{10} C_2 + \log_{10} 2\pi\right] \tag{4.115}$$

then

$$\frac{d\left(\log_{10} S_V\right)}{d\left(\log_{10} T\right)} = -1 \tag{4.116}$$

This result indicates that a line at -45° on a tripartite graph represents a constant relative displacement spectrum, S_D. Figure 4.31 illustrates, for different damping values, the response spectra of the El Centro earthquake (1940-05-18, comp. S00E). Figure 4.32 represents a normalized form of the same spectra. On this graph, all scales (displacement, velocity, and acceleration) are divided by the maximum ground components. As we can see, the response spectrum of an earthquake is very irregular. Nevertheless, the spectrum has a general trapezoidal (tent) shape. This shape is a characteristic of earthquake response spectra and has the following physical explanation:

- For low natural frequencies, the maximum relative displacement is equal to the maximum ground displacement and the maximum absolute acceleration tends toward zero.
- For intermediate natural frequencies, the relative displacement, relative velocity, and absolute acceleration are amplified.
- For high natural frequencies, the maximum absolute acceleration is equal to the maximum ground acceleration and the maximum relative displacement tends toward zero.

The results shown in figure 4.31 are typical of other base excitations.

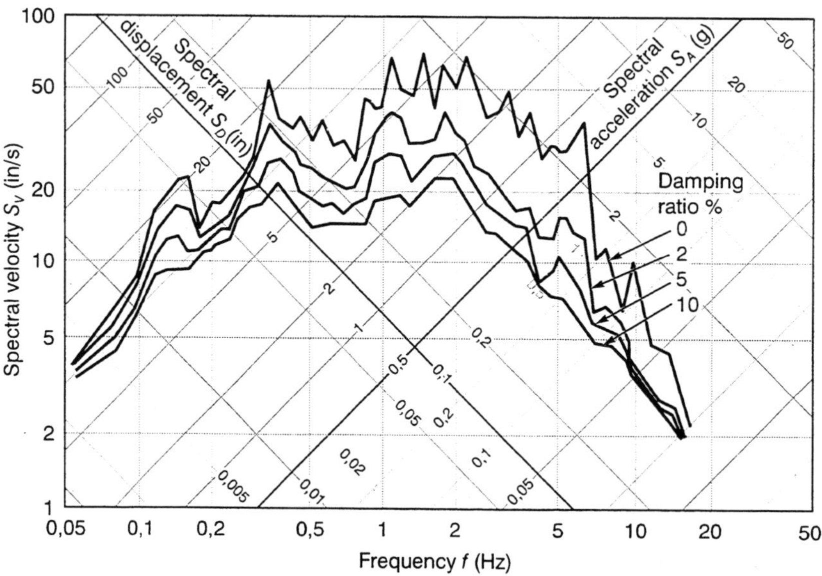

Figure 4.31   Response spectra of the El Centro earthquake (1940-05-18, comp. S00E).

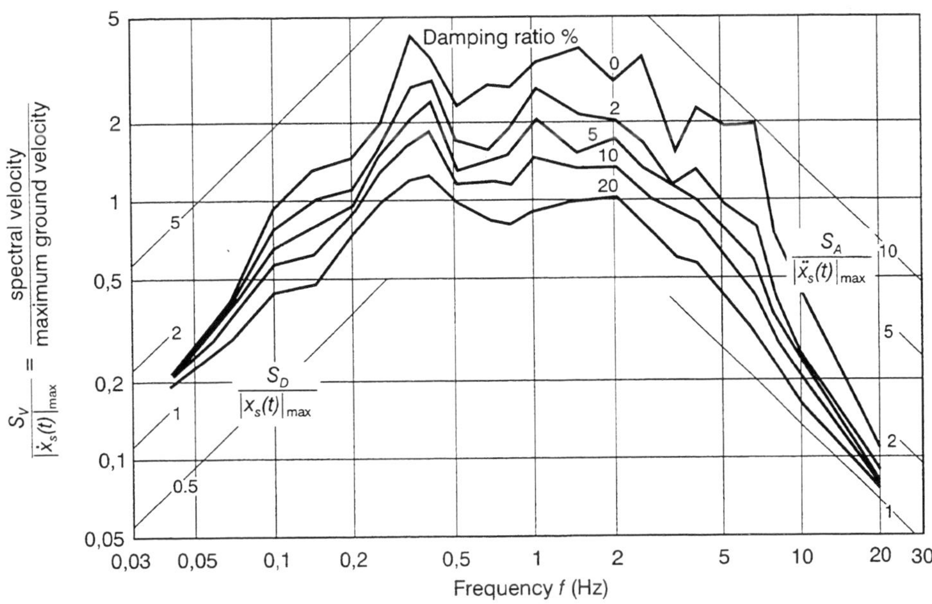

Figure 4.32   Normalized response spectra for the El Centro earthquake (1940-05-18, comp. S00E).

4.4   SIMPLIFIED DESIGN RESPONSE SPECTRA

4.4.1 Motivation

For the practical seismic design of structures, simplified response spectra are used. The different regions of the simplified spectra are represented by straight lines. The position of these lines – their amplitude – is a function of the seismic hazard of the region. Many different simplified design response spectra have been proposed. The most common are described below.

4.4.2 Housner's Response Spectra (1959)

Relying on response spectra obtained for four historical earthquakes in southern California, Housner (1959) proposed, for the first time, the average design spectrum shown in figure 4.33. This spectrum was calibrated for a maximum ground acceleration of 0,20g and for a probability of excedance of 50% – in other words, for the average of the historical spectral values. The values obtained from this spectrum are to be multiplied by a scale factor to take into account the seismic hazard of the region. For example, if the design earthquake of a given site is 0,15g, then the spectral acceleration will be:

$$S_A \text{ is equal to } \left(\frac{15}{20}\right) \text{ of Housner's } S_A \tag{4.117}$$

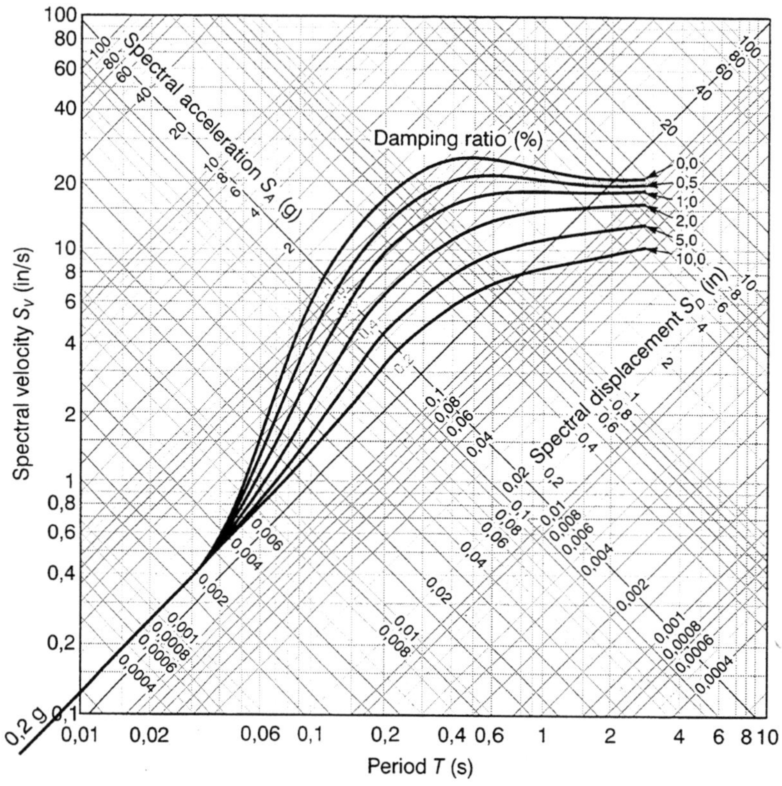

Figure 4.33   Housner's average response spectra (from Housner, 1970).

4.4.3 Newmark and Hall's Response Spectra (1969)

Figure 4.34 illustrates the simplified spectra proposed by Newmark and Hall (1969) based on standard ground-motion parameters. These parameters are:

- maximum ground acceleration: 0,50 g;
- maximum ground velocity: 61 cm/s (24 in/s);
- maximum ground displacement: 46 cm (18 in).

These values, based on many earthquake records, represent the average relation between different ground seismic parameters. For a given site, they are scaled directly to the maximum design acceleration, which is a function of the seismic hazard of the region. The simplified spectrum is obtained by multiplying each branch of the ground parameters by an amplification factor that depends on the damping coefficient of the structure, as shown in table 4.2.

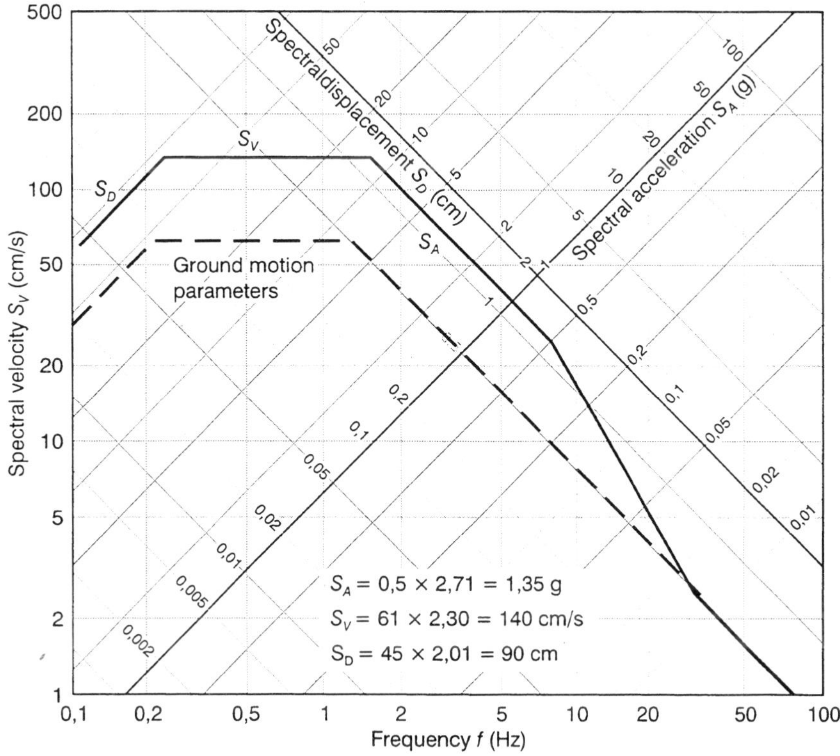

Figure 4.34 Newmark and Hall's design response spectra (1969).

Table 4.2 Amplification factors for Newmark and Hall's design response spectra (from Newmark and Hall, 1982).

% of critical damping	Probability of excedance of 16% (mean + 1 standard deviation)			Probability of excedance of 50% (mean value)		
	S_A	S_V	S_D	S_A	S_V	S_D
0,5	5,10	3,84	3,04	3,68	2,59	2,01
1,0	4,38	3,38	2,73	3,21	2,31	1,82
2,0	3,66	2,92	2,42	2,74	2,03	1,63
3,0	3,24	2,64	2,24	2,46	1,86	1,52
5,0	2,71	2,30	2,01	2,12	1,65	1,39
7,0	2,36	2,08	1,85	1,89	1,51	1,29
10,0	1,99	1,84	1,69	1,64	1,37	1,20
20,0	1,26	1,37	1,38	1,17	1,08	1,01

For the seismic design of nuclear plants, Newmark and Hall recommended the use of amplification factors corresponding to a probability of excedance of 16%. The procedure for plotting Newmark's simplified spectra on a tripartite graph is described in the following steps:

- *Step 1: Plot the ground-motion parameters*

The limits of the ground-motion parameters are linked by straight lines: maximum horizontal acceleration, maximum horizontal velocity, and maximum horizontal displacement. If the maximum horizontal acceleration is the only known parameter at the site, the standard ground motion parameters can be used with this maximum design acceleration. For example, if the maximum design acceleration of 0,33g, at the site, is the only known parameter, it yields:

$$\text{ground acceleration} = \left(\frac{0,33}{0,50}\right)(0,50\,g) = 0,33\,g$$

$$\text{ground velocity} = \left(\frac{0,33}{0,50}\right)(61\,\text{cm/s}) = 40,26\,\text{cm/s} \tag{4.118}$$

$$\text{ground displacement} = \left(\frac{0,33}{0,50}\right)(46\,\text{cm}) = 30,36\,\text{cm}$$

- *Step 2: Plot the different regions of the spectrum*

The amplification factors shown in table 4.2 are used to plot the different regions of the simplified spectrum. Table 4.3 shows recommended damping values to be used.

- *Step 3: Modify the spectral limits for high frequencies*

We find the corner frequency, f_1, which links the velocity branch to the acceleration branch. At a frequency of about $4f_1$, we start reducing linearly the acceleration branch of the spectrum until we reach the limit of the ground acceleration for a frequency of $10f_1$. Theoretically, the displacement branch should also be modified for low frequencies (below 0,1 Hz). But, as these low frequencies have little impact on civil-engineering structures, this modification is omitted.

Table 4.3 Recommended damping values.

Strain level	Types of structures and conditions	% of critical damping
Less than 50% of the elastic limit	welded steel, prestressed concrete, reinforced concrete with light cracking	2 to 3
	reinforced concrete with heavy cracking	3 to 5
	bolted or riveted steel, nailed or bolted timber	5 to 7
Close to or over the elastic limit	welded steel, prestressed concrete without complete loss of prestressing	5 to 7
	prestressed concrete with prestressing loss	7 to 10
	reinforced concrete	7 to 10
	bolted or riveted steel, bolted timber	10 to 15
	nailed timber	15 to 20

4.4.4 Design Response Spectra of the National Building Code of Canada (NBCC, 1995)

The 1995 edition of the NBCC proposes a design-response spectrum for the different regions of Canada when a structure requires spectral analysis. Figure 4.35 shows this spectrum based on 5% critical damping and maximum ground horizontal velocity of 1 m/s. To determine the design spectrum for a given site, each branch of the spectrum (fig. 4.35) must be scaled to the maximum horizontal design velocity for the site. The maximum horizontal design velocity, v, can be found for all regions of Canada, in appendix C of the NBCC or in the seismic zoning maps shown in figure 3.7. Figure 4.35 also shows that, for $Z_a = Z_v$, the acceleration branch is equal to 3 g, and the velocity branch is equal to 2 m/s. These values represent amplification factors of 3 for acceleration and 2 for velocity. In comparison with table 4.2, the code spectrum for $Z_a = Z_v$ is very similar to Newmark and Hall's spectrum for 5% damping and for a probability of excedance of 16%. In chapter 6, we will discuss further the use of the NBCC design-response spectra.

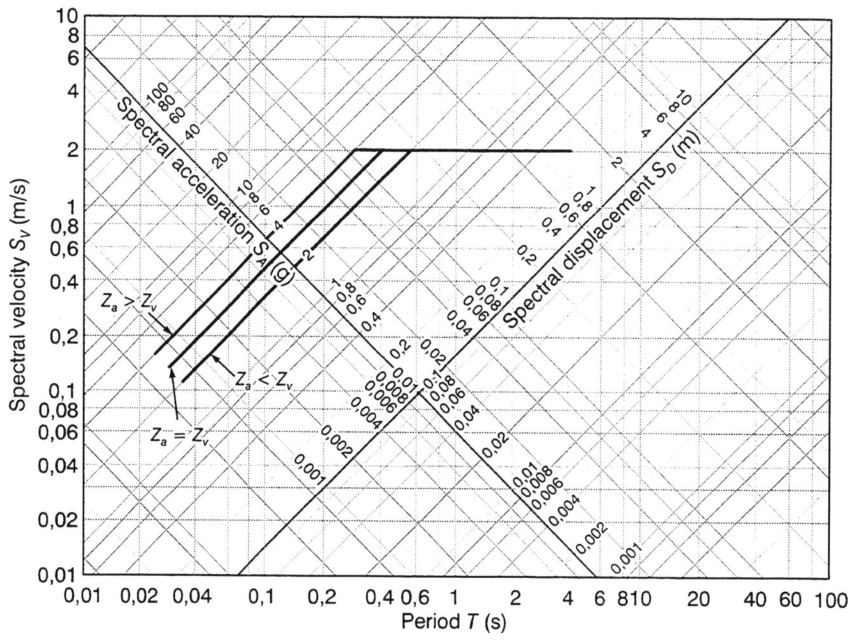

Figure 4.35 Design-response spectra (NBCC, 1995); maximum horizontal ground velocity: 1 m/s, maximum horizontal ground acceleration: 1 g.

4.4.5 Floor Response Spectra

The response spectra discussed above can be used to determine the maximum response of a SDOF structure subjected to a base motion. Similarly, the maximum response of equipment located in a building can be obtained using the response spectrum corresponding to the floor where the equipment is located. Clearly, the vibration of a complex building varies from storey to storey, thus creating a variation in the response spectra of the various floors.

The traditional technique used to generate a floor response spectrum is to calculate the historical horizontal acceleration of this floor, using techniques discussed below, and then use this accelerogram to construct a response spectrum. If a simplified design response spectrum of a floor is to be constructed, the procedure starts with a group of accelerograms at the base, and the resulting spectra are smoothed. Because of the large quantity of calculations required to generate a floor response spectrum, approximate methods have been proposed (Biggs and Roesset, 1970; Singh, 1975).

4.5   APPLICATION OF RESPONSE SPECTRA CONCEPTS IN THE NATIONAL BUILDING CODE OF CANADA (NBCC, 1995)

Figure 4.36 shows the equilibrium of horizontal forces (leaving aside damping) of a simple portal frame subjected to a seismic lateral load.

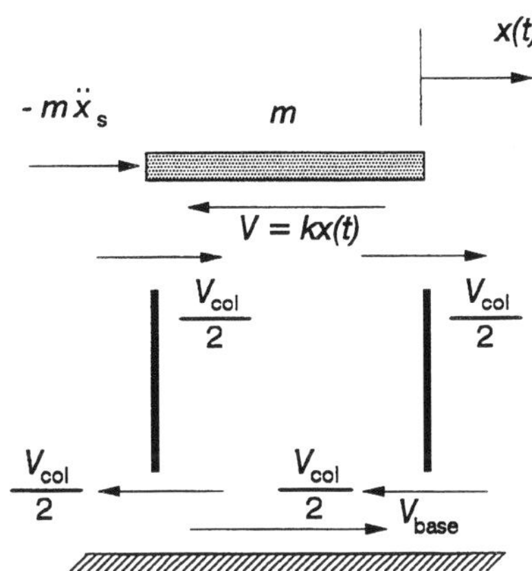

Figure 4.36   Equilibrium of horizontal forces for a portal frame subjected to a seismic lateral load.

The maximum base shear is obtained by:

$$V_{base}\big|_{max} = V_{col}\big|_{max} = k\,|x|_{max} = k S_D \tag{4.119}$$

or

$$V_{base} = \frac{k S_A}{\omega^2} = k S_A \frac{m}{k} = \frac{S_A}{g} W = C W \tag{4.120}$$

The factor C, called "seismic coefficient" by the NBCC, is a function of the fundamental period of the structure. As we will see in chapter 6, the formula included in the NBCC (1995) for the base shear of an elastic structure is:

$$V_{base} = V = v S I F W \tag{4.121}$$

The factor C is associated to the factors v (maximum horizontal velocity) and S (seismic coefficient) (equation 4.120). Factors I and F are included to take into account the importance of the building (I) and the soil amplification (F). It is important to note that the seismic coefficient, vS, is a response spectrum.

4.6 EXAMPLE OF APPLYING RESPONSE SPECTRA

The simple portal frame shown in figure 4.37 is subjected to an earthquake identical to the El Centro earthquake (1940-05-18, comp. S00E). We need to calculate the maximum displacement of the frame and the maximum base shear, supposing a damping of 2% critical.

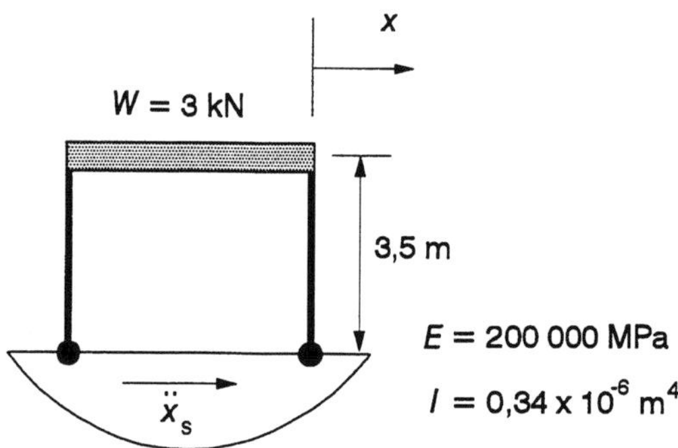

Figure 4.37 Simple portal frame subjected to the El Centro earthquake (1940-05-18, comp. S00E).

The lateral stiffness coefficient is obtained by:

$$k = 2\left[\frac{3EI}{L^3}\right] = \frac{6 \times 200 \times 10^6 \times 0,34 \times 10^{-6}}{(3,5)^3} = 9,52\,\text{kN/m} = 9520\,\text{N/m} \qquad (4.122)$$

The equivalent mass is:

$$m = \frac{3000\,\text{N}}{9,81\,\text{m/s}^2} = 305,81\,\frac{\text{N}-\text{s}^2}{\text{m}} \qquad (4.123)$$

The natural frequency looks like:

$$\omega = \sqrt{\frac{k}{m}} = \sqrt{\frac{9520}{305,81}} = 5,5\,\text{rad/s}$$

$$f = \frac{\omega}{2\pi} = 0,88\,\text{Hz} \qquad (4.124)$$

$$T = \frac{1}{f} = 1,14\text{s}$$

According to the response spectrum in figure 4.31, for a damping of 2% critical, the spectral values are:

$$S_V = 26 \text{ in/s} = 0,66 \text{ m/s}$$
$$S_A = 0,4 \text{ g} = 3,92 \text{ m/s}^2 \qquad (4.125)$$
$$S_D = 5 \text{ in} = 0,13 \text{ m}$$

The maximum relative displacement of the storey is equal to:

$$x_{\text{max}} = S_D = 5 \text{ in} = 0,13 \text{ m} \qquad (4.126)$$

The maximum base shear is obtained by:

$$V_{\text{base}}\big|_{\text{max}} = \frac{S_A}{g} W = \left[\frac{3,92}{9,81}\right] \times 3 = 1,2 \text{ kN} \qquad (4.127)$$

4.7 "EXACT" DYNAMIC ANALYSIS OF MULTI-DEGREE-OF-FREEDOM STRUCTURES

4.7.1 Introduction

In more complex structures, it is necessary to use a multi-degree-of-freedom (MDOF) system to obtain an adequate dynamic model. This section discusses the dynamic analysis of a linear MDOF system. In dynamic analysis, the term "exact" assumes that all of the structures' characteristics and the dynamic excitation (forces or ground motion) are precisely known. The responses obtained earlier for SDOF systems will be used and combined to obtain the total dynamic response of the MDOF system. Furthermore, this section focuses only on systems with concentrated masses; we assume that the concentrated masses attract the most important inertial forces and that the system has a finite number of DDOF. We will thus ignore systems with distributed masses that include an infinite number of DDOF; an excellent description of those systems can be found in Clough and Penzien (1993).

4.7.2 Equations of Motion for a Multi-Degree-of-Freedom Structure

There are N equations of motion for a system with N DDOF. These equations can be written as matrices of $N \times N$. As for the SDOF system, the dynamic excitation can either be an arbitrary dynamic loading or a base excitation.

Arbitrary dynamic loading. Consider an undamped system with two degrees of freedom, as shown in figure 4.38. The free-body diagrams for the two masses are also shown in figure 4.39. Applying Newton's second law of motion to each mass yields:

$$m_1 \ddot{x}_1 + k_1 x_1 - k_2 (x_2 - x_1) = F_1(t)$$
$$m_2 \ddot{x}_2 + k_2 (x_2 - x_1) = F_2(t) \qquad (4.128)$$

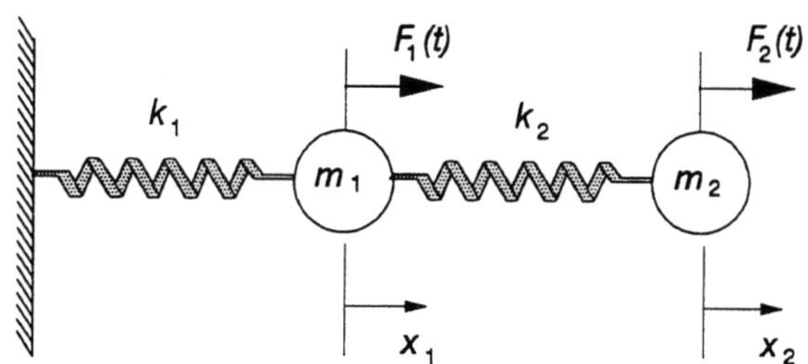

Figure 4.38   System with two degrees of freedom.

In matrix form, we write:

$$\begin{bmatrix} m_1 & 0 \\ 0 & m_2 \end{bmatrix}\begin{bmatrix} \ddot{x}_1 \\ \ddot{x}_2 \end{bmatrix} + \begin{bmatrix} (k_1+k_2) & -k_2 \\ -k_2 & k_2 \end{bmatrix}\begin{bmatrix} x_1 \\ x_2 \end{bmatrix} = \begin{bmatrix} F_1(t) \\ F_2(t) \end{bmatrix} \tag{4.129}$$

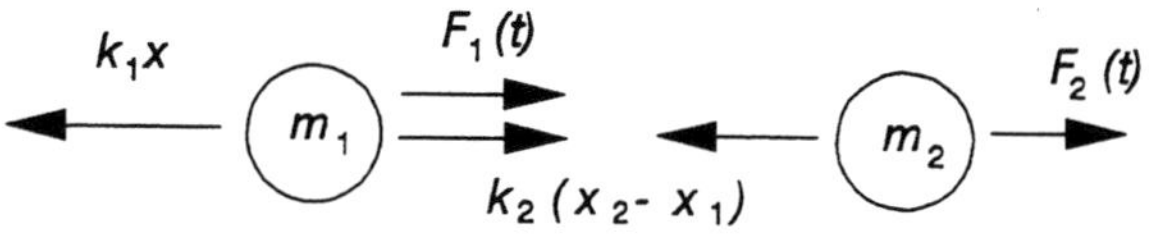

Figure 4.39   Free-body diagrams for a system with two degrees of freedom.

Equation 4.129 is also written:

$$[m](\ddot{x}) + [k](x) = (F(t)) \tag{4.130}$$

But if viscous damping is integrated into the model, a damping matrix $[c]$ is introduced and equation 4.130 becomes:

$$[m](\ddot{x}) + [c](\dot{x}) + [k](x) = (F(t)) \tag{4.131}$$

where $[m]$ =   global mass matrix (usually diagonal)
   $[c]$ =   global damping matrix (difficult to evaluate)
   $[k]$ =   global stiffness matrix
   $(F(t))$ =   dynamic loading vector
   (x), $(\dot{x})$, and $(\ddot{x})$ =   relative displacement, relative velocity, and relative acceleration vectors

Equation 4.131 represents a system of linear differential equations governing the movement of a MDOF system. This system of equations describes the fundamental relations to be solved for an arbitrary dynamic load vector, $[F(t)]$. It is important to note that equation 4.131 supposes that the properties of the system ($[m]$, $[k]$, $[c]$) do not change during the system's response (vibration). In other words, the system is linear. The elastic limit of the system's elements will never be exceeded; as a result, yielding is not considered.

Base excitation (earthquake problem). To determine the equations of motion, let us consider the same two-degrees-of-freedom system subjected to a base displacement, $x_s(t)$ (fig. 4.40). Applying Newton's second law to each mass yields:

$$\begin{aligned}
m_1(\ddot{x}_1 + \ddot{x}_s) + k_1 x_1 - k_2(x_2 - x_1) &= 0 \\
m_2(\ddot{x}_2 + \ddot{x}_s) + k_2(x_2 - x_1) &= 0
\end{aligned} \tag{4.132}$$

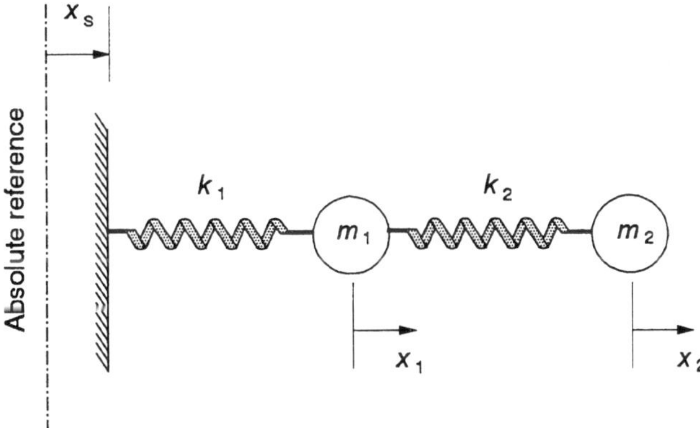

Figure 4.40   Two-degrees-of-freedom system subjected to a base displacement.

In matrix form, we write:

$$\begin{bmatrix} m_1 & 0 \\ 0 & m_2 \end{bmatrix} \begin{bmatrix} \ddot{x}_1 \\ \ddot{x}_2 \end{bmatrix} + \begin{bmatrix} (k_1+k_2) & -k_2 \\ -k_2 & k_2 \end{bmatrix} \begin{bmatrix} x_1 \\ x_2 \end{bmatrix} = -\begin{bmatrix} m_1 & 0 \\ 0 & m_2 \end{bmatrix} \begin{bmatrix} 1 \\ 1 \end{bmatrix} \ddot{x}_s \tag{4.133}$$

or

$$[m](\ddot{x}) + [k](x) = -[m](r)\ddot{x}_s \tag{4.134}$$

If viscous damping is introduced into this model, it yields:

$$[m](\ddot{x}) + [c](\dot{x}) + [k](x) = -[m](r)\ddot{x}_s \qquad (4.135)$$

where
$$
\begin{aligned}
[m] &= \text{global mass matrix (usually diagonal)} \\
[c] &= \text{global damping matrix (difficult to evaluate)} \\
[k] &= \text{global stiffness matrix} \\
(x),\ (\dot{x}),\ \text{and}\ (\ddot{x}) &= \text{relative displacement, relative velocity and relative} \\
&\quad\ \text{acceleration vectors} \\
\ddot{x}_s &= \text{absolute base acceleration} \\
(r) &= \text{dynamic coupling vector}
\end{aligned}
$$

The dynamic coupling vector, (r), links the direction of the base motion with the direction of each DDOF when the structure moves as a rigid body. In general, this vector is made of 1 and 0. For example, the dynamic coupling vector for the corresponding DDOF of figure 4.41 is:

$$(r) = \begin{bmatrix} 1 \\ 1 \\ 0 \end{bmatrix} \qquad (4.136)$$

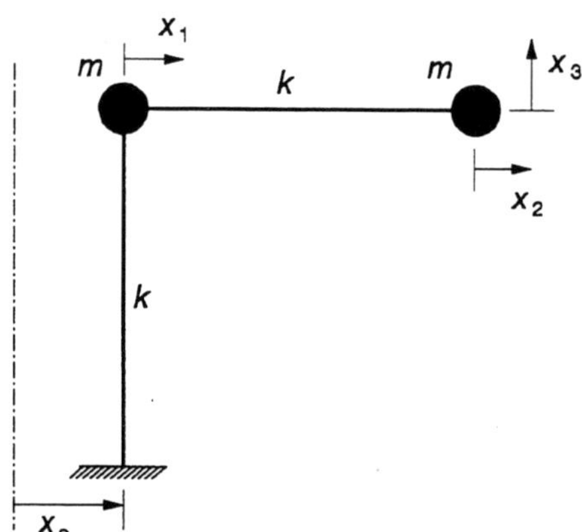

Figure 4.41   Three-degrees-of-freedom system subjected to horizontal base motion.

Note that equation 4.135 is identical to equation 4.131 except for the dynamic load vector, $(F(t))$, which is replaced by a fictitious dynamic load vector: $-(m)\,(r)\,\ddot{x}_s(t)$.

4.7.3 Stiffness Matrix Computation

The stiffness matrix, $[k]$, shown in equations 4.131 and 4.135, can be evaluated by standard structural analysis techniques (see, for example, Laursen, 1978). The elements of a stiffness matrix are called k_{ij} and are defined as the necessary force applied to the DDOF i to obtain a unit displacement at the DDOF j (all other DDOF have no displacement).

In the static problem, the stiffness matrix is defined as:

$$[k](x) = (F) \tag{4.137}$$

where $[k]$ = global stiffness matrix
(x) = static displacement vector
(F) = static load vector

In many instances, it is easier to evaluate the flexibility matrix of the system, $[f]$, and inverse it to obtain the stiffness matrix of the system, $[k]$. Using Betti-Maxwell's theorem, it can be shown (Laursen, 1978) that:

$$\begin{aligned}
[k] &= [f]^{-1} \\
[f] &= [k]^{-1} \\
[k] &= [k]^{T} \\
[f] &= [f]^{T}
\end{aligned} \tag{4.138}$$

The elements of a flexibility matrix are called f_{ij} and are defined as the displacement of the DDOF i resulting from a unit force applied to the DDOF j (all other DDOF have zero force). In the static problem, the flexibility matrix is defined as:

$$[f](F) = (x) \tag{4.139}$$

where $[f]$ = global flexibility matrix
(x) = static displacement vector
(F) = static loading vector

4.7.4 Mass Matrix Computation

The simplest method for evaluation of the mass matrix of a system lies in the idea that the totality of the mass is lumped at the DDOF's locations (nodes). The usual procedure considers the structure in sections between the points of the DDOF; therefore, the mass of each section is equally distributed at the nodes. Figure 4.2 illustrates the procedure used for a beam with three DDOF.

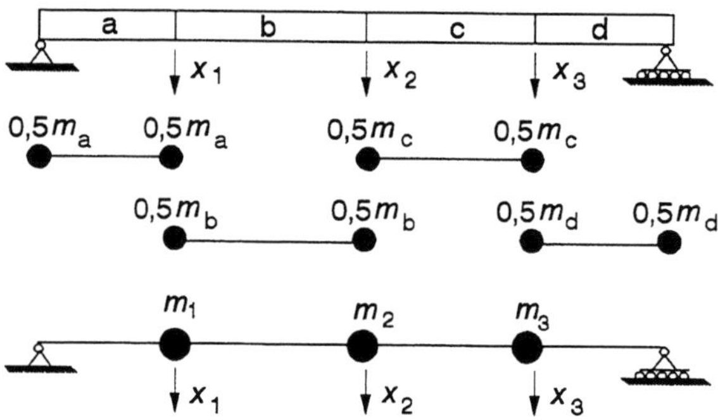

Figure 4.42 Lumped mass distribution for a beam with three DDOF.

For a system with N translating DDOF (no rotating DDOF), the mass matrix will be diagonal and on the order of $N \times N$:

$$[m] = \begin{bmatrix} m_1 & 0 & 0 & \cdots & 0 & \cdots & 0 \\ 0 & m_2 & 0 & \cdots & 0 & \cdots & 0 \\ 0 & 0 & m_3 & \cdots & 0 & \cdots & 0 \\ \cdots & \cdots & \cdots & \cdots & \cdots & \cdots & \cdots \\ 0 & 0 & 0 & \cdots & m_i & \cdots & 0 \\ \cdots & \cdots & \cdots & \cdots & \cdots & \cdots & \cdots \\ 0 & 0 & 0 & \cdots & 0 & \cdots & m_N \end{bmatrix} \qquad (4.140)$$

If more than one translating DDOF is specified for the same node (for example, vertical and horizontal), the same mass is associated with each DDOF. However, the mass associated with a rotating DDOF will be zero, as the masses are concentrated and have no rotating inertia. It is always possible, however, to calculate the mass moment of inertia of a segment of a system and to define it as a concentrated mass. Another method for evaluation of the mass matrix, based on energy concepts (Cook, 1981), results in a consistent mass matrix. This method is more complex but ensures a convergence of dynamic response in terms of energy.

4.7.5 Computation of Damping Matrix

The damping forces applied on a structure are impossible to evaluate precisely. So, because of its mathematical convenience, the simple model of viscous damping is adopted. According to this model, the coefficients of the damping matrix, $[c]$, are generally defined in terms of modal damping. As we will see below, the form of damping matrix is chosen in order to solve the system easily.

4.7.6 Examples

Example 1

Calculate the stiffness matrix and mass matrix of the spring-mass system with two degrees of freedom (figure 4.43).

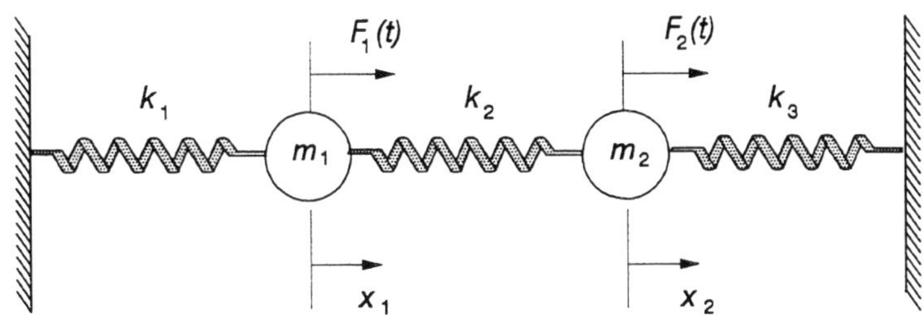

Figure 4.43   Spring-mass system with two degrees of freedom.

As the DDOF are only translating, the mass matrix will be diagonal:

$$[m] = \begin{bmatrix} m_1 & 0 \\ 0 & m_2 \end{bmatrix} \tag{4.141}$$

The stiffness coefficients, k_{ij} , are evaluated by imposing a unit displacement on each DDOF. To impose a unit displacement of only x_1, the necessary forces k_{11} and k_{21} at the DDOF x_1 and x_2 are:

$$\begin{aligned} k_{11} &= k_1 + k_2 \\ k_{21} &= -k_2 \end{aligned} \tag{4.142}$$

To induce a unit displacement of only x_2, the necessary forces k_{12} and k_{22} at the DDOF x_1 and x_2 are:

$$\begin{aligned} k_{12} &= -k_2 \\ k_{22} &= k_2 + k_3 \end{aligned} \tag{4.143}$$

The stiffness matrix is then:

$$[k] = \begin{bmatrix} (k_1 + k_2) & -k_2 \\ -k_2 & (k_2 + k_3) \end{bmatrix} \tag{4.144}$$

Example 2

Calculate the mass matrix and the stiffness matrix for a cable in tension, as shown in figure 4.44. The cable is brought to tension T_o and supports three masses. The mass matrix is diagonal:

$$[m] = \begin{bmatrix} m & 0 & 0 \\ 0 & 3m & 0 \\ 0 & 0 & 2m \end{bmatrix} \qquad (4.145)$$

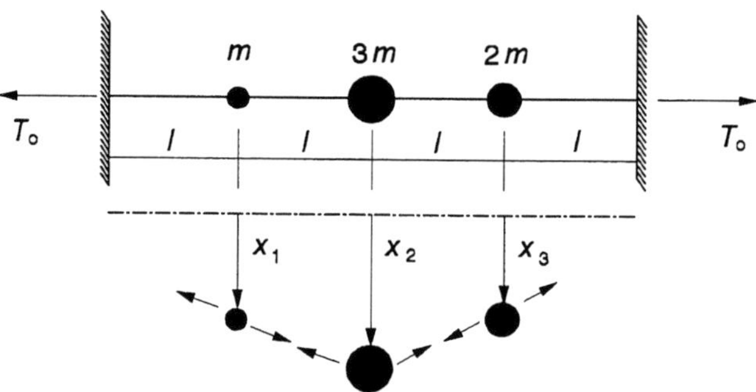

Figure 4.44 Cable under tension supporting three masses, represented by a system of three DDOF.

The free-body diagrams corresponding to unit displacements of each DDOF are shown in figure 4.45.

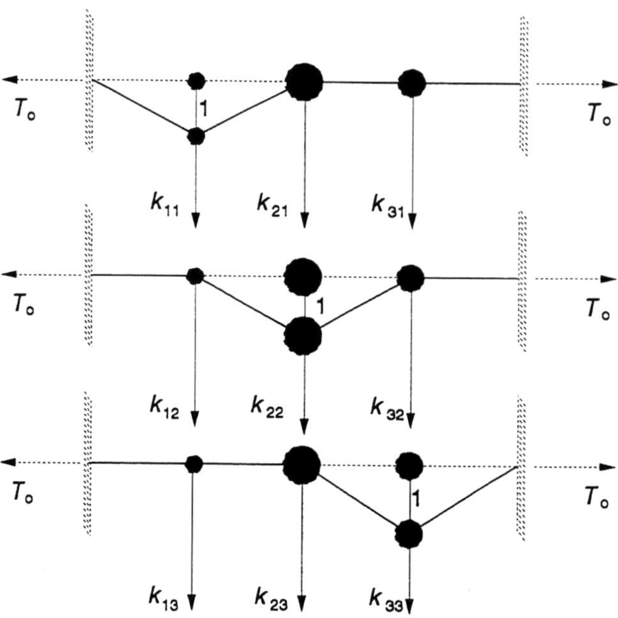

Figure 4.45 Computation of the stiffness coefficients for a cable in tension.

The stiffness coefficients are the following:

$$k_{11} = k_{22} = k_{33} = \frac{2T_o}{\sqrt{1+l^2}} \approx \frac{2T_o}{l}$$

$$k_{21} = k_{12} = k_{32} = k_{23} = \frac{-T_o}{\sqrt{1+l^2}} \approx \frac{-T_o}{l} \tag{4.146}$$

$$k_{31} = k_{13} = 0$$

The stiffness matrix is:

$$[k] = \frac{T_o}{l}\begin{bmatrix} 2 & -1 & 0 \\ -1 & 2 & -1 \\ 0 & -1 & 2 \end{bmatrix} \tag{4.147}$$

Example 3

Calculate the mass matrix and the stiffness matrix for the frame shown in figure 4.46.

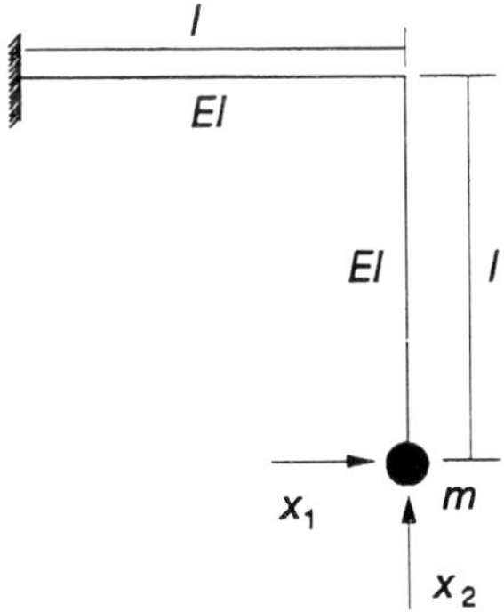

Figure 4.46   Frame represented by a two-degrees-of-freedom system.

The mass matrix is diagonal with the same mass assigned to each DDOF.

$$[m] = \begin{bmatrix} m & 0 \\ 0 & m \end{bmatrix} \tag{4.148}$$

In this system, it is easier to evaluate and invert the flexibility matrix, $[f]$, to obtain the stiffness matrix, $[k]$. To determine $[f]$, unit forces are applied in the x_1 and x_2 directions to get:

$$[f] = \frac{l^3}{6EI}\begin{bmatrix} 8 & 3 \\ 3 & 2 \end{bmatrix} \tag{4.149}$$

The inverse becomes the stiffness matrix:

$$[k] = \frac{6EI}{7l^3}\begin{bmatrix} 2 & -3 \\ -3 & 8 \end{bmatrix} \tag{4.150}$$

4.7.7 Free Vibrations of Undamped Systems

Natural frequencies. The equations of motion for an undamped MDOF system in free vibrations are written:

$$[m](\ddot{x}) + [k](x) = (0) \tag{4.151}$$

To solve equation 4.151, the method of separation of variables is used and the solution is:

$$\begin{aligned}
x_1 &= A_1 \sin(\omega t + \phi) \\
x_2 &= A_2 \sin(\omega t + \phi) \\
&\quad\vdots \qquad\qquad \vdots \\
x_N &= A_N \sin(\omega t + \phi)
\end{aligned} \tag{4.152}$$

where A_i corresponds to the amplitude of the vibrations of the ith DDOF and ϕ is the phase angle.

Equation 4.152 can also be written:

$$(x) = (A)\sin(\omega t + \phi) \tag{4.153}$$

Substituting equation 4.153 in equation 4.151, we obtain:

$$-\omega^2[m](A)\sin(\omega t + \phi) + [k](A)\sin(\omega t + \phi) = (0)$$

$$\big[[k] - \omega^2[m]\big](A)\sin(\omega t + \phi) = (0) \tag{4.154}$$

This equation must be satisfied at all times t; therefore:

$$\big[[k] - \omega^2[m]\big](A) = (0) \tag{4.155}$$

The values A_i are obtained by:

$$(A) = \big[[k] - \omega^2[m]\big]^{-1}(0) = \frac{\text{adjunct}\big[[k] - \omega^2[m]\big]}{\big|[k] - \omega^2[m]\big|}(0) \tag{4.156}$$

In order for the system to accept non-zero values for A_i, the determinant of equation 4.155 must be equal to zero.

$$\big|[k] - \omega^2[m]\big| = (0) \tag{4.157}$$

If the mass matrix is diagonal, it yields:

$$\begin{vmatrix} (k_{11}-\omega^2 m_1) & k_{12} & \cdots & \cdots & k_{1N} \\ k_{21} & (k_{22}-\omega^2 m_2) & \cdots & \cdots & k_{2N} \\ \cdots & \cdots & \cdots & \cdots & \cdots \\ k_{N1} & k_{N2} & \cdots & \cdots & (k_{NN}-\omega^2 m_N) \end{vmatrix} = 0 \tag{4.158}$$

The values for k_{ij} and m_i are known; therefore, expansion of the determinant gives an Nth degree equation in ω^2. This frequency equation gives N real roots corresponding to N natural frequencies of the system. The lowest frequency of the system is called the fundamental frequency of the system; the corresponding period (the longest) is called the fundamental period of the system.

Examples

Example 1
Calculate the natural frequencies of a two-degrees-of-freedom system, as shown in figure 4.43. The mass and stiffness matrices have already been calculated in equations 4.141 and 4.144. Using equation 4.158 yields:

$$\begin{vmatrix} (k_1+k_2-\omega^2 m_1) & -k_2 \\ -k_2 & (k_2+k_3-\omega^2 m_2) \end{vmatrix} = 0 \tag{4.159}$$

Expansion of the determinant takes the form of:

$$(k_1+k_2-\omega^2 m_1)(k_2+k_3-\omega^2 m_2) - k_2^2 = 0 \tag{4.160}$$

or

$$(\omega^2)^2 - \left(\frac{k_1 + k_2}{m_1} + \frac{k_2 + k_3}{m_2} \right) \omega^2 + \left(\frac{k_1 k_2 + k_2 k_3 + k_1 k_3}{m_1 m_2} \right) = 0 \tag{4.161}$$

The solution of this quadratic equation gives two real values for ω^2. To simplify the computations, we assume:

$$\begin{aligned} k_1 &= k_2 = k_3 = k \\ m_1 &= m_2 = m \end{aligned} \tag{4.162}$$

Equation 4.161 then becomes:

$$(\omega^2)^2 - \left(\frac{4k}{m} \right) \omega^2 + \left(\frac{3k^2}{m^2} \right) = 0 \tag{4.163}$$

The roots of equation 4.163 are obtained by:

$$\omega^2 = \frac{\frac{4k}{m} \pm \sqrt{ \left(-\frac{4k}{m} \right)^2 - 4(1) \left(\frac{3k^2}{m^2} \right) }}{2} \tag{4.164}$$

Therefore:

$$\omega_1^2 = \frac{k}{m} \ ; \ f_1 = \frac{1}{2\pi} \sqrt{\frac{k}{m}} \ ; \ T_1 = 2\pi \sqrt{\frac{m}{k}}$$

$$\tag{4.165}$$

$$\omega_2^2 = \frac{3k}{m} \ ; \ f_2 = \frac{1}{2\pi} \sqrt{\frac{3k}{m}} \ ; \ T_2 = 2\pi \sqrt{\frac{m}{3k}}$$

Example 2
Calculate the natural frequencies for the cable under tension shown in figure 4.44. The stiffness and mass matrices were determined by equations 4.145 and 4.147. The natural frequencies are obtained from a zero determinant:

$$\begin{vmatrix} \left(\dfrac{2T_o}{l} - m\,\omega^2\right) & -\dfrac{T_o}{l} & 0 \\[2ex] -\dfrac{T_o}{l} & \left(\dfrac{2T_o}{l} - 3m\,\omega^2\right) & -\dfrac{T_o}{l} \\[2ex] 0 & -\dfrac{T_o}{l} & \left(\dfrac{2T_o}{l} - 2m\,\omega^2\right) \end{vmatrix} = 0 \qquad (4.166)$$

Define:

$$\xi = \left(\frac{lm}{T_o}\right)\omega^2 \qquad (4.167)$$

Expansion of the determinant gives a cubic equation:

$$F(\xi) = 6\xi^3 - 22\xi^2 + 21\xi - 4 = 0 \qquad (4.168)$$

Newton-Raphson's iterative method (Cook, 1981) can be used to determine quickly the roots of equation 4.168:

$$\xi_j = \xi_{j-1} - \frac{F(\xi_{j-1})}{\left.\dfrac{dF(\xi)}{d\xi}\right|_{\xi=\xi_{j-1}}} \qquad (4.169)$$

The roots of equation 4.168 are:

$$\xi_1 = 0{,}25282 \qquad \xi_2 = 1{,}181 \qquad \xi_3 = 2{,}2329 \qquad (4.170)$$

Substituting these values into equation 4.167 yields:

$$\omega_1 = 0,503\sqrt{\frac{T_o}{lm}} \;;\; f_1 = 0,080\sqrt{\frac{T_o}{lm}} \;;\; T_1 = 12,496\sqrt{\frac{lm}{T_o}}$$

$$\omega_2 = 1,087\sqrt{\frac{T_o}{lm}} \;;\; f_2 = 0,173\sqrt{\frac{T_o}{lm}} \;;\; T_2 = 5,780\sqrt{\frac{lm}{T_o}} \tag{4.171}$$

$$\omega_3 = 1,494\sqrt{\frac{T_o}{lm}} \;;\; f_3 = 0,238\sqrt{\frac{T_o}{lm}} \;;\; T_3 = 4,206\sqrt{\frac{lm}{T_o}}$$

Example 3

Calculate the natural frequencies for the frame shown in figure 4.46. The mass and stiffness matrices have been calculated in equations 4.148 and 4.150. The natural frequencies are obtained from a zero determinant:

$$\begin{vmatrix} (2-\xi) & -3 \\ -3 & (8-\xi) \end{vmatrix} = 0 \tag{4.172}$$

where

$$\xi = \left(\frac{7ml^3}{6EI}\right)\omega^2 \tag{4.173}$$

Expansion of the determinant gives a quadratic equation:

$$\xi^2 - 10\xi + 7 = 0 \tag{4.174}$$

The roots of equation 4.174 are:

$$\xi_1 = 0,757 \qquad \xi_2 = 9,245 \tag{4.175}$$

Substituting in equation 4.173 yields:

$$\omega_1 = 0,806\sqrt{\frac{EI}{ml^3}} \;;\; f_1 = 0,128\sqrt{\frac{EI}{ml^3}} \;;\; T_1 = 7,796\sqrt{\frac{ml^3}{EI}}$$

$$\tag{4.176}$$

$$\omega_2 = 2,815\sqrt{\frac{EI}{ml^3}} \;;\; f_2 = 0,448\sqrt{\frac{EI}{ml^3}} \;;\; T_2 = 2,232\sqrt{\frac{ml^3}{EI}}$$

Modes of vibration (mode shapes). When the natural frequencies are determined, they can be substituted one by one into equation 4.155 to solve for each mode of vibration, $(A^{(i)})$.

$$\left[[k] - \omega_i^2 [m] \right] (A^{(i)}) = (0) \tag{4.177}$$

It is important to note that each vector $(A^{(i)})$ does not have an absolute value because only its shape has been determined. In fact, natural frequencies represent eigenvalues and modes of vibration represent eigenvectors of equation 4.177.

Examples

Example 1

Calculate the mode shapes for the two-degrees-of-freedom system shown in figure 4.43. Using the simplifications of equation 4.162, we find:

$$\omega_1^2 = \frac{k}{m}; \qquad \omega_2^2 = \frac{3k}{m} \tag{4.178}$$

Substituting for the first mode in equation 4.155 yields:

$$\begin{bmatrix} 2k - \dfrac{k}{m}m & -k \\ -k & 2k - \dfrac{k}{m}m \end{bmatrix} \begin{pmatrix} A_1^{(1)} \\ A_2^{(1)} \end{pmatrix} = (0) \tag{4.179}$$

Let us suppose:

$$A_1^{(1)} = 1 \tag{4.180}$$

then

$$A_2^{(1)} = 1 \tag{4.181}$$

Substituting for the second mode in equation 4.155 yields:

$$\begin{bmatrix} 2k - \dfrac{3k}{m}m & -k \\ -k & 2k - \dfrac{3k}{m}m \end{bmatrix} \begin{pmatrix} A_1^{(2)} \\ A_2^{(2)} \end{pmatrix} = (0) \tag{4.182}$$

Suppose:

$$A_1^{(2)} = 1 \tag{4.183}$$

then

$$A_2^{(2)} = -1 \tag{4.184}$$

Figure 4.47 illustrates these two modes of vibration.

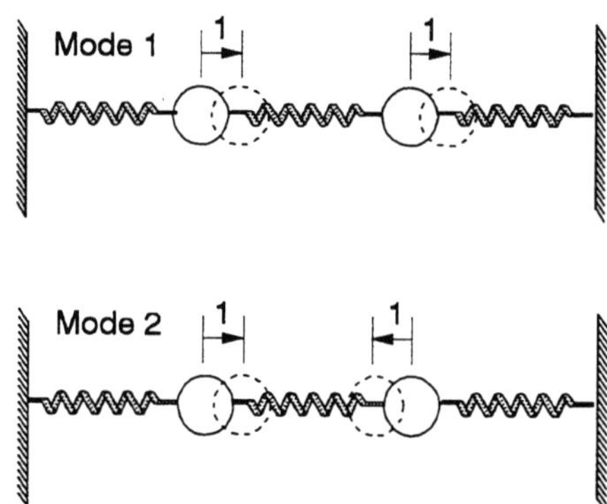

Figure 4.47   Modes of vibration of a two-degrees-of-freedom system.

Example 2
Calculate the mode shapes for the cable under tension shown in figure 4.44. The stiffness
matrix and mass matrix were determined by equations 4.145 and 4.147. As well, equation
4.171 yields:

$$\omega_1^2 = 0{,}253 \frac{T_o}{lm}$$

$$\omega_2^2 = 1{,}182 \frac{T_o}{lm} \tag{4.185}$$

$$\omega_3^2 = 2{,}232 \frac{T_o}{lm}$$

Substituting for the first mode in equation 4.155 yields:

$$\frac{T_o}{l}\begin{bmatrix} 2-0,253 & -1 & 0 \\ -1 & 2-0,759 & -1 \\ 0 & -1 & 2-0,506 \end{bmatrix}\begin{pmatrix} A_1^{(1)} \\ A_2^{(1)} \\ A_3^{(1)} \end{pmatrix} = (0) \qquad (4.186)$$

Let us suppose:

$$A_1^{(1)} = 1 \qquad (4.187)$$

then

$$A_2^{(1)} = 1,747$$
$$A_3^{(1)} = 1,169 \qquad (4.188)$$

Substituting for the second mode in equation 4.155 yields:

$$\frac{T_o}{l}\begin{bmatrix} 2-1,182 & -1 & 0 \\ -1 & 2-3,545 & -1 \\ 0 & -1 & 2-2,363 \end{bmatrix}\begin{pmatrix} A_1^{(2)} \\ A_2^{(2)} \\ A_3^{(2)} \end{pmatrix} = (0) \qquad (4.189)$$

Let us suppose:

$$A_1^{(2)} = 1 \qquad (4.190)$$

then

$$A_2^{(2)} = 0,818$$
$$A_3^{(2)} = -2,253 \qquad (4.191)$$

Substituting for the third mode in equation 4.155 yields:

$$\frac{T_o}{l}\begin{bmatrix} 2-2,232 & -1 & 0 \\ -1 & 2-6,696 & -1 \\ 0 & -1 & 2-4,464 \end{bmatrix}\begin{pmatrix} A_1^{(3)} \\ A_2^{(3)} \\ A_3^{(3)} \end{pmatrix} = (0) \qquad (4.192)$$

Let us suppose:

$$A_1^{(3)} = 1 \tag{4.193}$$

then

$$A_2^{(3)} = -0,232$$
$$A_3^{(3)} = 0,094 \tag{4.194}$$

These three modes of vibration are shown in figure 4.48.

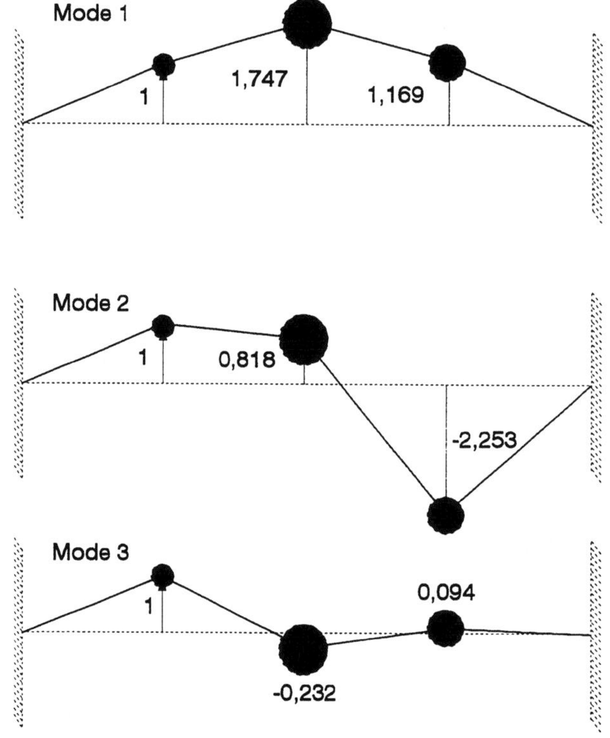

Figure 4.48 Modes of vibration of a cable under tension.

Example 3
Calculate the mode shapes for the frame shown in figure 4.46. The mass and stiffness matrices were calculated by equations 4.148 and 4.150. It was also determined by equation 4.176 that:

$$\omega_1^2 = 0{,}650 \frac{EI}{ml^3}$$
$$\omega_2^2 = 7{,}924 \frac{EI}{ml^3} \tag{4.195}$$

Substituting for the first mode in equation 4.155 yields:

$$\begin{bmatrix} 2-0{,}758 & -3 \\ -3 & 8-0{,}758 \end{bmatrix} \begin{pmatrix} A_1^{(1)} \\ A_2^{(1)} \end{pmatrix} = (0) \tag{4.196}$$

Let us suppose:

$$A_1^{(1)} = 1 \tag{4.197}$$

then

$$A_2^{(1)} = 0{,}414 \tag{4.198}$$

Substituting for the second mode in equation 4.155 yields:

$$\begin{bmatrix} 2-9{,}245 & -3 \\ -3 & 8-9{,}245 \end{bmatrix} \begin{pmatrix} A_1^{(2)} \\ A_2^{(2)} \end{pmatrix} = (0) \tag{4.199}$$

Let us suppose:

$$A_1^{(2)} = 1 \tag{4.200}$$

then

$$A_2^{(2)} = -2{,}415 \tag{4.201}$$

The modes of vibration are illustrated in figure 4.49.

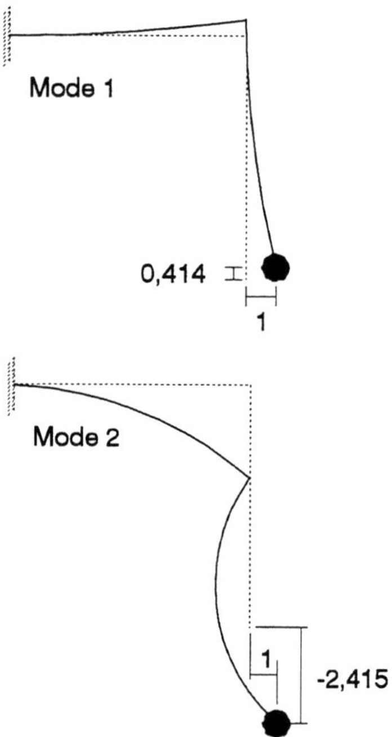

Figure 4.49 Modes of vibration of a frame represented by a
two-degrees-of-freedom system.

Frequency and modal matrices. Once the natural frequencies of a system are calculated, they
can be positioned in ascending order:

$$\omega_1^2 < \omega_2^2 < ... < \omega_N^2 \tag{4.202}$$

A frequency matrix can then be constructed:

$$[\omega^2] = \begin{bmatrix} \omega_1^2 & 0 & 0 & ... & 0 & ... & 0 \\ 0 & \omega_2^2 & 0 & ... & 0 & ... & 0 \\ 0 & 0 & \omega_3^2 & ... & 0 & ... & 0 \\ ... & ... & ... & ... & ... & ... & ... \\ 0 & 0 & 0 & ... & \omega_i^2 & ... & 0 \\ ... & ... & ... & ... & ... & ... & ... \\ 0 & 0 & 0 & ... & 0 & ... & \omega_N^2 \end{bmatrix} \tag{4.203}$$

$$[A] = \left[(A^{(1)})(A^{(2)})\dots(A^{(N)}) \right] \tag{4.204}$$

or

$$[A] = \begin{bmatrix} A_1^{(1)} & A_1^{(2)} & A_1^{(3)} & \dots & A_1^{(i)} & \dots & A_1^{(N)} \\ A_2^{(1)} & A_2^{(2)} & A_2^{(3)} & \dots & A_2^{(i)} & \dots & A_2^{(N)} \\ A_3^{(1)} & A_3^{(2)} & A_3^{(3)} & \dots & A_3^{(i)} & \dots & A_3^{(N)} \\ \dots & \dots & \dots & \dots & \dots & \dots & \dots \\ A_i^{(1)} & A_i^{(2)} & A_i^{(3)} & \dots & A_i^{(i)} & \dots & A_i^{(N)} \\ \dots & \dots & \dots & \dots & \dots & \dots & \dots \\ A_N^{(1)} & A_N^{(2)} & A_N^{(3)} & \dots & A_N^{(i)} & \dots & A_N^{(N)} \end{bmatrix} \tag{4.205}$$

As we will see below, the modal matrix plays a strategic role for solving the dynamic response of a MDOF system.

Mode shape normalization. As the mode shapes do not have absolute values, they always have to be normalized. We can suppose that the first (or the last) element of each mode shape has a unit value, as illustrated in the examples in section 4.7.7. Another method for normalization of a mode of vibration, $(A^{(i)})$, is by the mass matrix. Each element is normalized thus:

$$(A^{(i)})^T [m] (A^{(i)}) = 1 \tag{4.206}$$

For example, if the mass matrix is used to normalize the first mode of vibration for the cable under tension shown in figure 4.44:

$$(A^{(1)}) = \begin{pmatrix} 1 \\ 1{,}747 \\ 1{,}169 \end{pmatrix} \tag{4.207}$$

then

$$(A^{(1)})^T [m] (A^{(1)}) = \begin{pmatrix} 1 & 1{,}747 & 1{,}169 \end{pmatrix} \begin{bmatrix} m & 0 & 0 \\ 0 & 3m & 0 \\ 0 & 0 & 2m \end{bmatrix} \begin{pmatrix} 1 \\ 1{,}747 \\ 1{,}169 \end{pmatrix} = 12{,}889 \ m \tag{4.208}$$

To normalize this mode, each element is divided by $\sqrt{12{,}889 \ m}$.

$$(A^{(1)}) = \frac{1}{\sqrt{m}} \begin{pmatrix} 0{,}279 \\ 0{,}487 \\ 0{,}326 \end{pmatrix} \tag{4.209}$$

This method of normalizing the modes of vibration simplifies the computations for obtaining the dynamic response of a MDOF system.

Orthogonality conditions of mode shapes. For each mode of vibration $(A^{(i)})$, we have:

$$\left[[k] - \omega_i^2 [m] \right] (A^{(i)}) = (0) \tag{4.210}$$

or

$$[k](A^{(i)}) = \omega_i^2 [m](A^{(i)}) \tag{4.211}$$

Let us write this equation for two different modes, $(A^{(r)})$ and $(A^{(s)})$ $(r \neq s)$.

$$[k](A^{(r)}) = \omega_r^2 [m](A^{(r)}) \tag{4.212}$$

$$[k](A^{(s)}) = \omega_s^2 [m](A^{(s)}) \tag{4.213}$$

Equation 4.212 can be premultiplied by $(A^{(s)})^T$. Equation 4.213 can be transposed and postmultiplied by $(A^{(r)})$. It yields:

$$(A^{(s)})^T [k](A^{(r)}) = \omega_r^2 (A^{(s)})^T [m](A^{(r)}) \tag{4.214}$$

$$(A^{(s)})^T [k]^T (A^{(r)}) = \omega_s^2 (A^{(s)})^T [m]^T (A^{(r)}) \tag{4.215}$$

But the mass matrix and the stiffness matrix are symmetrical:

$$\begin{aligned} [k] &= [k]^T \\ [m] &= [m]^T \end{aligned} \tag{4.216}$$

Subtracting equation 4.215 from equation 4.214 yields:

$$0 = (\omega_r^2 - \omega_s^2)\left[(A^{(s)})^T [m](A^{(r)}) \right] \tag{4.217}$$

However, as the modes are different:

$$\omega_r \neq \omega_s \tag{4.218}$$

we must conclude that for $r \neq s$:

$$(A^{(s)})^T [m] (A^{(r)}) = 0 \tag{4.219}$$

and

$$(A^{(s)})^T [k] (A^{(r)}) = 0 \tag{4.220}$$

Equations 4.219 and 4.220 represent orthogonality conditions of mode shapes. The mode shapes are then said to be normal (or orthogonal) with respect to the mass matrix and the stiffness matrix. These orthogonality conditions are very useful for:

- simplifying the modal analysis, as we will see below;
- verifying computation of mode shapes;
- simplifying mode-shape computation by other techniques (Clough and Penzien, 1993).

Moreover, if all mode shapes are normalized with respect to the mass matrix (equation 4.206), it yields:

$$[A]^T [m] [A] = [I] \tag{4.221}$$

where $[I]$ is the identity matrix.
It also yields:

$$[A]^T [k] [A] = [\omega^2] \tag{4.222}$$

If equations 4.221 and 4.222 are satisfied, the modes are said to be orthonormal.

Rayleigh's method. Rayleigh (1945) developed a procedure for analyzing vibrating systems based on the conservation of energy. This technique is used mostly in earthquake engineering to estimate the fundamental period of vibration of a structure. As we will see in chapter 6, the NBCC (1995) recommends this approach.

To develop Rayleigh's formula, we use the equations of motion of an undamped MDOF system in free vibration (equation 4.151). This equation is premultiplied by the transpose of the velocity vector $(\dot{x})^T$ and is integrated with respect to time.

$$\int_0^t (\dot{x})^T [m] (\ddot{x}) \, dt + \int_0^t (\dot{x})^T [k] (x) \, dt = \text{constant} \tag{4.223}$$

Now, differential equations are used between the displacement, the velocity, and the acceleration.

$$(\dot{x}) = \frac{(dx)}{dt}$$

$$(\ddot{x}) = \frac{(d\dot{x})}{dt} \tag{4.224}$$

Substituting equation 4.224 into equation 4.223 yields:

$$\int_0^{\dot{x}} (\dot{x})^T [m] (d\dot{x}) + \int_0^x (x)^T [k] (dx) = \text{constant} \tag{4.225}$$

The two terms on the left-hand side of this equation can be integrated.

$$\frac{1}{2}(\dot{x})^T [m](\dot{x}) + \frac{1}{2}(x)^T [k](x) = \text{constant} \tag{4.226}$$

The first term on the left-hand side represents the kinetic energy of the system at time t. The second term on the left-hand side represents the elastic strain energy of the system at time t. Equation 4.226 reflects the law of energy conservation of an elastic, undamped system. At all times, the energy in the system is equal to a constant.

Now, let us suppose that a system vibrates in a specific mode $(A^{(i)})$ with a corresponding natural frequency ω. The system's response can be expressed by:

$$(x) = (A^{(i)}) \sin(\omega_i t + \phi)$$

$$(\dot{x}) = (A^{(i)}) \omega_i \cos(\omega_i t + \phi) \tag{4.227}$$

Substituting equation 4.227 into equation 4.226 yields:

$$\frac{1}{2}\omega_i^2 \cos^2(\omega_i t + \phi)(A^{(i)})^T [m](A^{(i)}) + \frac{1}{2}\sin^2(\omega_i t + \phi)(A^{(i)})^T [k](A^{(i)}) = \text{constant} \tag{4.228}$$

In equation 4.228, we notice that the strain energy is zero ($\cos^2(\omega_i t + \phi) = 0$) when the kinetic energy is maximum ($\sin^2(\omega_i t + \phi) = 1$) and that the strain energy is maximum ($\cos^2(\omega_i t + \phi) = 1$) when the kinetic energy is zero ($\sin^2(\omega_i t + \phi) = 0$). As total energy in the system remains constant, the maximum kinetic energy must be equal to the maximum strain energy.

$$\frac{1}{2}\omega_i^2 (A^{(i)})^T [m](A^{(i)}) = \frac{1}{2}(A^{(i)})^T [k](A^{(i)}) \tag{4.229}$$

The natural frequency can be isolated and an equation for the natural period of vibration can be found.

$$\omega_i = \sqrt{\frac{(A^{(i)})^T [k] (A^{(i)})}{(A^{(i)})^T [m] (A^{(i)})}} \tag{4.230}$$

$$T_i = 2\pi \sqrt{\frac{(A^{(i)})^T [m] (A^{(i)})}{(A^{(i)})^T [k] (A^{(i)})}} \tag{4.231}$$

When the exact mode of vibration $(A^{(i)})$ is known, the exact corresponding natural period, T_i, can be calculated.

However, if the mode shape is only approximated, then the period of vibration is an estimate. Generally, the first mode of vibration of a system is estimated by static deformations, (δ), generated by a careful selection of the static loads, (F).

$$\begin{aligned}(F) &= [k](\delta) \\ (A^{(i)}) &\approx (\delta)\end{aligned} \tag{4.232}$$

The estimate of the fundamental period becomes:

$$T_1 \approx 2\pi \sqrt{\frac{(\delta)^T [m] (\delta)}{(\delta)^T (F)}} \tag{4.233}$$

For a diagonal matrix, this estimate can be written as:

$$T_1 \approx 2\pi \sqrt{\frac{\displaystyle\sum_{j=1}^{N} W_j \delta_j^2}{g \displaystyle\sum_{j=1}^{N} F_j \delta_j}} \tag{4.234}$$

where W_j = weight associated to the jth degree of freedom
$\quad\quad\;\; N$ = number of degrees of freedom
$\quad\quad\;\; g$ = acceleration of gravity

Often, the fundamental mode shape can be estimated, using static loads equal to the weight associated with the degrees of freedom $(F_j = W_j)$.

$$T_1 \approx 2\pi \sqrt{\dfrac{\sum\limits_{j=1}^{N} W_j \delta_j^2}{g \sum\limits_{j=1}^{N} W_j \delta_j}} \tag{4.235}$$

Equation 4.235 is Rayleigh's formula included in the NBCC (1995).

Example

Estimate the fundamental period of the frame, shown in figure 4.46, using Rayleigh's formula. The mass and stiffness matrices have been calculated by equations 4.148 and 4.150. We use a static load vector corresponding to the weight associated with the degrees of freedom.

$$(F) = \begin{pmatrix} mg \\ mg \end{pmatrix} \tag{4.236}$$

The fundamental mode of vibration corresponds to the static deformations under this static load vector.

$$(A^{(1)}) \approx (\delta) = [k]^{-1}(F) = [f](F)$$

$$(\delta) = \dfrac{l^3}{6EI}\begin{bmatrix} 8 & 3 \\ 3 & 2 \end{bmatrix}\begin{pmatrix} mg \\ mg \end{pmatrix} = \dfrac{mgl^3}{6EI}\begin{pmatrix} 11 \\ 5 \end{pmatrix} \tag{4.237}$$

Substituting Rayleigh's formula yields:

$$T_1 \approx 2\pi \sqrt{\dfrac{\left[\dfrac{mgl^3}{6EI}\right]^2 (mg)\left[(11)^2 + (5)^2\right]}{g\left[\dfrac{mgl^3}{6EI}\right](mg)(11+5)}} \tag{4.238}$$

$$T_1 \approx 7{,}749 \sqrt{\dfrac{ml^3}{EI}}$$

We note that the difference between the estimate of Rayleigh's formula and the exact fundamental period calculated in equation 4.176 is only 0,6%.

4.7.8 Modal Analysis: the Modal Superimposition Method

Strategy. The modal analysis (or the modal superimposition method) is a very useful technique for determining the dynamic response of a linear MDOF system subjected to an arbitrary dynamic load (or a base acceleration). The governing system of differential equations is written as:

$$[m](\ddot{x}) + [c](\dot{x}) + [k](x) = (F(t)) \tag{4.239}$$

where $[m]$　　　　$=$　global mass matrix (usually diagonal)
　　　　$[c]$　　　　$=$　global damping matrix (difficult to evaluate)
　　　　$[k]$　　　　$=$　global stiffness matrix
　　　　$(F(t))$　　　$=$　dynamic load vector
　　　　$(x), (\dot{x}),$ and $(\ddot{x}) =$　relative displacement, velocity, and acceleration vectors

In general, the differential equations can be coupled by $[m]$, $[c]$, or $[k]$. In other words, these matrices are not necessarily diagonal matrices. If k_{ij} exists for $i \neq j$, the system is said to be statistically coupled. If m_{ij} exists for $i \neq j$, the system is said to be dynamically coupled. The basic strategy of the modal analysis is to introduce a linear transformation of the variables, using the modal matrix, $[A]$, in order to uncouple the equations of motion.

Normal coordinates. To uncouple the equations of motion, a linear transformation is introduced using the modal matrix, $[A]$. This transformation converts the differential equations from the real coordinates $(x(t))$ (or degrees of freedom) into a new system of normal coordinates $(u(t))$.

$$(x(t)) = [A](u(t)) \tag{4.240}$$

Uncoupling the equations of motion. Equation 4.240 is substituted into equation 4.239.

$$[m][A](\ddot{u}) + [c][A](\dot{u}) + [k][A](u) = (F(t)) \tag{4.241}$$

Equation 4.241 is premultiplied by the transpose of a specific mode shape $(A^{(i)})^T$:

$$(A^{(i)})^T[m][A](\ddot{u}) + (A^{(i)})^T[c][A](\dot{u}) + (A^{(i)})^T[k][A](u) = (A^{(i)})^T(F(t)) \tag{4.242}$$

Remembering the orthogonality conditions of the mode shapes, for $i \neq j$:

$$(A^{(i)})^T[m](A^{(j)}) = (A^{(i)})^T[k](A^{(j)}) = 0 \tag{4.243}$$

and also assuming for the moment for $i \neq j$:

$$(A^{(i)})^T [c] (A^{(j)}) = 0 \tag{4.244}$$

then, only one relation in u_i remains in equation 4.242.

$$M_i \ddot{u}_i + C_i \dot{u}_i + K_i u_i = P_i(t) \tag{4.245}$$

where

$$M_i = (A^{(i)})^T [m] (A^{(i)}) = \text{generalized mass in mode } i$$
$$C_i = (A^{(i)})^T [c] (A^{(i)}) = \text{generalized damping coefficient in mode } i$$
$$K_i = (A^{(i)})^T [k] (A^{(i)}) = \text{generalized stiffness coefficient in mode } i \tag{4.246}$$
$$P_i(t) = (A^{(i)})^T (F(t)) = \text{generalized load in mode } i$$

Equation 4.245 has the same form as equation 4.64 for a SDOF system. Equation 4.245 was already solved by Duhamel's integral (equation 4.69).

$$u_i(t) = \frac{1}{M_i \omega_{d_i}} \int_0^t P_i(\tau) e^{-\zeta_i \omega_i (t-\tau)} \sin \omega_{d_i}(t-\tau) \, d\tau \tag{4.247}$$

where

$$\omega_i^2 = \frac{K_i}{M_i}$$
$$\zeta_i = \frac{C_i}{2 \omega_i M_i} \tag{4.248}$$
$$\omega_{d_i} = \omega_i \sqrt{1 - \zeta_i^2}$$

After solving for each normal coordinate u_i $(i=1$ to $N)$, the response of each degree of freedom can be determined using the same linear transformation.

$$(x(t)) = [A](u(t)) \tag{4.249}$$

The response of a specific DDOF, $x_i(t)$, is obtained by:

$$x_i(t) = \sum_{j=1}^{N_m} A_i^{(j)} u_j(t)$$

$$x_i(t) = \sum_{j=1}^{N_m} \left[\frac{A_i^{(j)}}{M_j \omega_j} \int_0^t P_j(\tau) e^{-\zeta_j \omega_j(t-\tau)} \sin \omega_{d_j}(t-\tau) d\tau \right]$$

(4.250)

with N_m representing the number of modes considered in the modal analysis. In general, it is not necessary to consider all the modes of vibration and $N_m < N$ (section 4.8.3) in the modal analysis of a civil-engineering structure.

Conditions needed for damping matrix. In the above section, the assumption was made that the modes of vibration were orthogonal with respect to the damping matrix (equation 4.244). The modes of vibration that satisfy equation 4.244 are called classic normal modes. These modes ensure that the equations of motion are uncoupled and that the damped modes of vibration are identical to the undamped modes of vibration. In this section, we study the conditions that the damping matrix must satisfy in order to ensure the existence of classic normal modes.

Rayleigh's Damping

As the mode shapes are already orthogonal to the mass matrix and the stiffness matrix (equations 4.219 and 4.220), the existence of normal modes is ensured by defining the damping matrix as a linear combination of the mass matrix and the stiffness matrix.

$$[c] = \alpha[m] + \beta[k]$$

(4.251)

This type of damping is called Rayleigh's damping. In this case, the damping ratio in an arbitrary mode i is obtained by substituting equation 4.251 into equation 4.248.

$$\zeta_i = \frac{C_i}{2\omega_i M_i} = \frac{(A^{(i)})^T [c](A^{(i)})}{2\omega_i (A^{(i)})^T [m](A^{(i)})}$$

$$\zeta_i = \alpha \frac{(A^{(i)})^T [m](A^{(i)})}{2\omega_i (A^{(i)})^T [m](A^{(i)})} + \beta \frac{(A^{(i)})^T [k](A^{(i)})}{2\omega_i (A^{(i)})^T [m](A^{(i)})}$$

(4.252)

Simplifying the terms yields:

$$\zeta_i = \frac{\alpha}{2\omega_i} + \frac{\beta \omega_i}{2}$$

(4.253)

In practice, the coefficients α and β are obtained by specifying (or measuring) the damping ratios in two distinct modes of vibration i and j.

$$\alpha = 2\omega_i\omega_j \left[\frac{\zeta_j\omega_i - \zeta_i\omega_j}{\omega_i^2 - \omega_j^2}\right]$$

$$\beta = 2\left[\frac{\zeta_i\omega_i - \zeta_j\omega_j}{\omega_i^2 - \omega_j^2}\right]$$

$$(4.254)$$

It is important to note that if the damping matrix is proportional only to the mass matrix ($\beta = 0$), the damping ratio is inversely proportional to the natural frequency. In other words, modes with high frequencies will have very little damping. However, if the damping matrix is only proportional to the stiffness matrix ($\alpha = 0$), the damping ratio is directly proportional to the natural frequency. In this case, the modes with high frequencies will be very damped and will have little influence on the response. In general, high-frequency modes show high damping (fig. 4.50) when Rayleigh's damping is used.

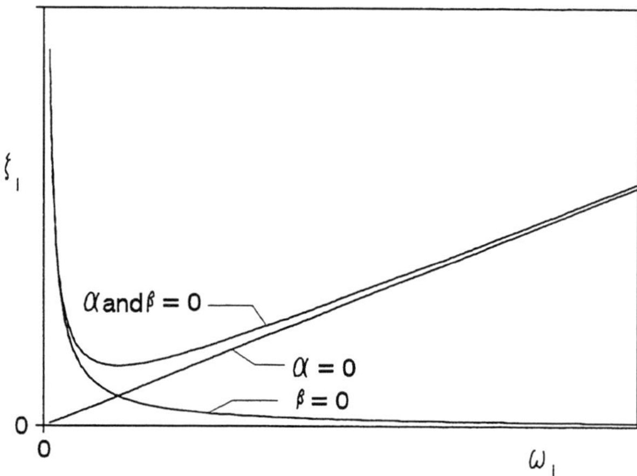

Figure 4.50   Variation of Rayleigh's damping with natural frequencies.

In some cases, we note that Rayleigh's damping is not physically attainable and may even be unsafe.

Modal Damping

Another procedure to ensure the existence of classical normal modes is to specify directly the damping ratio in each mode. In other words, we want:

$$[A]^T[c][A] = \begin{bmatrix} 2\zeta_1\omega_1 & 0 & 0 & \ldots & 0 & \ldots & 0 \\ 0 & 2\zeta_2\omega_2 & 0 & \ldots & 0 & \ldots & 0 \\ 0 & 0 & 2\zeta_3\omega_3 & \ldots & 0 & \ldots & 0 \\ \ldots & \ldots & \ldots & \ldots & \ldots & \ldots & \ldots \\ 0 & 0 & 0 & \ldots & 2\zeta_i\omega_i & \ldots & 0 \\ \ldots & \ldots & \ldots & \ldots & \ldots & \ldots & \ldots \\ 0 & 0 & 0 & \ldots & 0 & \ldots & 2\zeta_N\omega_N \end{bmatrix} \tag{4.255}$$

The necessary condition that the matrix $[c]$ must meet is obtained by premultiplying equation 4.255 by $[m]\,[A]$ and by postmultiplying it by $[A]^T\,[m]$.

$$[m][A][A]^T[c][A][A]^T[m] =$$

$$[m][A] \begin{bmatrix} 2\zeta_1\omega_1 & 0 & 0 & .. & 0 & .. & 0 \\ 0 & 2\zeta_2\omega_2 & 0 & .. & 0 & .. & 0 \\ 0 & 0 & 2\zeta_3\omega_3 & .. & 0 & .. & 0 \\ .. & .. & .. & .. & .. & .. & .. \\ 0 & 0 & 0 & .. & 2\zeta_i\omega_i & .. & 0 \\ .. & .. & .. & .. & .. & .. & .. \\ 0 & 0 & 0 & .. & 0 & .. & 2\zeta_N\omega_N \end{bmatrix} [A]^T[m] \tag{4.256}$$

Let us suppose that the mode shapes are orthonormal:

$$[A]^T[m][A] = [I] \tag{4.257}$$

then

$$[m][A] = \left[[A]^T\right]^{-1}$$

$$[A]^T[m] = [A]^{-1} \tag{4.258}$$

Substituting equation 4.258 into equation 4.256 yields:

$$[c] = [m][A] \begin{bmatrix} 2\zeta_1\omega_1 & 0 & 0 & \dots & 0 & \dots & 0 \\ 0 & 2\zeta_2\omega_2 & 0 & \dots & 0 & \dots & 0 \\ 0 & 0 & 2\zeta_3\omega_3 & \dots & 0 & \dots & 0 \\ \dots & \dots & \dots & \dots & \dots & \dots & \dots \\ 0 & 0 & 0 & \dots & 2\zeta_i\omega_i & \dots & 0 \\ \dots & \dots & \dots & \dots & \dots & \dots & \dots \\ 0 & 0 & 0 & \dots & 0 & \dots & 2\zeta_N\omega_N \end{bmatrix} [A]^T[m] \tag{4.259}$$

Equation 4.259 defines the components of the damping matrix, $[c]$, to ensure that the equations of motion become uncoupled by the linear transformation using the modal matrix. We note that the damping matrix depends on the damping ratio assigned to each mode. If matrix $[c]$ is constructed according to equation 4.259, the damped modes of vibration will be classical normal modes; therefore, they will be identical to the undamped normal modes.

Effects of initial conditions. The response of each normal coordinate (equation 4.247) is valid when the system is initially at rest. The total response must include the effects of the initial conditions. Referring to the total response of a SDOF system (equation 4.70):

$$u_i(t) = e^{-\zeta_i\omega_i t}\left[\frac{\dot{u}_{i_0} + \zeta_i\omega_i u_{i_0}}{\omega_{d_i}}\sin\omega_{d_i}t + u_{i_0}\cos\omega_{d_i}t\right] + $$
$$\frac{1}{M_i\omega_{d_i}}\int_0^t P_i(\tau)e^{-\zeta_i\omega_i(t-\tau)}\sin\omega_{d_i}(t-\tau)\,d\tau \tag{4.260}$$

where

$$u_{i_o} = \text{initial displacement of } u_i$$
$$\dot{u}_{i_o} = \text{initial velocity of } \dot{u}_i \tag{4.261}$$

Before solving each modal coordinate, $u_i(t)$, the initial conditions must be known. The initial conditions of the system are specified in the real coordinates (degrees of freedom). In theory, these initial conditions can be transformed into normal coordinates by inversing the linear transformation.

$$(u_o) = [A]^{-1}(x_o)$$
$$(\dot{u}_o) = [A]^{-1}(\dot{x}_o) \tag{4.262}$$

However, use of equation 4.262 requires the inversion of the modal matrix, which translates into long computations. Therefore, another approach is suggested: the linear transformation given by equation 4.224 is premultiplied by $[A]^T[m]$:

$$[A]^T[m](x_o) = [A]^T[m][A](u_o)$$

$$[A]^T[m](\dot{x}_o) = [A]^T[m][A](\dot{u}_o)$$

$$(4.263)$$

Each term i becomes:

$$u_{i_o} = \frac{(A^{(i)})^T[m](x_o)}{(A^{(i)})^T[m](A^{(i)})} = \frac{(A^{(i)})^T[m](x_o)}{M_i}$$

$$\dot{u}_{i_o} = \frac{(A^{(i)})^T[m](\dot{x}_o)}{(A^{(i)})^T[m](A^{(i)})} = \frac{(A^{(i)})^T[m](\dot{x}_o)}{M_i}$$

$$(4.264)$$

4.7.9 Specialization of Modal Analysis for Earthquake Response

In computation of the earthquake response of a MDOF system subjected to a base acceleration, equation 4.247 takes the form:

$$u_i(t) = \frac{-(A^{(i)})^T[m](I)}{M_i \omega_{d_i}} \int_0^t \ddot{x}_s(\tau) e^{-\zeta_i \omega_i (t-\tau)} \sin \omega_{d_i}(t-\tau) d\tau \qquad (4.265)$$

or

$$u_i(t) = \frac{-\alpha_i}{\omega_{di}} S^{(i)}(t) \qquad (4.266)$$

where

$$\alpha_i = \frac{(A^{(i)})^T[m](I)}{M_i}$$

$$S^{(i)}(t) = \int_0^t \ddot{x}_s e^{-\zeta_i \omega_i (t-\tau)} \sin\left(\omega_{di}(t-\tau)\right) d\tau$$

$$(4.267)$$

The factor α is called the modal participation factor for mode i. This factor represents the relative contribution of mode i (assuming that the ratio $S^{(i)}(t)/\omega_{di}$ is constant for each mode) to the total response of the system. Note that this factor can be positive or negative. As seen above, when the solution of each normal coordinate is found $u_i(t)$ *(i=1 to N)*, the response of each DDOF can be determined using the same linear transformation.

$$(x(t)) = [A](u(t)) \qquad (4.268)$$

To illustrate these points, let us consider the multi-storey structure shown in figure 4.51. We assume only one horizontal DDOF per storey, x_i *(t)*, and only one concentrated mass per storey, m_i, as well as a base acceleration, $\ddot{x}_s$ *(t)*. Having solved for all degrees of freedom, *(x(t))*, a set of equivalent seismic loads is calculated, *(Q(t))*.

$$(Q(t)) = [k](x(t)) \qquad (4.269)$$

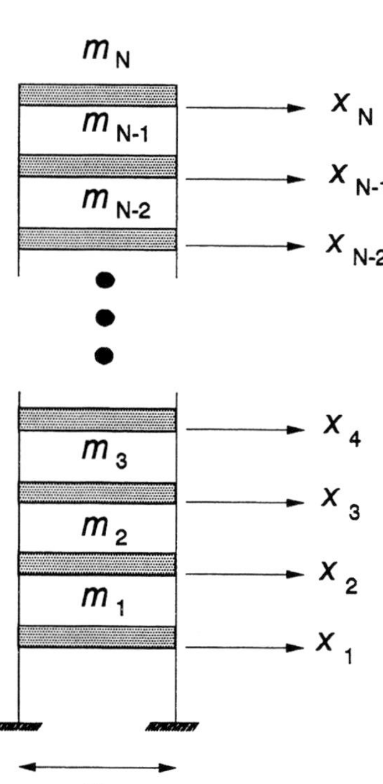

Figure 4.51 Multi-storey structure subjected to a base excitation.

This set of seismic loads, *(Q(t))* (fig. 4.52) represents the lateral static loads that must be applied to the structure at time *t* to cause the same displacements, *(x(t))*. When considering only one DDOF per storey (fig. 4.51), the shear, $V_i(t)$, and overturning moment, $M_i(t)$, can be calculated for each storey.

$$\begin{pmatrix} V_1 \\ V_2 \\ \bullet \\ \bullet \\ \bullet \\ V_N \end{pmatrix} = \begin{bmatrix} 1 & 1 & 1 & \dots & 1 \\ 0 & 1 & 1 & \dots & 1 \\ 0 & 0 & 1 & \dots & 1 \\ \dots & \dots & \dots & \dots & \dots \\ 0 & 0 & 0 & \dots & 1 \end{bmatrix} \begin{pmatrix} Q_1 \\ Q_2 \\ \bullet \\ \bullet \\ \bullet \\ Q_N \end{pmatrix} \qquad (4.270)$$

$$(V(t)) = [S](Q(t))$$

$$\begin{pmatrix} M_0 \\ M_1 \\ \bullet \\ \bullet \\ \bullet \\ M_{N-1} \end{pmatrix} = \begin{bmatrix} h_1 & h_2 & h_3 & \dots & h_N \\ 0 & h_2 & h_3 & \dots & h_N \\ 0 & 0 & h_3 & \dots & h_N \\ \dots & \dots & \dots & \dots & \dots \\ 0 & 0 & 0 & \dots & h_N \end{bmatrix} \begin{pmatrix} V_1 \\ V_2 \\ \bullet \\ \bullet \\ \bullet \\ V_N \end{pmatrix} \qquad (4.271)$$

$$(M(t)) = [h](V(t))$$

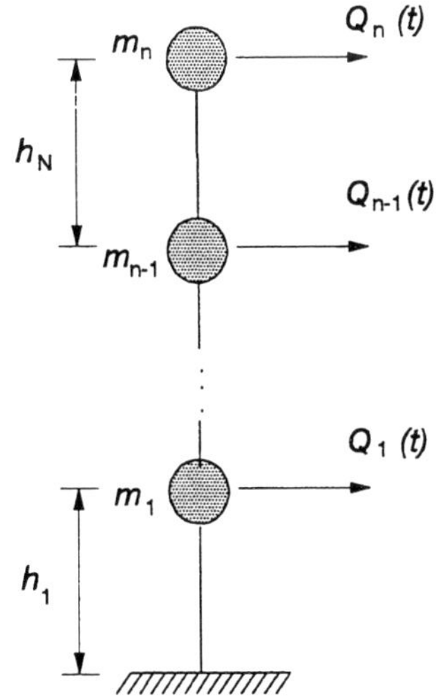

Figure 4.52 Equivalent seismic loads.

4.7.10   Summary of Modal Analysis

The modal analysis (or modal superimposition method) is a powerful tool for analysis of any linear structure with lumped mass subjected to an arbitrary dynamic loading represented by a set of dynamic loads or by a base accelerogram. The following summary represents the steps of the modal analysis.

Step 1: Selection of the degrees of freedom and computation of mass and stiffness matrices
- Selection of the most important DDOF.
- Construction of the mass and stiffness matrices.
- Each mode is assigned a damping ratio.

Step 2: Calculation of the natural frequencies and mode shapes
The eigenvalues and eigenvectors problem is written as:

$$\left[[k] - \omega_i^2 [m]\right](A^{(i)}) = (0) \tag{4.272}$$

Step 3: Calculation of the generalized masses and dynamic loads
For each mode i, the generalized mass, M_i, and the generalized dynamic load, $P_i(t)$, are calculated.

$$M_i = (A^{(i)})^T [m](A^{(i)})$$
$$P_i(t) = (A^{(i)})^T (F(t)) \tag{4.273}$$

For the earthquake problem, we have:

$$P_i(t) = -(A^{(i)})^T [m](r)\ddot{x}_s(t) \tag{4.274}$$

Step 4: Initial conditions calculation (if necessary)
If the structure is not initially at rest, the initial conditions are calculated:

$$u_{i_o} = \frac{(A^{(i)})^T [m](x_o)}{(A^{(i)})^T [m](A^{(i)})} = \frac{(A^{(i)})^T [m](x_o)}{M_i}$$

$$\dot{u}_{i_o} = \frac{(A^{(i)})^T [m](\dot{x}_o)}{(A^{(i)})^T [m](A^{(i)})} = \frac{(A^{(i)})^T [m](\dot{x}_o)}{M_i} \tag{4.275}$$

Step 5: Modal response calculation
The response in each mode is as follows:

$$u_i(t) = e^{-\zeta_i \omega_i t}\left[\frac{\dot{u}_{i_0} + \zeta_i \omega_i u_{i_0}}{\omega_{d_i}}\sin \omega_{d_i} t + u_{i_0}\cos \omega_{d_i} t\right] +$$
$$\frac{1}{M_i \omega_{d_i}}\int_0^t P_i(\tau) e^{-\zeta_i \omega_i (t-\tau)}\sin \omega_{d_i}(t-\tau)\,d\tau \tag{4.276}$$

For the earthquake problem, we have:

$$u_i(t) = \frac{-\alpha_i}{\omega_{di}}S^{(i)}(t) \tag{4.277}$$

where

$$\alpha_i = \frac{(A^{(i)})^T[m](I)}{M_i}$$

$$S^{(i)}(t) = \int_0^t \ddot{x}_s e^{-\zeta_i \omega_i (t-\tau)}\sin\left(\omega_{di}(t-\tau)\right)d\tau \tag{4.278}$$

Step 6: Response calculation in real coordinates
The modal responses are superimposed to solve for the response of each DDOF, $x_i(t)$.

$$(x(t)) = [A](u(t))$$

$$x_i(t) = \sum_{j=1}^{N_m} A_i^{(j)} u_j(t) \tag{4.279}$$

For the earthquake problem, a set of equivalent static loads, $(Q(t))$, is estimated.

$$(Q(t)) = [k](x(t)) \tag{4.280}$$

The internal forces (shear, moment, etc.) can be calculated from the equivalent seismic forces, as in the static analysis.

4.7.11   Example

To illustrate the modal analysis procedure, let us consider a three-storey structure modelled as a three-degrees-of-freedom system. Let us calculate the displacements of each storey caused by a constant load suddenly applied to the third floor, as shown in figure 4.53. Note that damping is neglected. Figure 4.53 also shows the mass and lateral stiffness of each storey.

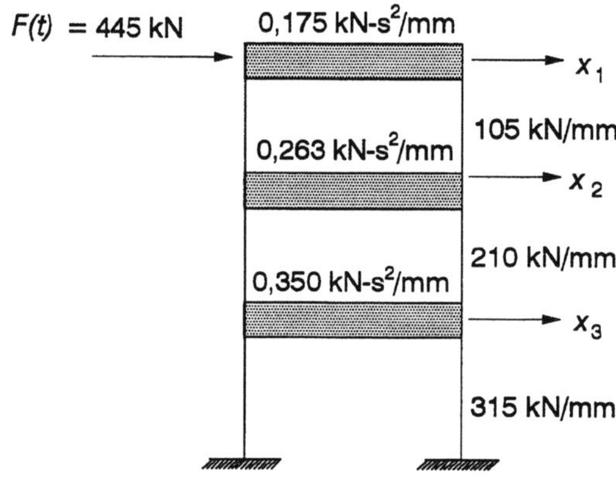

Figure 4.53   Three-storey structure modelled as a three-degrees-of-freedom system subjected to a constant load on the third floor.

Step 1
The mass and stiffness matrices are given. The reader can verify these results using standard methods of structural analysis.

$$[m] = \begin{bmatrix} 0,175 & 0 & 0 \\ 0 & 0,263 & 0 \\ 0 & 0 & 0,350 \end{bmatrix} \frac{kN{\bullet}s^2}{mm}$$

$$[k] = 105 \begin{bmatrix} 1 & -1 & 0 \\ -1 & 3 & -2 \\ 0 & -2 & 5 \end{bmatrix} \frac{kN}{mm}$$

(4.281)

Step 2

The natural frequencies and mode shapes are given. The reader can verify these results using the methods shown above.

$$\omega_1 = 14,5 \text{ rad/s}; \quad \omega_2 = 31,1 \text{ rad/s}; \quad \omega_3 = 46,1 \text{ rad/s}$$

$$(A^{(1)}) = \begin{pmatrix} 1,000 \\ 0,644 \\ 0,300 \end{pmatrix}; \quad (A^{(2)}) = \begin{pmatrix} 1,000 \\ -0,601 \\ -0,676 \end{pmatrix}; \quad (A^{(3)}) = \begin{pmatrix} 1,000 \\ -2,570 \\ 2,470 \end{pmatrix} \tag{4.282}$$

Step 3

The generalized masses and loads are calculated.

$$M_1 = (1,000 \quad 0,644 \quad 0,300) \begin{bmatrix} 0,175 & 0 & 0 \\ 0 & 0,263 & 0 \\ 0 & 0 & 0,350 \end{bmatrix} \begin{pmatrix} 1,000 \\ 0,644 \\ 0,300 \end{pmatrix} = 0,315 \; \frac{\text{kN}\bullet\text{s}^2}{\text{mm}}$$

$$M_2 = (1,000 \quad -0,601 \quad -0,676) \begin{bmatrix} 0,175 & 0 & 0 \\ 0 & 0,263 & 0 \\ 0 & 0 & 0,350 \end{bmatrix} \begin{pmatrix} 1,000 \\ -0,601 \\ -0,676 \end{pmatrix} = 0,430 \; \frac{\text{kN}\bullet\text{s}^2}{\text{mm}} \tag{4.283}$$

$$M_3 = (1,000 \quad -2,570 \quad 2,470) \begin{bmatrix} 0,175 & 0 & 0 \\ 0 & 0,263 & 0 \\ 0 & 0 & 0,350 \end{bmatrix} \begin{pmatrix} 1,000 \\ -2,570 \\ 2,470 \end{pmatrix} = 4,047 \; \frac{\text{kN}\bullet\text{s}^2}{\text{mm}}$$

$$P_1 = (1,000 \quad 0,644 \quad 0,300) \begin{pmatrix} 445 \\ 0 \\ 0 \end{pmatrix} = 445 \text{ kN}$$

$$P_2 = (1,000 \quad -0,601 \quad -0,676) \begin{pmatrix} 445 \\ 0 \\ 0 \end{pmatrix} = 445 \text{ kN} \tag{4.284}$$

$$P_3 = (1,000 \quad -2,570 \quad 2,470) \begin{pmatrix} 445 \\ 0 \\ 0 \end{pmatrix} = 445 \text{ kN}$$

Step 4
There is no initial condition as the system was initially at rest.

Step 5
The response of a SDOF system subjected to a constant load has already been calculated (equation 4.72). The response in each mode is then:

$$u_i(t) = \frac{P_i}{M_i\omega_i^2}(1 - \cos\omega_i t) \qquad (4.285)$$

$$u_1(t) = \frac{445}{0,315(14,5)^2}(1 - \cos 14,5t) = 6,719(1 - \cos 14,5t) \text{ mm}$$

$$u_2(t) = \frac{445}{0,430(31,1)^2}(1 - \cos 31,1t) = 1,070(1 - \cos 31,1t) \text{ mm} \qquad (4.286)$$

$$u_3(t) = \frac{445}{4,047(46,1)^2}(1 - \cos 46,1t) = 0,052(1 - \cos 46,1t) \text{ mm}$$

Step 6
The modal responses can now be superimposed to calculate the response of each DDOF (storey).

$$\begin{pmatrix} x_1(t) \\ x_2(t) \\ x_3(t) \end{pmatrix} = \begin{bmatrix} 1,000 & 1,000 & 1,000 \\ 0,644 & -0,601 & -2,570 \\ 0,300 & -0,676 & 2,470 \end{bmatrix} \begin{pmatrix} 6,719(1 - \cos 14,5t) \\ 1,070(1 - \cos 31,1t) \\ 0,052(1 - \cos 46,1t) \end{pmatrix} \qquad (4.287)$$

$$x_1(t) = 6,719(1 - \cos 14,5t) + 1,070(1 - \cos 31,1t) + 0,052(1 - \cos 46,1t)$$
$$x_2(t) = 4,327(1 - \cos 14,5t) - 0,643(1 - \cos 31,1t) - 0,134(1 - \cos 46,1t) \qquad (4.288)$$
$$x_3(t) = 2,016(1 - \cos 14,5t) - 0,723(1 - \cos 31,1t) + 0,128(1 - \cos 46,1t)$$

Figure 4.54 shows the time-histories of the displacements at each floor. We notice that the effect of the first mode is dominant. This is frequently the case for multi-storey buildings in which the focus is strictly on displacements. However, the effect of higher modes can be accentuated when calculating the internal forces in a structure.

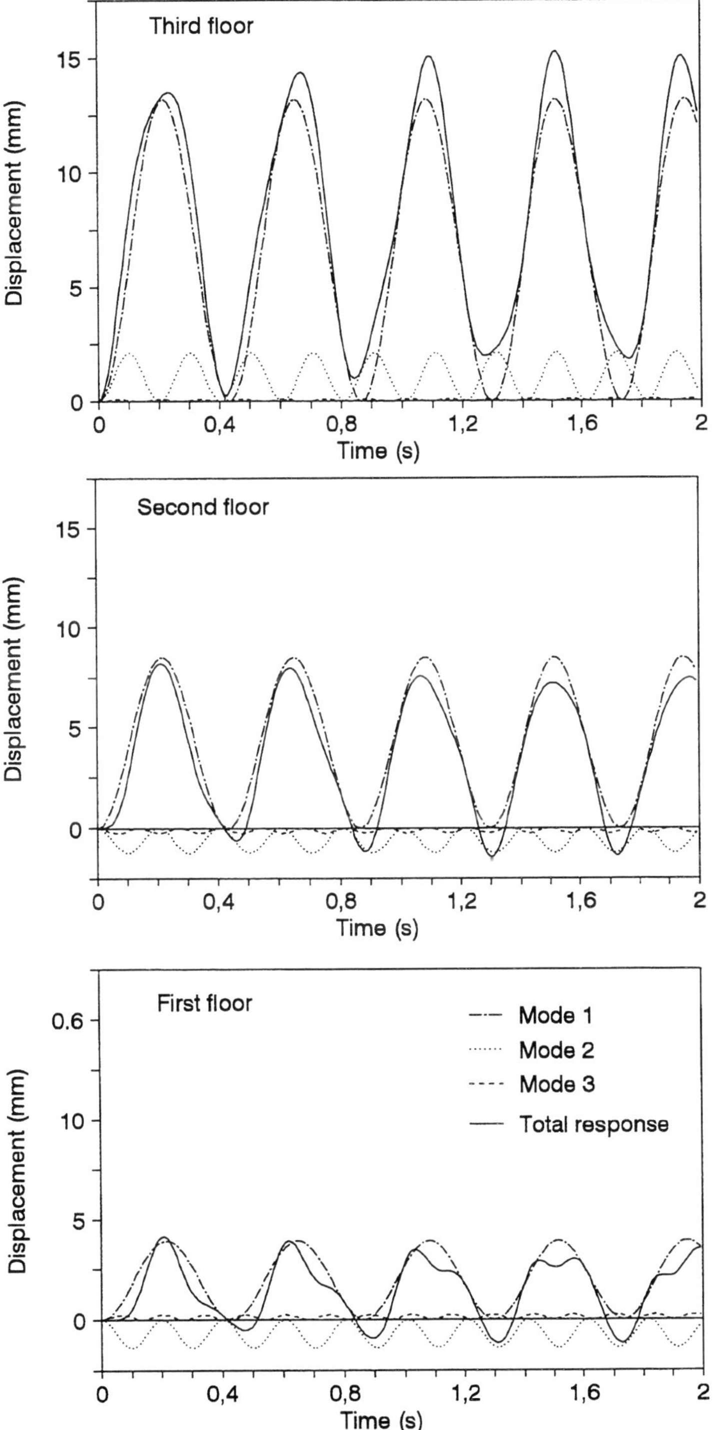

Figure 4.54 Displacement time-histories for each floor of a three-storey structure subjected to a constant load on the third floor.

4.8 APPROXIMATE EARTHQUAKE ANALYSIS: THE SPECTRAL METHOD

4.8.1 Motivation

In the earthquake response of a structure modelled as a MDOF system, an approximate analysis method can be used. Called the spectral method, it is much faster than modal analysis. This method uses the concept of elastic seismic response spectra to estimate maximum responses (displacements, velocities, internal forces, etc.) of the structure for a given base accelerogram.

4.8.2 Maximum Modal Responses

According to equation 4.266, the absolute value of the maximum response in mode i is:

$$\mu_{i\,max} = |u_i(t)|_{max} = \frac{\alpha_i}{\omega_{di}} |S^{(i)}(t)|_{max} \qquad (4.289)$$

where

$$\alpha_i = \frac{(A^{(i)})^T [m] (I)}{M_i}$$

$$\qquad (4.290)$$

$$|S^{(i)}(t)|_{max} = \left| \int_0^t \ddot{x}_s e^{-\zeta_i \omega_i (t-\tau)} \sin\left(\omega_{di}(t-\tau)\right) d\tau \right|_{max}$$

Comparing equation 4.290 with equations 4.105 and 4.106, we note:

$$|S^{(i)}|_{max} = S_V^{(i)} = \text{relative velocity spectrum} \qquad (4.291)$$

for $\ddot{x}_s(t)$, ω_i, and ζ_i
then

$$u_{i\,max} = \frac{\alpha_i}{\omega_{di}} S_V^{(i)} = \alpha_i S_D^{(i)} \qquad (4.292)$$

As seen above, the response in real coordinates is obtained by the following equation.

$$(x(t)) = [A](u(t)) \qquad (4.293)$$

The response of each degree of freedom, $x_i(t)$, can then be calculated.

$$x_i(t) = \sum_{j=1}^{N_m} A_i^{(j)} u_j(t) = \sum_{j=1}^{N_m} x_i^{(j)} \tag{4.294}$$

with N_m representing the number of modes considered in the analysis $(N_m \leq N)$.

It is important to note that $x_i^{(j)}$ represents the contribution of mode j to the response of the DDOF i. The maximum response of this modal contribution can be determined as $x^{(j)}{}_{imax}$.

$$x_i^{(j)}{}_{max} = A_i u_{jmax} = A_i^{(j)} \alpha_j S_D^{(j)} \tag{4.295}$$

where $S_D^{(j)}$ represents the relative displacement response spectrum corresponding to mode j, meaning for ω_j and ζ_j . All maximum relative displacements in mode j occurred at the same time and can be written in the following matrix form:

$$\left(x_{max}^{(j)} \right) = \alpha_j S_D^{(j)} \left(A^{(j)} \right) \tag{4.296}$$

The lateral maximum forces in mode i are then written:

$$\left(Q_{max}^{(i)} \right) = [k] \left(x_{max}^{(i)} \right) = \omega_i^2 [m] \, \alpha_i S_D^{(i)} \left(A^{(i)} \right) = \alpha_i S_A^{(i)} [m] \left(A^{(i)} \right) \tag{4.297}$$

For only one DDOF per storey, the maximum base shear in mode i can be written as:

$$V_{max}^{(i)} = \sum_{j=1}^{N} Q_{j_{max}}^{(i)} = \left(Q_{max}^{(i)} \right)^T (I) \tag{4.298}$$

Equation 4.297 is substituted in equation 4.298.

$$V_{max}^{(i)} = \left(A^{(i)} \right)^T [m]^T (I) \, \alpha_i S_A^{(i)} \tag{4.299}$$

Referring to the definition of the modal participation factor (equation 4.267), equation 4.299 can be rewritten.

$$V_{max}^{(i)} = M_i \alpha_i^2 S_A^{(i)} = M_i^* S_A^{(i)} \tag{4.300}$$

The term M_i^* represents the modal mass for mode i. The sum of all modal masses from $i = 1$ to N must be equal to the total mass of the structure.

4.8.3 Number of Modes to Consider

Obviously, it is not necessary to consider the effect of all modes of vibration to obtain an acceptable level of accuracy for civil-engineering structures. To model a three-dimensional building, the number of modes required for dynamic modal analysis or for spectral analysis depends on the direction of the base excitation and the degree of coupling between the translational and torsional modes. Modal mass can be used (equation 4.300) as an indicator of the number of modes to consider in the analysis. The number of modes to be used must represent at least 90% of the total mass of the structure (Carr, 1994).

4.8.4 Statistical Combinations of Maximum Modal Responses

Motivation. To determine the maximum response of a DDOF, $x_{i\,max}$, maximum modal responses cannot be directly added. These maximum values do not occur at the same time and summing these absolute values is too conservative. In practice, a statistical combination of maximum modal responses must be used to estimate the maximum response of each DDOF. The most common combinations used are briefly described below.

   Square root of the sum of the squares (SRSS) combination. This combination is often used in practice.

$$x_{i\,max} = \left| x_i(t) \right|_{max} \approx \sqrt{\left| x_i^{(1)} \right|^2_{max} + \left| x_i^{(2)} \right|^2_{max} + \dots \left| x_i^{(N_m)} \right|^2_{max}} \qquad (4.301)$$

   This combination method, also called probable response, generally gives good results for systems with well-separated natural frequencies. Originally, this method was proposed for the two-dimensional analysis of structures; therefore, the lateral natural frequencies are not close together. In a three-dimensional analysis, however, certain modes in different directions may have similar natural frequencies. Consequently, the SRSS combination may produce unconservative results.

   The complete quadratic combination (CQC). When a system has closely spaced natural frequencies, the CQC combination, proposed by Wilson, Der Kiureghian, and Bayo (1981), provides a better estimate. This method is based on the random vibration theory and gives exact results if the accelerogram is represented by white noise. If the frequencies are well defined, the results converge toward the SRSS combination.

$$x_{i\,max} \approx \sqrt{\sum_{j=1}^{N_m} \sum_{k=1}^{N_m} x_{i\,max}^{(j)} \rho_{jk} x_{i\,max}^{(k)}} \qquad (4.302)$$

where $x_{i\,max}^{(j)}$ = maximum modal response in mode j (+ or -)

$\quad\quad\quad x_{i\,max}^{(k)}$ = maximum modal response in mode k (+ or -)

$\quad\quad\quad \rho_{jk}$ = correlation coefficient between modes j and k

$$\rho_{jk} = \frac{8\sqrt{\zeta_j \zeta_k}\left(\zeta_j + r\,\zeta_k\right) r^{\frac{3}{2}}}{\left(1 - r^2\right)^2 + 4\,\zeta_j \zeta_k r\left(1 + r^2\right) + 4\left(\zeta_j^2 + \zeta_k^2\right)r^2} \tag{4.303}$$

and

$$r = \frac{\omega_k}{\omega_j} = \text{frequency ratio}$$
$$\zeta_j = \text{critical damping ratio in mode } j \tag{4.304}$$

The double-sum combination (DSC). The double-sum combination, proposed by Rosenblueth and Elorduy (1969), takes the same form as equation 4.302 except for the correlation coefficients.

$$\rho_{jk} = \frac{1}{\left[1 + \left(\dfrac{\left(\omega_{dj} - \omega_{dk}\right)}{\left(\zeta_j^* \omega_j + \zeta_k^* \omega_k\right)}\right)^2\right]} \tag{4.305}$$

where

$$\omega_{dj} = \omega_j \sqrt{1 - \left(\zeta_j^*\right)^2}$$
$$\zeta_j^* = \zeta_j + \frac{2}{S\,\omega_j} \tag{4.306}$$

where S represents the duration of the central portion of the accelerogram showing the characteristics of stationary white noise. For historical earthquakes, S is associated with the strong motion duration (section 2.12.3).

Humar's combination. For the torsional earthquake response of asymmetric buildings, Humar (1984) proposed to use the DSC combination by making S equal to infinity and using the same damping ratio, ζ, for each mode of vibration (the assumption generally used in practice).

$$\rho_{jk} = \frac{1}{\left[1 + \left(\dfrac{\sqrt{1 - \zeta^2}}{\zeta}\left(\dfrac{\omega_j - \omega_k}{\omega_j + \omega_k}\right)\right)^2\right]} \tag{4.307}$$

Gupta's combination. Gupta (1990) proposed to modify the correlation coefficients of the DSC combination in the following manner:

$$\rho_{jk} = \cfrac{1}{\left[1 + \left(\cfrac{\sqrt{1-\zeta^2}\left(\omega_j - \omega_k\right)}{\left(\zeta\omega_j + \zeta\omega_k + c_{jk}\right)}\right)^2\right]} \tag{4.308}$$

where

$$c_{jk} = \left(0,16 - 0,5\,\zeta\right)\left(1,4 - \left|\omega_j^2 - \omega_k^2\right|\right) \geq 0 \tag{4.309}$$

This combination produces greater values for the correlation coefficients when two frequencies are very close together. When the frequency difference increases, however, the combination degenerates to a CQC combination and a Humar's combination.

Final remarks on statistical combinations. Carr (1994) compared the correlation coefficients generated by the different statistical combinations of the maximum modal responses. He concluded that when two frequencies, associated with two different modes, have a variation exceeding 20%, the coefficients are negligible and all combinations degenerate to the SRSS combination.

It is extremely important to understand that whatever statistical combination is used, it must be applied to the maximum parameter to be calculated (displacement, force, etc.). First, the maximum value of this parameter must be calculated in each mode of vibration, and only then can the statistical combination be used on these modal responses.

4.8.5 Summary of the Spectral Method

Spectral analysis is a powerful tool for estimating maximum dynamic response of any linear structure with lumped mass subjected to an arbitrary seismic load, which is represented by its elastic earthquake response spectrum. Here is a summary of the steps in spectral analysis.

Step 1: Selection of degrees of freedom and computation of mass and stiffness matrices
- Selection of the most important DDOF.
- Construction of the mass and stiffness matrices.
- Assigning a damping ratio to each mode.

Step 2: Selection of the earthquake response spectrum
Selection of the response spectrum generally depends on the seismicity of the region. The choice can be a historical earthquake spectrum or a simplified design response spectrum (section 4.4).

Step 3: Calculation of natural frequencies and mode shapes
The eigenvalues and eigenvectors are solved for each mode i.

$$\left[[k] - \omega_i^2[m]\right](A^{(i)}) = (0) \tag{4.310}$$

Step 4: Generalized masses calculation
The generalized mass, M_i, is calculated for each mode i.

$$M_i = (A^{(i)})^T[m](A^{(i)}) \tag{4.311}$$

Step 5: Calculation of the modal participation factors.
The modal participation factor is calculated for each mode i.

$$\alpha_i = \frac{(A^{(i)})^T[m](I)}{M_i} \tag{4.312}$$

Step 6: Calculation of spectral responses
The relative displacement spectral response, $S_D^{(i)}$, is calculated for each mode corresponding to ω_i and ζ_i.

Step 7: Calculation of maximum modal responses
The maximum response for DDOF i is calculated in each mode j.

$$x_{i\,\mathrm{max}}^{(j)} = A_i^{(j)} \alpha_j S_D^{(j)} \tag{4.313}$$

Step 8: Estimation of maximum responses by statistical combination of the maximum modal responses
Maximum modal responses are combined with one of the statistical methods presented in section 4.8.4 to estimate the maximum response of each DDOF. In practice, the SRSS method is often chosen for its simplicity.

$$x_{i\,\mathrm{max}} = |x_i(t)|_{\mathrm{max}} \approx \sqrt{\left|x_i^{(1)}\right|_{\mathrm{max}}^2 + \left|x_i^{(2)}\right|_{\mathrm{max}}^2 + \dots \left|x_i^{(N_m)}\right|_{\mathrm{max}}^2} \tag{4.314}$$

In this section, the word "response" was associated with the relative displacements of each DDOF and the SRSS combination was used on displacements. As stated above, when estimating the maximum value of another parameter (shear at each floor, for example), the maximum value in each mode must be found and the statistical combination applied on this parameter and not on the displacements. The following example demonstrates this point, which proves to be important when using the spectral method.

4.8.6 Example

Let us consider that the three-storey structure shown in figure 4.53 is subjected to the El Centro earthquake (1940-05-18, comp. S00E). Estimate the probable maximum base shear using the SRSS combination. Suppose a damping of 2% critical for each mode of vibration.

Step 1

The mass and stiffness matrices are given. The reader may verify these results using standard structural analysis methods.

$$[m] = \begin{bmatrix} 0{,}175 & 0 & 0 \\ 0 & 0{,}263 & 0 \\ 0 & 0 & 0{,}350 \end{bmatrix} \frac{kN \bullet s^2}{mm}$$

$$[k] = 105 \begin{bmatrix} 1 & -1 & 0 \\ -1 & 3 & -2 \\ 0 & -2 & 5 \end{bmatrix} \frac{kN}{mm}$$

$$(4.315)$$

Step 2

Figure 4.55 illustrates the response spectrum of the El Centro earthquake.

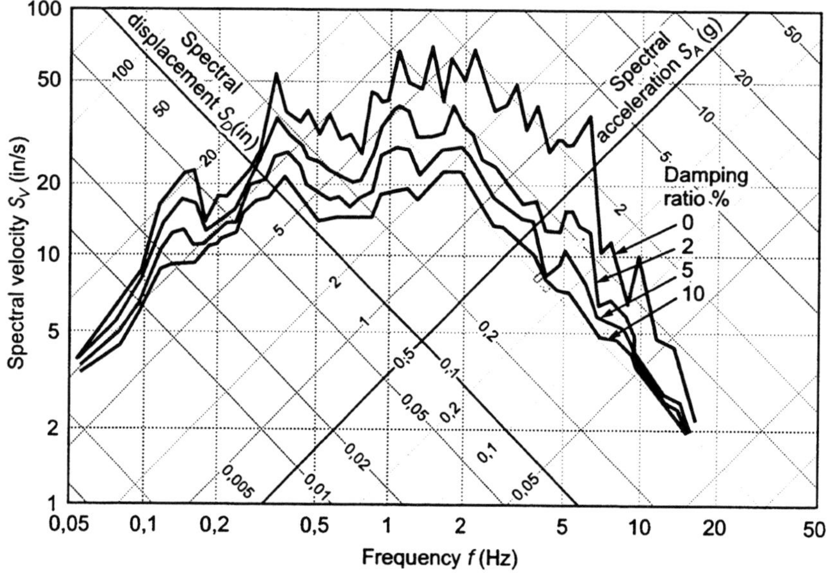

Figure 4.55   Response spectrum of the El Centro earthquake (1940-05-18, comp. S00E).

Step 3

The natural frequencies and the mode shapes are given. The reader may verify these results using methods discussed above.

$$\omega_1 = 14,5\,\text{rad/s}; \quad \omega_2 = 31,1\,\text{rad/s}; \quad \omega_3 = 46,1\,\text{rad/s}$$

$$(A^{(1)}) = \begin{pmatrix} 1,000 \\ 0,644 \\ 0,300 \end{pmatrix}; \quad (A^{(2)}) = \begin{pmatrix} 1,000 \\ -0,601 \\ -0,676 \end{pmatrix}; \quad (A^{(3)}) = \begin{pmatrix} 1,000 \\ -2,570 \\ 2,470 \end{pmatrix} \tag{4.316}$$

Step 4

The generalized masses are calculated.

$$M_1 = (1,000 \quad 0,644 \quad 0,300) \begin{bmatrix} 0,175 & 0 & 0 \\ 0 & 0,263 & 0 \\ 0 & 0 & 0,350 \end{bmatrix} \begin{pmatrix} 1,000 \\ 0,644 \\ 0,300 \end{pmatrix} = 0,315\ \frac{\text{kN}\cdot\text{s}^2}{\text{mm}}$$

$$M_2 = (1,000 \quad -0,601 \quad -0,676) \begin{bmatrix} 0,175 & 0 & 0 \\ 0 & 0,263 & 0 \\ 0 & 0 & 0,350 \end{bmatrix} \begin{pmatrix} 1,000 \\ -0,601 \\ -0,676 \end{pmatrix} = 0,430\ \frac{\text{kN}\cdot\text{s}^2}{\text{mm}} \tag{4.317}$$

$$M_3 = (1,000 \quad -2,570 \quad 2,470) \begin{bmatrix} 0,175 & 0 & 0 \\ 0 & 0,263 & 0 \\ 0 & 0 & 0,350 \end{bmatrix} \begin{pmatrix} 1,000 \\ -2,570 \\ 2,470 \end{pmatrix} = 4,407\ \frac{\text{kN}\cdot\text{s}^2}{\text{mm}}$$

Step 5

The modal participation factors are calculated.

$$\alpha_1 = \frac{1}{0,315}(1,000 \quad 0,644 \quad 0,300)\begin{bmatrix} 0,175 & 0 & 0 \\ 0 & 0,2630 & \\ 0 & 0 & 0,350 \end{bmatrix}\begin{pmatrix} 1 \\ 1 \\ 1 \end{pmatrix} = 1,425$$

$$\alpha_2 = \frac{1}{0,430}(1,000 \quad -0,601 \quad -0,676)\begin{bmatrix} 0,175 & 0 & 0 \\ 0 & 0,263 & 0 \\ 0 & 0 & 0,350 \end{bmatrix}\begin{pmatrix} 1 \\ 1 \\ 1 \end{pmatrix} = -0,511 \qquad (4.318)$$

$$\alpha_3 = \frac{1}{4,407}(1,000 \quad -2,570 \quad 2,470)\begin{bmatrix} 0,175 & 0 & 0 \\ 0 & 0,263 & 0 \\ 0 & 0 & 0,350 \end{bmatrix}\begin{pmatrix} 1 \\ 1 \\ 1 \end{pmatrix} = 0,090$$

Step 6

Using the response spectrum shown in figure 4.55, we find the spectral responses in each mode at 2% critical damping.

- First mode, $f_1 = 2,308$ Hz

$$S_D^{(1)} = 2,00 \text{ in} = 50,8 \text{ mm} \qquad (4.319)$$

- Second mode, $f_2 = 4,950$ Hz

$$S_D^{(2)} = 0,46 \text{ in} = 11,7 \text{ mm} \qquad (4.320)$$

- Third mode, $f_3 = 7,337$ Hz

$$S_D^{(3)} = 0,20 \text{ in} = 5,1 \text{ mm} \qquad (4.321)$$

Step 7

Now, the maximum modal responses are calculated in each mode, for each DDOF.

- First mode

$$
\begin{aligned}
x_{1^{(1)}_{max}} &= (1{,}000)(1{,}425)(50{,}8) = 72{,}4 \ \text{mm} \\
x_{2^{(1)}_{max}} &= (0{,}644)(1{,}425)(50{,}8) = 46{,}6 \ \text{mm} \\
x_{3^{(1)}_{max}} &= (0{,}300)(1{,}425)(50{,}8) = 21{,}7 \ \text{mm}
\end{aligned}
\tag{4.322}
$$

- Second mode

$$
\begin{aligned}
x_{1^{(2)}_{max}} &= (1{,}00)(-0{,}511)(11{,}7) = -6{,}0 \ \text{mm} \\
x_{2^{(2)}_{max}} &= (-0{,}601)(-0{,}511)(11{,}7) = 3{,}6 \ \text{mm} \\
x_{3^{(2)}_{max}} &= (-0{,}676)(-0{,}511)(11{,}7) = 4{,}0 \ \text{mm}
\end{aligned}
\tag{4.323}
$$

- Third mode

$$
\begin{aligned}
x_{1^{(3)}_{max}} &= (1{,}00)(0{,}090)(5{,}1) = 0{,}5 \ \text{mm} \\
x_{2^{(3)}_{max}} &= (-2{,}57)(0{,}090)(5{,}1) = -1{,}2 \ \text{mm} \\
x_{3^{(3)}_{max}} &= (2{,}47)(0{,}090)(5{,}1) = 1{,}1 \ \text{mm}
\end{aligned}
\tag{4.324}
$$

Step 8

Now, the maximum equivalent seismic loads can be calculated for each mode by multiplying the maximum modal responses by the stiffness matrix, $[k]$.

$$
(Q^{(i)})_{max} = [k](x^{(i)})_{max}
\tag{4.325}
$$

- First mode

$$
\begin{pmatrix} Q_{1^{(1)}_{max}} \\ Q_{2^{(1)}_{max}} \\ Q_{3^{(1)}_{max}} \end{pmatrix}
= 105 \begin{bmatrix} 1 & -1 & 0 \\ -1 & 3 & -2 \\ 0 & -2 & 5 \end{bmatrix}
\begin{pmatrix} 72{,}4 \\ 46{,}6 \\ 21{,}7 \end{pmatrix}
= \begin{pmatrix} 2\ 709 \\ 2\ 520 \\ 1\ 607 \end{pmatrix} \ \text{kN}
\tag{4.326}
$$

- Second mode

$$\begin{pmatrix} Q_{1^{(2)}_{max}} \\ Q_{2^{(2)}_{max}} \\ Q_{3^{(2)}_{max}} \end{pmatrix} = 105 \begin{bmatrix} 1 & -1 & 0 \\ -1 & 3 & -2 \\ 0 & -2 & 5 \end{bmatrix} \begin{pmatrix} -6,0 \\ 3,6 \\ 4,0 \end{pmatrix} = \begin{pmatrix} -1\,008 \\ 924 \\ 1\,344 \end{pmatrix} \text{ kN} \tag{4.327}$$

- Third mode

$$\begin{pmatrix} Q_{1^{(3)}_{max}} \\ Q_{2^{(3)}_{max}} \\ Q_{3^{(3)}_{max}} \end{pmatrix} = 105 \begin{bmatrix} 1 & -1 & 0 \\ -1 & 3 & -2 \\ 0 & -2 & 5 \end{bmatrix} \begin{pmatrix} 0,5 \\ -1,2 \\ 1,1 \end{pmatrix} = \begin{pmatrix} 179 \\ -662 \\ 830 \end{pmatrix} \text{ kN} \tag{4.328}$$

Figure 4.56 illustrates these maximum equivalent seismic loads for each mode.

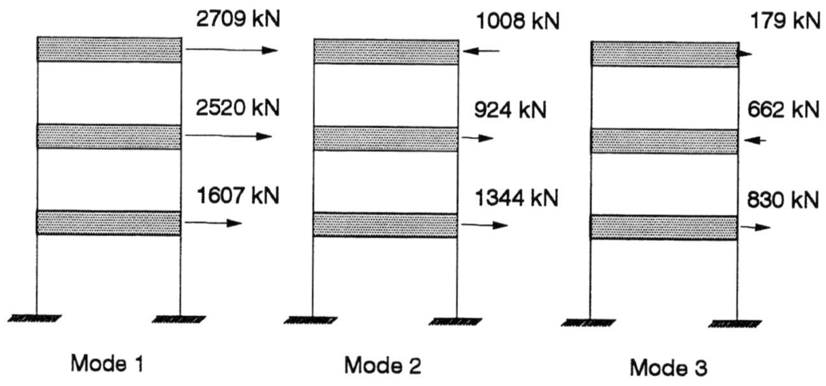

Figure 4.56 Maximum equivalent seismic loads for each mode of a three-storey structure.

Step 9

The maximum base shear for each mode is obtained by algebraically adding the maximum seismic loads.

- First mode

$$V^{(1)}_{max} = 2709 + 2520 + 1607 = 6836 \text{ kN} \tag{4.329}$$

- Second mode

$$V^{(2)}_{\text{max}} = -1008 + 924 + 1344 = 1260 \text{ kN} \qquad (4.330)$$

- Third mode

$$V^{(3)}_{\text{max}} = 179 - 662 + 830 = 347 \text{ kN} \qquad (4.331)$$

Step 10
The maximum probable base shear can be estimated by combining the maximum modal shears by the SRSS method.

$$|V_{\text{base}}|_{\text{max}} \approx \sqrt{|6836|^2 + |1260|^2 + |347|^2} = 6960 \text{ kN} \qquad (4.332)$$

The first mode is dominant in this base shear calculation. In this first dominant mode, the maximum seismic loads are more or less distributed as an inverted triangle (fig. 4.56). The seismic-load distribution of the static method included in the NBCC (1995) is based on these results (see chapter 6).

4.9 INTRODUCTION TO NONLINEAR ANALYSIS OF MULTI-DEGREE-OF-FREEDOM STRUCTURES

4.9.1 Basic Concept

In practice, and especially during earthquake response of civil engineering structures, the physical properties of a structure vary during vibrations. As a result, the accuracy of the linear analysis becomes uncertain. A nonlinear dynamic analysis must then be performed to obtain a realistic solution. A few examples requiring this type of nonlinear analysis are:

- If the elastic limit of the material is reached, therefore causing yielding of certain members, the global stiffness of the structure will be modified during the dynamic response.
- If the structure is slender, the axial loads in the columns can induce important second order (P-Delta) effects, once again causing a reduction of the structure's stiffness during the dynamic response.
- If the content of a reservoir is evacuated during the vibrations, then its mass will vary.

If the properties of the structure change during the dynamic response, a nonlinear analysis must be used. In this case, the coupled equations of motion are integrated directly. This approach is called a time-step (or piece-wise) integration procedure. The complete response of the system is divided into short time increments (time-steps). The system's response is obtained at each time-step by assuming that the system is linear based on the calculated properties at the beginning of the time-step. At the end of each time-step, the system's properties are modified according to the levels of stress and strains in the members. In other words, the nonlinear response of the system is estimated by a succession of linear analyses with variable dynamic properties.

4.9.2 Incremental Equations of Motion

As an example, let us consider the multi-storey nonlinear structure shown in figure 4.57. We assume a single concentrated mass at each floor, m_i, only one horizontal DDOF per floor, x_i, and a base acceleration, $\ddot{x}_s(t)$.

The equation of motion for mass i at time t is:

$$m_i\ddot{x}_i(t) + F_{si}(t) + F_{Di}(t) = -m_i\ddot{x}_s(t) \tag{4.333}$$

where $F_{si}(t)$ = total nonlinear restoring force applied on mass i at time t
  $F_{di}(t)$ = total nonlinear damping force applied on mass i at time t
  $x_i(t), \dot{x}_i(t), \ddot{x}_i(t)$ = relative displacement, relative velocity, and relative acceleration of mass i
  $\ddot{x}_s(t)$ = base acceleration

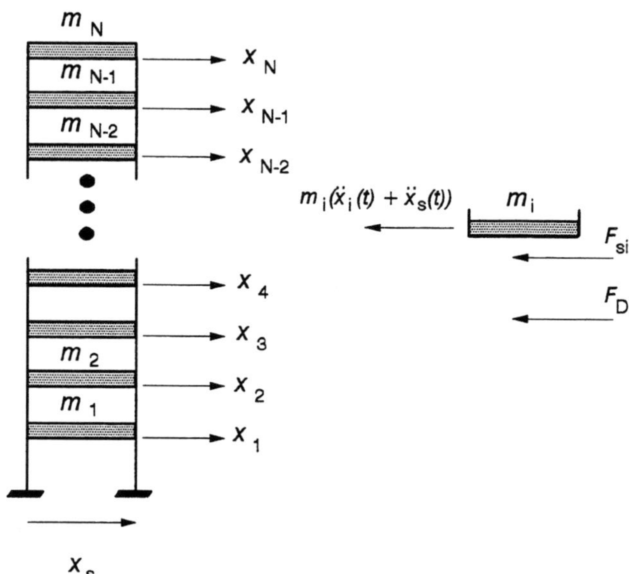

Figure 4.57 Multi-storey nonlinear structure subjected to a base excitation.

The restoring force, $F_{si}(t)$, and the damping force, $F_{Di}(t)$, represent the influence of all elements of the system on mass i and are written as:

$$F_{si}(t) = \sum_{j=1}^{N} f_{sij}(t)$$

$$F_{Di}(t) = \sum_{j=1}^{N} f_{Dij}(t)$$

(4.334)

where $f_{sij}(t)$ = nonlinear force applied to mass i at time t to induce a unit displacement of mass j, $x_j(t)$, at time t (all other masses have zero displacement)

$f_{Dij}(t)$ = nonlinear force applied to mass i at time t to induce a unit velocity in mass j, $\dot{x}_j(t)$, at time t (all other masses have zero velocity)

We assume that the nonlinear properties of the system and the characteristics of the base accelerogram are known (fig. 4.58). Now, the equations of motion for mass i are written for one time increment later $t+\Delta t$.

$$m_i \ddot{x}_i(t+\Delta t) + F_{si}(t+\Delta t) + F_{Di}(t+\Delta t) = -m_i \ddot{x}_s(t+\Delta t)$$

(4.335)

Equation 4.334 is subtracted from equation 4.335.

$$m_i\left[\ddot{x}_i(t+\Delta t) - \ddot{x}_i(t)\right] + \left[F_{si}(t+\Delta t) - F_{si}(t)\right] + \left[F_{Di}(t+\Delta t) - F_{Di}(t)\right]$$

$$= -m_i\left[\ddot{x}_s(t+\Delta t) - \ddot{x}_s(t)\right]$$

(4.336)

or

$$m_i \Delta \ddot{x}_i(t) + \Delta F_{si}(t) + \Delta F_{Di}(t) = -m_i \Delta \ddot{x}_s(t)$$

(4.337)

These equations can be combined for all N masses of the system and can be written in a matrix format.

$$[m]_{N\text{x}N}\left(\Delta\ddot{x}(t)\right)_{N\text{x}1} + \left(\Delta F_s(t)\right)_{N\text{x}1} + \left(\Delta F_D(t)\right)_{N\text{x}1}$$

$$= -[m]_{N\text{x}N}(r)_{N\text{x}1}\,\Delta\ddot{x}_s(t)$$

(4.338)

where (fig. 4.58)

$$\left(\Delta F_s(t)\right)_{N\text{x}1} = \left[k(t)\right]_{N\text{x}N}\left(\Delta x(t)\right)_{N\text{x}1}$$

$$\left(\Delta F_D(t)\right)_{N\text{x}1} = \left[c(t)\right]_{N\text{x}N}\left(\Delta\dot{x}(t)\right)_{N\text{x}1}$$

(4.339)

$$\Delta\ddot{x}_s(t) = \ddot{x}_s(t+\Delta t) - \ddot{x}_s(t)$$

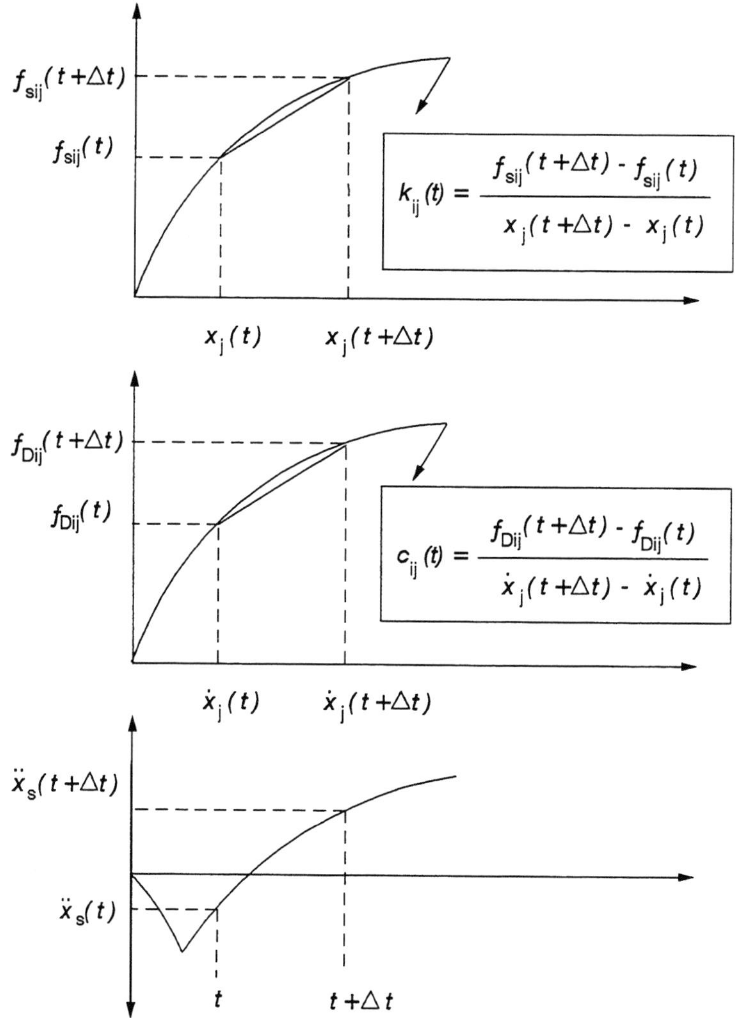

Figure 4.58 Properties of a nonlinear system.

The elements of matrices $[k(t)]$ and $[c(t)]$ are calculated for each time-step using the secants shown in figure 4.58. This procedure requires an iteration at each step because, to calculate the secants, it is necessary to know the displacement and velocity vectors at the end of each step. To reduce the amount of time required for the calculation, the tangent can be used at the beginning of the integration step to solve for the elements of these matrices.

$$c_{ij}(t) \approx \left(\frac{df_{dij}}{d\dot{x}_j} \right)_t$$

$$k_{ij}(t) \approx \left(\frac{df_{sij}}{dx_j} \right)_t \tag{4.340}$$

Equation 4.339 is substituted in equation 4.338 to obtain the incremental equations of motion.

$$[m]\big(\Delta\ddot{x}(t)\big) + \big[c(t)\big]\big(\Delta\dot{x}(t)\big) + \big[k(t)\big]\big(\Delta x(t)\big)$$
$$= -[m](r)\,\Delta\ddot{x}_s(t) \tag{4.341}$$

The assumption that the mass matrix is constant proves to be arbitrary and is therefore unnecessary. However, this assumption is reasonable when a civil-engineering structure is subjected to an earthquake. There are many numerical techniques for solution of equation 4.341. In this chapter, we will study two of these techniques: the average constant acceleration method and the linear acceleration method.

4.9.3 Average Constant Acceleration Method

The basic assumption of this method is that the relative acceleration of each DDOF is constant during a time-step and that the properties of the system do not change during this time. This assumption is illustrated in figure 4.59.

The relative acceleration of the DDOF i during one time-step is written as follows:

$$\ddot{x}_i(\tau) = \frac{1}{2}\Big[\ddot{x}_i(t) + \ddot{x}_i(t+\Delta t)\Big] \quad t \le \tau \le t+\Delta t \tag{4.342}$$

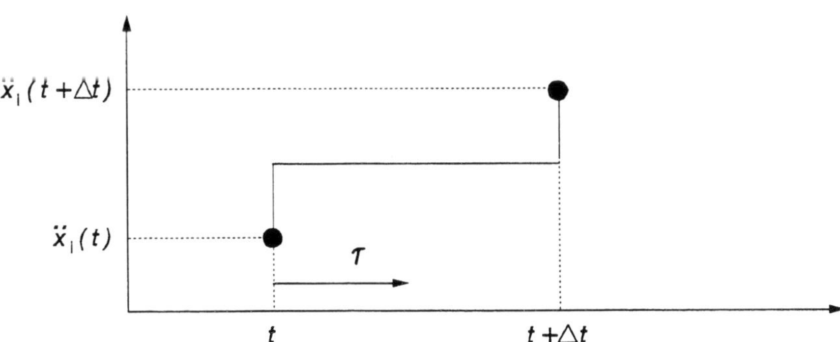

Figure 4.59   Assumption of average constant acceleration method.

The relative velocity of the DDOF i during one time-step is obtained by integrating equation 4.342.

$$\dot{x}_i(\tau) = \dot{x}_i(t) + \int_t^\tau \ddot{x}_i(\tau)\,d\tau \tag{4.343}$$

$$\dot{x}_i(\tau) = \dot{x}_i(t) + \frac{1}{2}\Big[\ddot{x}_i(t) + \ddot{x}_i(t+\Delta t)\Big]\big[\tau - t\big] \tag{4.344}$$

The velocity at the end of the time-step is:

$$\dot{x}_i(t+\Delta t) = \dot{x}_i(t) + \frac{1}{2}\Delta t\left[\ddot{x}_i(t)+\ddot{x}_i(t+\Delta t)\right]$$

$$\dot{x}_i(t+\Delta t) = \dot{x}_i(t) + \frac{1}{2}\Delta t\left[2\ddot{x}_i(t)+\Delta\ddot{x}_i(t)\right] \qquad (4.345)$$

The increment of the velocity during the time-step is:

$$\Delta\dot{x}_i(t) = \dot{x}_i(t+\Delta t) - \dot{x}_i(t) = \frac{1}{2}\Delta t\left[2\ddot{x}_i(t)+\Delta\ddot{x}_i(t)\right] \qquad (4.346)$$

Combining all DDOF, we can write an equation for the vector of incremental velocities.

$$\left(\Delta\dot{x}(t)\right) = \frac{1}{2}\Delta t\left(2\left(\ddot{x}(t)\right)+\left(\Delta\ddot{x}(t)\right)\right) \qquad (4.347)$$

Similarly, the relative displacement of the DDOF *i* during one time-step is written as follows:

$$x_i(\tau) = x_i(t) + \int_t^\tau \dot{x}_i(\tau)d\tau \qquad (4.348)$$

Equation 4.344 is replaced in equation 4.348.

$$x_i(\tau) = x_i(t) + (\tau-t)\dot{x}_i(t) + \frac{1}{4}(\tau-t)^2\left[\ddot{x}_i(t)+\ddot{x}_i(t+\Delta t)\right] \qquad (4.349)$$

The displacement increment during the time-step is then:

$$\Delta x_i(t) = x_i(t+\Delta t) - x_i(t) = (\Delta t)\dot{x}_i(t) + (\Delta t)^2\left[\frac{1}{2}\ddot{x}_i(t)+\frac{1}{4}\Delta\ddot{x}_i(t)\right] \qquad (4.350)$$

By combining all DDOFs, an equation for the vector of incremental displacements can be found.

$$\left(\Delta x(t)\right) = \Delta t\left(\dot{x}(t)\right) + [\Delta t]^2\left(\frac{1}{2}\left(\ddot{x}(t)\right)+\frac{1}{4}\left(\Delta\ddot{x}(t)\right)\right) \qquad (4.351)$$

4.9.4 Linear Acceleration Method

The basic assumption of this method is that the relative acceleration for each DDOF varies linearly during one time-step and that the properties of the system remain constant throughout the time-step (fig. 4.60).

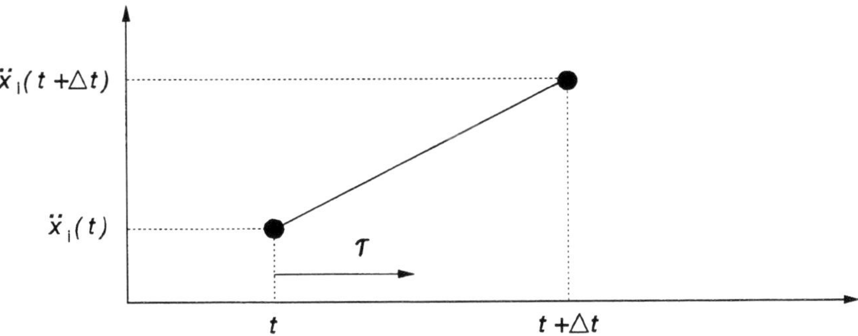

Figure 4.60 Assumption of linear acceleration method.

The relative acceleration of the DDOF *i* during one time-step is written as:

$$\ddot{x}_i(\tau) \;=\; \ddot{x}_i(t) \;+\; \left[\frac{(\tau-t)}{\Delta t}\right]\left[\ddot{x}_i(t+\Delta t)-\ddot{x}_i(t)\right] \quad t\le \tau\le t+\Delta t \tag{4.352}$$

The relative velocity of the DDOF *i* during one time-step is obtained by integrating equation 4.352.

$$\dot{x}_i(\tau) \;=\; \dot{x}_i(t) \;+\; (\tau-t)\ddot{x}_i(t) \;+\; \left[\frac{(\tau-t)^2}{2\Delta t}\right]\left[\ddot{x}_i(t+\Delta t)-\ddot{x}_i(t)\right] \tag{4.353}$$

The velocity at the end of the time-step is:

$$\dot{x}_i(t+\Delta t) \;=\; \dot{x}_i(t) \;+\; \Delta t\,\ddot{x}_i(t) \;+\;\frac{1}{2}\Delta t\,\Delta\ddot{x}_i(t) \tag{4.354}$$

The velocity increment during the time-step is obtained by:

$$\Delta\dot{x}_i(t) \;=\; \Delta t\,\ddot{x}_i(t) \;+\;\frac{1}{2}\Delta t\,\Delta\ddot{x}_i(t) \tag{4.355}$$

Combining all DDOF, we find an equation for the vector of incremental velocities.

$$\bigl(\Delta\dot{x}(t)\bigr) \;=\; \Delta t\bigl(\ddot{x}(t)\bigr)+\frac{1}{2}\Delta t\bigl(\Delta\ddot{x}(t)\bigr) \tag{4.356}$$

Similarly, the relative displacement of the DDOF *i* during one time-step is:

$$x_i(\tau) \;=\; x_i(t) \;+\; \dot{x}_i(t)(\tau-t) \;+\; \frac{1}{2}(\tau-t)^2\ddot{x}_i(t) \;+\; \left[\frac{(\tau-t)^3}{6\Delta t}\right]\Delta\ddot{x}_i(t) \quad t\le \tau\le t+\Delta t \tag{4.357}$$

The displacement increment during the time-step is obtained by the following equation:

$$\Delta x_i(t) = x_i(t + \Delta t) - x_i(t) = (\Delta t)\dot{x}_i(t) + (\Delta t)^2\left[\frac{1}{2}\ddot{x}_i(t) + \frac{1}{6}\Delta\ddot{x}_i(t)\right] \tag{4.358}$$

By combining all DDOF, we find an equation for the vector of incremental displacements.

$$(\Delta x(t)) = \Delta t(\dot{x}(t)) + [\Delta t]^2\left(\frac{1}{2}(\ddot{x}(t)) + \frac{1}{6}(\Delta\ddot{x}(t))\right) \tag{4.359}$$

4.9.5 Newmark-Beta's Algorithm

Newmark-Beta's algorithm, which is very useful in programming, combines the constant acceleration method and the linear acceleration method. The vectors of incremental velocities and incremental displacements are expressed by the following general equation:

$$(\Delta\dot{x}(t)) = \Delta t(\ddot{x}(t)) + \Gamma\Delta t(\Delta\ddot{x}(t)) \tag{4.360}$$

$$(\Delta x(t)) = \Delta t(\dot{x}(t)) + [\Delta t]^2\left(\frac{1}{2}(\ddot{x}(t)) + \beta(\Delta\ddot{x}(t))\right) \tag{4.361}$$

- If $\Gamma = 1/2$ and $\beta = 1/4$

Equations 4.360 and 4.361 are identical to equations 4.347 and 4.351 and represent the constant acceleration method.

- If $\Gamma = 1/2$ and $\beta = 1/6$

Equations 4.360 and 4.361 are identical to equations 4.356 and 4.359 and represent the linear acceleration method.

4.9.6 Integration of Equations of Motion

To solve the displacement, velocity, and acceleration vectors for each time-step, the displacement vector, $(x(t))$, is considered the basic variable. Equation 4.361 is first written as a function of the vector of incremental accelerations.

$$(\Delta\ddot{x}(t)) = \left[\frac{1}{\beta[\Delta t]^2}\right](\Delta x(t)) - \left[\frac{1}{\beta\Delta t}\right](\dot{x}(t)) - \left[\frac{1}{2\beta}\right](\ddot{x}(t)) \tag{4.362}$$

Equation 4.362 is substituted in equation 4.360.

$$(\Delta\dot{x}(t)) = \Delta t(\ddot{x}(t)) + \left[\frac{\Gamma}{\beta\Delta t}\right](\Delta x(t)) - \left[\frac{\Gamma}{\beta}\right](\dot{x}(t)) - \left[\frac{\Gamma\Delta t}{2\beta}\right](\ddot{x}(t)) \tag{4.363}$$

In order to keep only one variable for this problem, $(x(t))$, equations 4.362 and 4.363 are substituted into equation 4.341 (incremental equations of motion).

$$[m]\left(\left[\frac{1}{\beta[\Delta t]^2}\right](\Delta x(t)) - \left[\frac{1}{\beta \Delta t}\right](\dot{x}(t)) - \left[\frac{1}{2\beta}\right](\ddot{x}(t))\right)$$

$$+$$

$$[c(t)]\left(\left[\frac{\Gamma}{\beta \Delta t}\right](\Delta x(t)) - \left[\frac{\Gamma}{\beta}\right](\dot{x}(t)) - \left[\left[\frac{\Gamma}{2\beta}\right] - 1\right]\Delta t(\ddot{x}(t))\right) \tag{4.364}$$

$$+$$

$$[k(t)](\Delta x(t)) = -[m](r)\Delta \ddot{x}_s(t)$$

or

$$[\tilde{k}(t)](\Delta x(t)) = (\Delta \tilde{P}(t)) \tag{4.365}$$

$$[\tilde{k}(t)] = [k(t)] + \left[\frac{1}{\beta[\Delta t]^2}\right][m] + \left[\frac{\Gamma}{\beta \Delta t}\right][c(t)] \tag{4.366}$$

$$(\Delta \tilde{P}(t)) = -[m](r)\Delta \ddot{x}_s(t)$$
$$+$$
$$[m]\left(\left[\frac{1}{\beta \Delta t}\right](\dot{x}(t)) + \left[\frac{1}{2\beta}\right](\ddot{x}(t))\right)$$
$$+ \tag{4.367}$$
$$[c(t)]\left(\left[\frac{\Gamma}{\beta}\right](\dot{x}(t)) + \left[\left[\frac{\Gamma}{2\beta}\right] - 1\right]\Delta t(\ddot{x}(t))\right)$$

Equation 4.365 is a system of linear equations that can be solved for the vector of incremental displacements. Then, the vector of incremental velocities can be obtained (equation. 4.363). From these results, we can solve for the displacement and velocity vectors at the beginning of the next time-step.

$$(x(t+\Delta t)) = (x(t)) + (\Delta x(t))$$
$$(\dot{x}(t+\Delta t)) = (\dot{x}(t)) + (\Delta \dot{x}(t)) \tag{4.368}$$

The vector of incremental accelerations can be solved by equation 4.362. However, this could create a series of numerical errors due to the fact that the tangent stiffness and damping matrices are used instead of the secant values. To avoid this problem, we compute the acceleration vector for the next time-step by imposing the dynamic equilibrium equation directly.

$$\left(\ddot{x}(t+\Delta t)\right) = [m]^{-1}\left(\left(-[m](r)\ddot{x}_s(t+\Delta t)\right) - \left(F_D(t+\Delta t)\right) - \left(F_s(t+\Delta t)\right)\right) \tag{4.369}$$

where

$$\left(F_s(t+\Delta t)\right) = \left(F_s(t)\right) + [k(t)](\Delta x(t))$$

$$\left(F_D(t+\Delta t)\right) = \left(F_D(t)\right) + [c(t)](\Delta \dot{x}(t)) \tag{4.370}$$

4.9.7 Summary of Nonlinear Time-Step Dynamic Analysis

A nonlinear time-step dynamic analysis of a MDOF structure involves the following steps:

Step 1
Obtain the displacement and velocity vectors, $(x(t))$ and $(\dot{x}(t))$, from the initial conditions of the problem or from the results of the previous time-step.

Step 2
From the displacement and velocity vectors, calculate the tangent stiffness and damping matrices, $[k(t)]$ and $[c(t)]$.

Step 3
Calculate the acceleration vector at the beginning of the time-step, $(\ddot{x}(t))$, using equation 4.369.

Step 4
Solve for the vector of incremental displacements, $(\Delta x(T))$, using equation 4.365.

Step 5
Solve for the vector of incremental velocities, $(\Delta \dot{x}(t))$, using equation 4.363.

Step 6
Calculate the new initial conditions for the next time-step, using equation 4.368.

Step 7
Repeat this process until the end of the analysis.

We can see that the nonlinear analysis of a MDOF structure is much more complex and time consuming than a linear analysis (modal or spectral) of the same structure. For each time-step, an equivalent static problem needs to be solved (equation 4.465).

4.9.8 Stability and Accuracy of Numerical Results

In a linear system, the average constant acceleration method ($\Gamma = 1/2$ and $\beta = 1/4$) is unconditionally stable no matter what time-step is used. To obtain reasonable accuracy for a nonlinear system, try a first run with a time-step that corresponds to the following equation:

$$\frac{\Delta t}{T_n} \leq 0{,}02 \tag{4.371}$$

where T_n represents the shortest natural period that we wish to consider in the analysis.

It is necessary to try at least two runs, with two different time-steps, in order to ensure reasonable accuracy of the results.

In a linear system, the linear acceleration method is conditionally stable. The method is stable if the time-step corresponds to the following inequality:

$$\frac{\Delta t}{T_n} < 0{,}55 \tag{4.372}$$

For a seismic analysis, the time-step must be much smaller than the value given by equation 4.372, in order to represent the accelerogram accurately. The linear acceleration method can be made unconditionally stable by using the Wilson-θ's algorithm (Bathe and Wilson, 1976).

4.10 PROBLEMS

4.1　a) Evaluate Duhamel's integral for an undamped SDOF system with a natural period, T, and excited by a constant ground acceleration, a_s. Derive an equation for the maximum relative displacement, S_D.

　b) Use the RESAS program available on the web site http://www.polymtl.ca/pub to calculate the relative displacement spectrum, S_D, for a constant ground acceleration of 1 m/s². Assume zero damping and natural periods of 0,1 to 5 s. Represent the ground acceleration by increments of 0,01s for a total duration of 5 s. On graph paper, compare your results with the equation found in a). Discuss these results.

　c) Use the RESAS program to calculate and plot the relative displacement, relative velocity, and absolute acceleration response spectra for the S00E component of the accelerogram recorded at El Centro (1940-05-18). Use 5% and 20% critical damping and periods between 0,01 and 0,5 s. The El Centro accelerogram is also on the web site http://www.polymtl.ca/pub (file EC.ACC).

　d) Compare graphically the relative velocity response spectrum and the absolute acceleration response spectrum obtained in c), with the corresponding pseudo-spectral values. Discuss the accuracy of the pseudo-spectra.

4.2   A water reservoir is located on a site influenced mainly by two seismic sources (fig. 4.61). The two sources have a small surface area and can be considered point sources. Figure 4.61 illustrates the hypocentral distances, d.

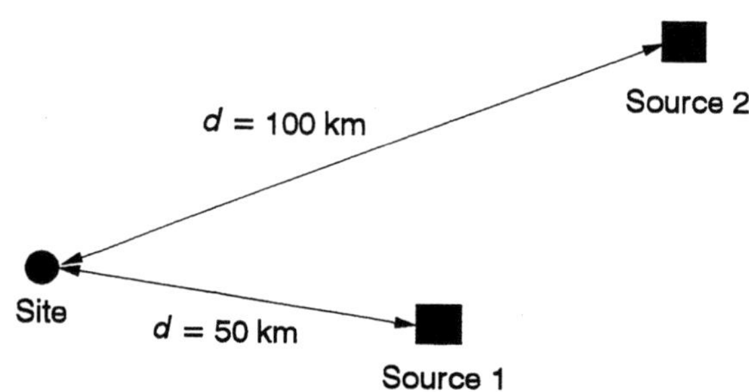

Figure 4.61   Site influenced by two seismic sources.

The seismicity model for each source is:

$$\log_{10} N_i = A_i - b_i M \quad i = 1, 2$$

where N_i = Number of earthquakes per annum occurring in source i with magnitude greater or equal to M

A_1 = 5,05
A_2 = 5,94
b_1 = 1,09
b_2 = 1,14
M = local magnitude

The attenuation model for the region is defined by the following equation:

$$a_{max} = 0{,}01 \, e^{1{,}3M} \, d^{-1{,}5}$$

where a_{max} = maximum ground horizontal acceleration in g
M = local magnitude
d = hypocentral distance in km

a)   Calculate the maximum acceleration at the site of the reservoir, which has a return period of 475 years.

b)   If the standard event at the site has a return period of 475 years, plot Newmark's response spectrum on tripartite graph paper (fig. 4.71). Use the parameters of Newmark's standard event with 2% critical damping.
The reservoir is modelled as a SDOF system (fig. 4.62). The weight of a full reservoir is 68,5 kN, an empty reservoir weighs 31,1 kN, and the lateral stiffness of the reservoir's column is 10,8 kN/mm.

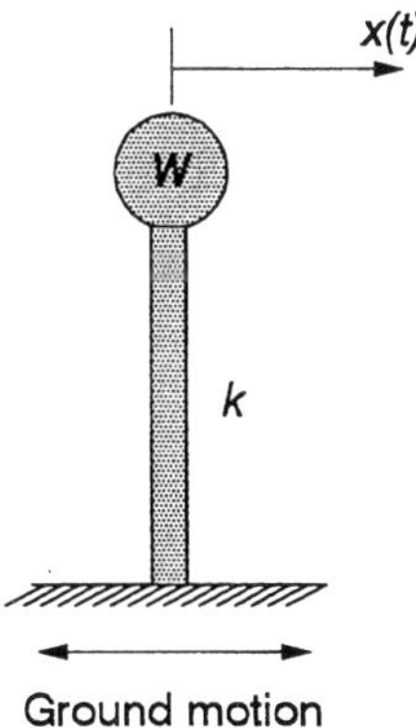

Figure 4.62   Water reservoir.

c)   Calculate the expected maximum lateral displacement at the top of the reservoir under the design earthquake. Compare the results for the full reservoir with the results for the empty reservoir.

d)   If the column must remain elastic during the design earthquake, calculate the base shear that must be sustained by the column.

e)   Find the value of the equivalent seismic coefficient for the base shear calculated in d).

f)   Repeat steps c), d), and e) using the El Centro earthquake spectrum obtained from the RESAS program in problem 4.1. (Note that the maximum acceleration for this earthquake is 0,34 g.) Compare and discuss your results.

4.3   A one-storey structure has a mass of 560 N-s^2/mm, lateral stiffness of 8 580 N/mm, and 10% critical damping.

a)   Calculate the maximum base shear and the maximum relative displacement using Housner's response spectrum with a design peak ground acceleration of 0,33 g.

b)   Calculate the same quantities, if the lateral stiffness of the structure is multiplied by 4. Discuss the usefulness of only increasing the stiffness of a structure to improve its earthquake resistance.

4.4   The cooling tower shown in figure 4.63 has a height of 40 m and can be modelled as a SDOF system. The weight of the tower is 8 000 kN and its lateral stiffness is 100 kN/mm. The S00E component of the El Centro earthquake is used as the design earthquake for the site (fig. 4.64).

a)   Assuming that the tower must remain elastic at all times, estimate the maximum base shear and the maximum base overturning moment that the column must sustain. Assume 5% critical damping.

b)   Estimate the maximum relative lateral displacement at the top of the tower for the El Centro earthquake.

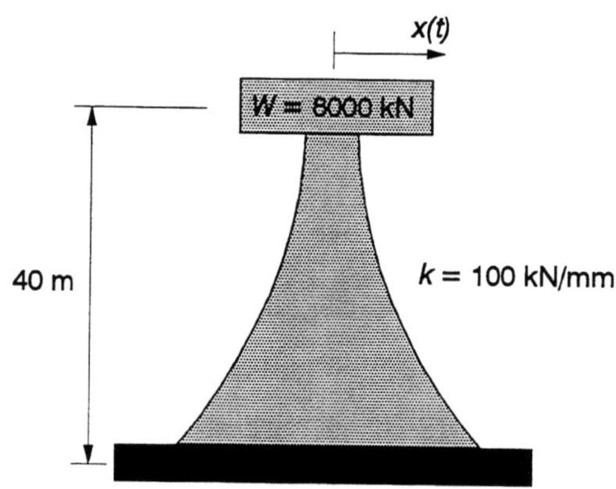

Figure 4.63 Cooling tower.

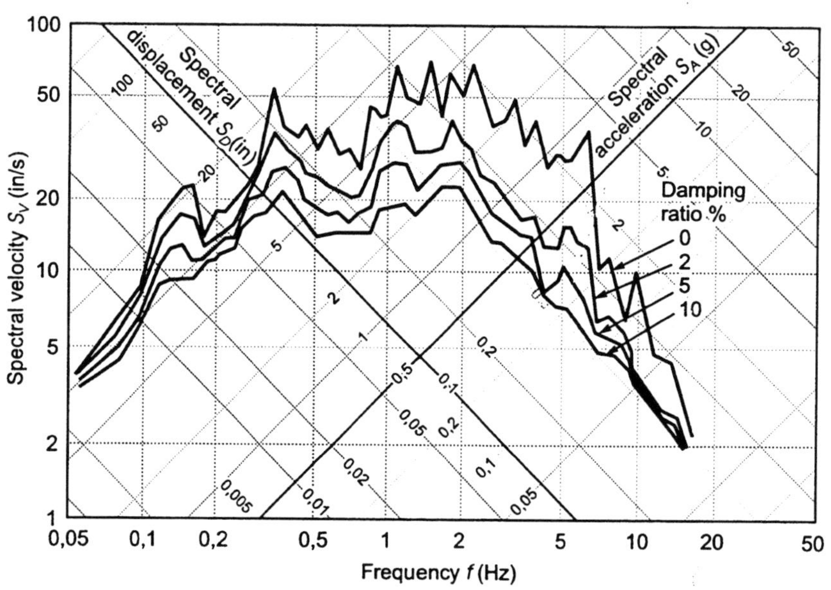

Figure 4.64 El Centro earthquake response spectrum (1940-05-18, comp. S00E).

4.5 A two-storey building was modelled as a simple two-degrees-of-freedom system (fig. 4.65). The equations of motion of this system in free vibration (ignoring damping) are written as:

$$\begin{bmatrix} 0,0385 & 0 \\ 0 & 0,0245 \end{bmatrix} \frac{kN-s^2}{mm} \begin{pmatrix} \ddot{x}_1 \\ \ddot{x}_2 \end{pmatrix} + \begin{bmatrix} 6,30 & -4,38 \\ -4,38 & 6,30 \end{bmatrix} \frac{kN}{mm} \begin{pmatrix} x_1 \\ x_2 \end{pmatrix} = \begin{pmatrix} 0 \\ 0 \end{pmatrix}$$

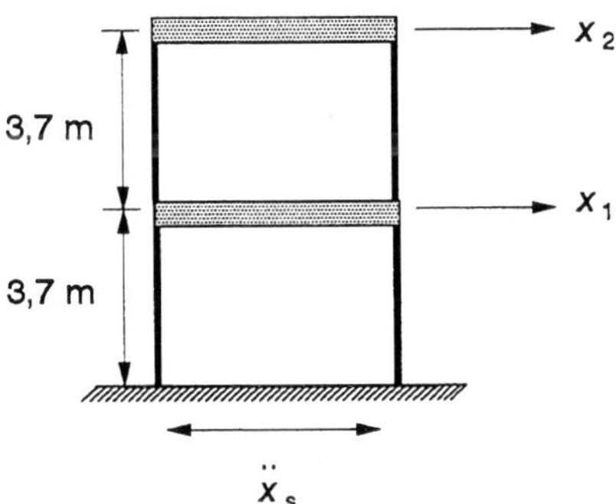

Figure 4.65 Two-storey building.

For each statistical combination presented in this chapter, estimate the maximum inter-storey shear and displacements for each floor. Assume that the structure must remain elastic for a design earthquake corresponding to a seismic zone in Canada with a peak ground horizontal acceleration of 0,2 g and a peak ground horizontal velocity of 0,2 m/s. Use the response spectrum of the NBCC, shown in figure 4.66, with 5% damping in each mode. Assume a strong motion duration equal to 20 s.

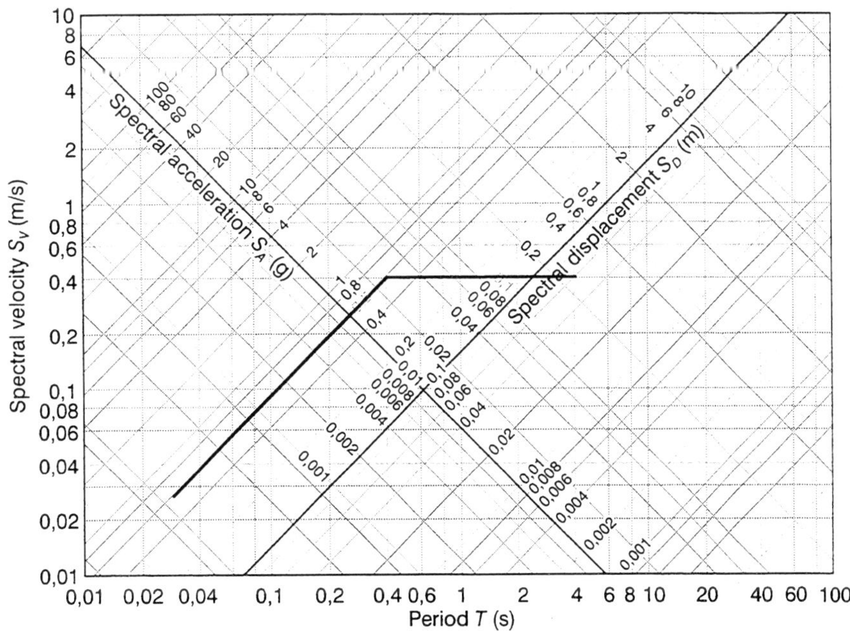

Figure 4.66 NBCC response spectrum for a seismic zone of peak ground horizontal acceleration of 0,2 g and peak ground horizontal velocity of 0,2 m/s.

4.6 Shown below are the first two natural periods and mode shapes of a four-storey building. The weight on each floor is 143,2 kN.

$$T_1 \quad = \quad 0,63 \text{ s}$$
$$T_2 \quad = \quad 0,32 \text{ s}$$
$$(A^{(1)})^T = \quad (1,00,\ 0,60,\ 0,30,\ 0,06)$$
$$(A^{(2)})^T = \quad (-0,31,\ -0,05,\ 1,00,\ 0,70)$$

a) The building must remain elastic when subjected to a design earthquake with the following ground-motion parameters: 0,16g, 20 cm/s, and 15 cm. Prepare Newmark's response spectrum for this design earthquake. Use amplification factors for acceleration, velocity, and displacements of 2, 1,5, and 1,2, respectively.

b) Calculate, based on purely elastic behaviour, the equivalent seismic forces of the first two modes of the structure. Then, calculate the maximum probable (using the SRSS combination) and possible inter-storey shears developed at each floor.

c) Calculate the probable and possible maximum relative displacements on the fourth floor for the design earthquake.

4.7 A simple one-storey steel portal frame has the properties shown in figure 4.67. The structure has a lateral elasto-plastic behaviour. The yield shear force, V_y, represents the necessary lateral force at the floor to simultaneously yield all columns of the portal frame. The frame is subjected to the S00E component of the accelerogram recorded at El Centro (file EC.ACC of the accompanying disk).

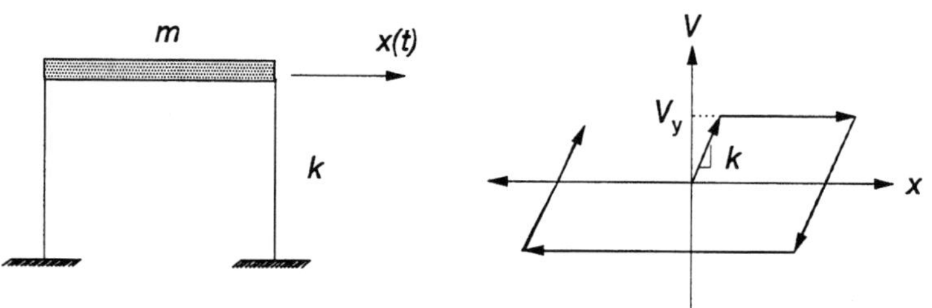

Figure 4.67 Simple steel portal frame with lateral elasto-plastic behaviour.

where $m =$ mass at the floor level $= 1$ kN - s²/mm
       $k =$ lateral stiffness $= 39,5$ kN/mm
       $\xi =$ critical damping ratio $= 0,05$

Plot the first 5 seconds of the lateral displacement time-history at the floor level, $x(t)$, for $V_y = 490$ kN and $V_y = \alpha$ (infinite). Use the average constant acceleration method with a time-step of 0,01s.

Suggestion: The direct integration method is easily programmable on a spreadsheet such as Lotus, Excel, or Quatro.

4.8   Figure 4.68 shows a rigid bar supported by a column of negligible mass.

a)   Using the indicated DDOF, evaluate the mass and stiffness matrices of the system.
b)   Using the matrices obtained in a), calculate the natural frequencies and mode shapes of the system.
c)   Normalize the modes with respect to the mass matrix.

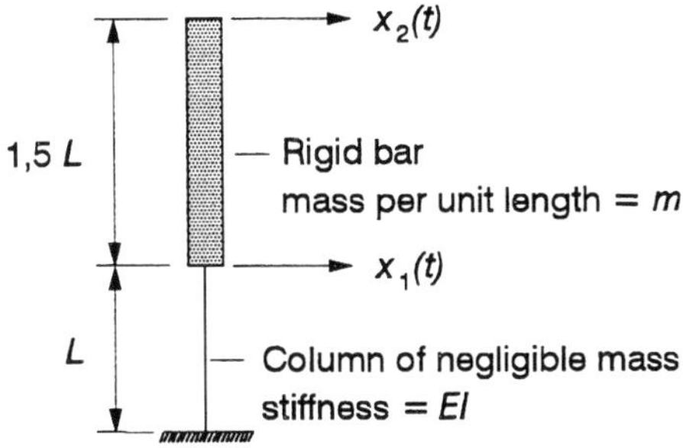

Figure 4.68   Rigid bar supported by a column of negligible mass.

4.9   The two-storey frame shown in figure 4.69 has rigid floors and lumped masses at the floor levels. The lateral-load-resisting system is made of a single diagonal bracing member in the first storey with fixed-ended columns in both storeys. Assuming only one DDOF per floor:

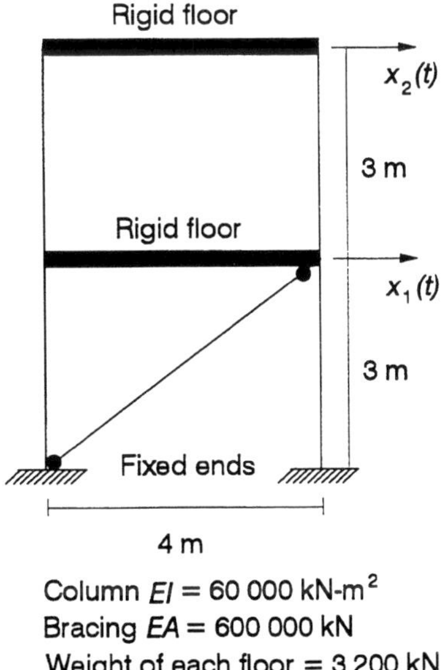

Figure 4.69   Two-storey frame.

a)   Find the mass matrix of the system.
b)   Find the stiffness matrix of the system.
c)   Find the frequency matrix.
d)   Find the modal matrix.
e)   Calculate the generalized mass matrix for each mode of vibration.
f)   Calculate the modal participation factor for each mode of vibration.

The structure, initially at rest, is subjected to the earthquake with the response spectrum shown in figure 4.31. Assuming that the structure remains elastic during the earthquake and 5% critical damping for each mode:

g)   Calculate the maximum possible base shear.
h)   Using the SRSS combination, calculate the maximum probable base shear.

4.10 A cantilever beam of length $2L$ and stiffness EI has a lumped mass of $2m$ at mid-span and another lumped mass m at the end (fig. 4.70). A dynamic force $F(t)$ is applied at mid-span of the cantilever.

Find the natural frequencies, mode shapes, and differential equations of motion in normal coordinates.

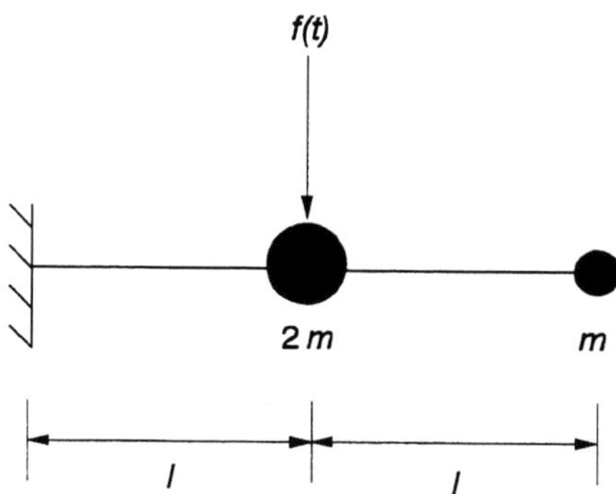

Figure 4.70 Cantilever beam modelled as a two-degrees-of freedom system.

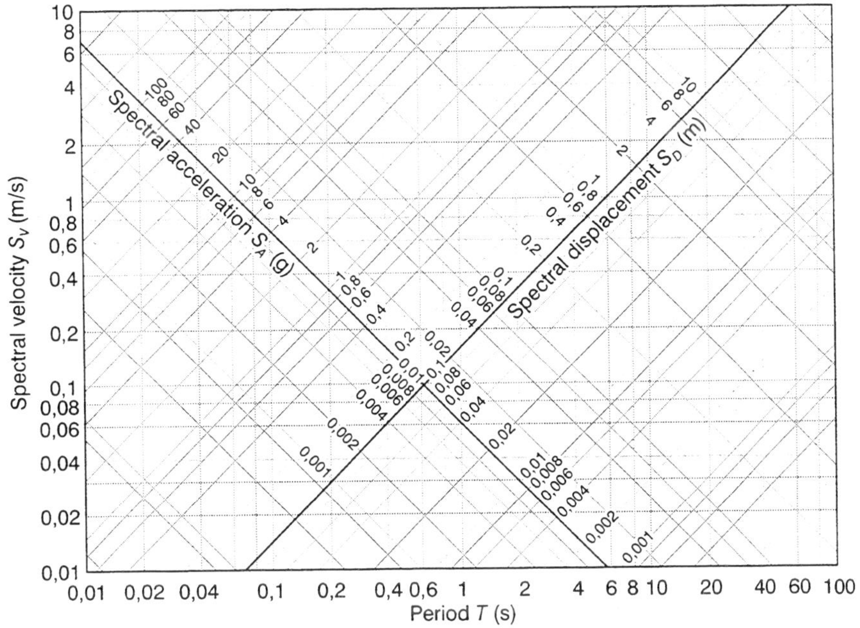

Figure 4.71 Tripartite graph.

4.11 REFERENCES

Anderson, J.-C. (1989). "Dynamic Response of Buildings." In F. Naeim (ed.), *The Seismic Design Handbook*. New York: Van Nostrand Reinhold.

Bathe, K.-J., and Wilson, E.L. (1976). *Numerical Methods in Finite Element Analysis*. Englewood Cliffs, NJ: Prentice-Hall.

Biggs, J.M., and Roesset, J.M. (1970). "Seismic Analysis of Equipment Mounted on a Massive Structure." In R.J. Hudson (ed.), *Seismic Design of Nuclear Power Plants*. Cambridge, MA: MIT Press.

Carr, A.J. (1994). "Dynamic Analysis of Structures." *Bulletin of the New Zealand National Society for Earthquake Engineering*, 27(2): 129–46.

Clough, R.W., and Penzien, J. (1993). *Dynamics of Structures*. 2nd ed. New York: McGraw-Hill.

Cook, R.D. (1981). *Concepts and Applications of Finite Element Analysis*. 2nd ed. New York: John Wiley.

Gupta, A.K. (1990). *Response Spectrum Method in Seismic Analysis and Design of Structures*. Cambridge, MA: Blackwell Scientific.

Housner, G. (1959). "Behavior of Structures During Earthquakes." ASCE, *Journal of the Engineering Mechanics Division*, 85(EW4): 109–29.

—(1970). "Design Spectrum." In R.L. Wiegel (ed.), *Earthquake Engineering*. Englewood Cliffs, NJ: Prentice-Hall.

Humar, J.L. (1984). "Design for Seismic Torsional Forces." *Canadian Journal of Civil Engineering,* 11(2): 150–63.

Laursen, H.I. (1978). *Structural Analysis.* 2nd ed. New York: McGraw-Hill.

National Building Code of Canada (1995). 18th ed. Ottawa: National Research Council of Canada.

Newmark, N.M., and Hall, W.J. (1969). "Seismic Design Criteria for Nuclear Reactor Facilities." *Proceedings of the Fourth World Conference on Earthquake Engineering*, Santiago, Chili, vol. 2, B-4, pp. 37–50.

—(1982). *Earthquake Spectra and Design.* Earthquake Engineering Research Institute Monograph Series.

Rayleigh, Lord (1945). *Theory of Sound.* New York: Dover.

Rosenblueth, E., and Elorduy, J. (1969). "Responses of Linear Systems to Certain Transient Disturbances." *Proceedings of the Fourth World Conference on Earthquake Engineering*, Santiago, Chili, vol. I, A-1, pp. 185–96.

Singh, M.P. (1975). "Generation of Seismic Floor Spectra." ASCE, *Journal of the Engineering Mechanics Division*, 101(EM5): 393–607.

Wilson, E.L., Der Kiureghian, A., and Bayo, E.P. (1981). "A Replacement for the SRSS Method in Seismic Analysis." *Earthquake Engineering and Structural Dynamics*, 9(2): 187–94.

Wilson, J.-C., and Liu, T. (1991). "Ambient Vibration Measurements on a Cable-Stayed Bridge." *Earthquake Engineering and Structural Dynamics*, 20(8): 723–47.

CHAPTER 5

Elements of Soil Dynamics

5.1 INTRODUCTION

As seen in chapter 3, the design earthquake is usually specified for a given region in Canada as occurring on rock sites or on firm ground, and specific local soil conditions at the site are not taken into account. In this chapter, we look at a structure situated on a soil deposit, as shown in figure 5.1, and investigate the following questions: Does a soil deposit influence the ground motion at the base of a structure? How is the dynamic response of a structure affected by the presence of a soil deposit?

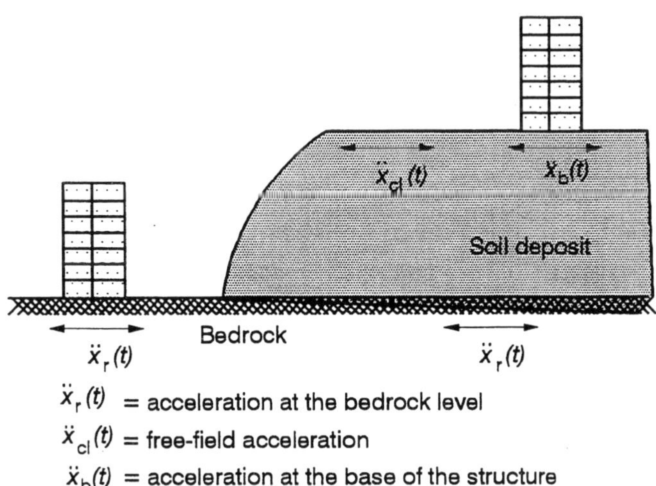

$\ddot{x}_{r}(t)$ = acceleration at the bedrock level

$\ddot{x}_{cl}(t)$ = free-field acceleration

$\ddot{x}_{b}(t)$ = acceleration at the base of the structure

Figure 5.1   Effect of a soil deposit during an earthquake.

When a structure rests on a soil deposit, three problems can be foreseen:

1. Seismic loads on a structure situated on a soil deposit may be greater than those on the same structure standing on bedrock.
2. Following violent vibrations, the soil may totally lose its load-bearing capacity and liquefy.
3. Following vibrations of large amplitude, the soil may collapse and cause differential settlement of the structure's foundation.

In order to understand better the causes of these problems, we will first consider the dynamic behaviour of a soil deposit during an earthquake.

5.2   SOIL DYNAMIC RESPONSE: LUMPED-MASS ANALYSIS

5.2.1 Description

In many cases, the surface motion of a soil deposit originates mainly from the vertical propagation of shear waves (S-waves) from the bedrock to the ground surface. In general, a soil deposit is made of many layers (clay, silt, sand, etc.) which exhibit different mechanical properties. If the rock layer or the different soil layers are on an incline, finite-element techniques must be used to compute the dynamic response. However, if the rock or the interfaces between the different soil layers are horizontal, as a first approximation, each layer can be considered linear and elastic and a lumped-mass model can be developed to analyze the dynamic response of the soil deposit during an earthquake. A simple mechanical model of a soil deposit (fig. 5.2) is presented as a semi-infinite medium with a unit width and an infinite length. The mass of the soil is assumed to be concentrated equally over and under each layer.

$$m_1 = \frac{\rho_1 h_1}{2} \tag{5.1}$$

$$m_i = \frac{\left(\rho_{i-1} h_{i-1} + \rho_i h_i\right)}{2} \quad i = 2,3,...,N \tag{5.2}$$

where m_i = lumped mass over soil layer i
      ρ_i = density (mass per unit of volume) of soil layer i
      h_i = thickness of soil layer i

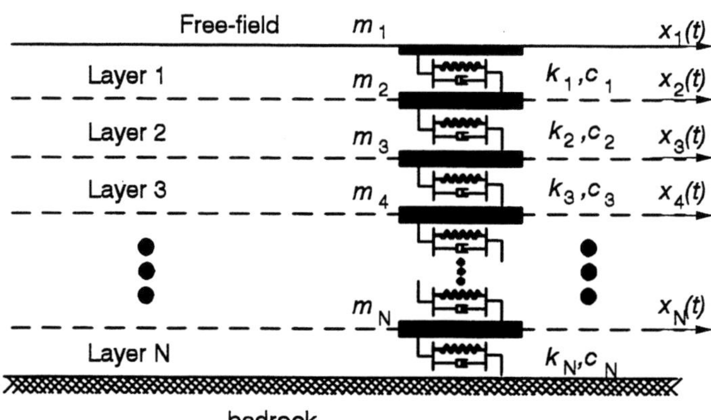

Figure 5.2 Lumped-mass model for the seismic analysis of a soil deposit.

These lumped masses are linked by spring elements and viscous dampers that can resist the lateral deformations of the soil deposit. The stiffness constant of layer i, k_i, is obtained by considering the shear strain of this soil layer (figure 5.3). This layer has a unit cross-section (width × length). In this case, the horizontal force, F_i, needed to cause a horizontal displacement, x_i, is the following:

$$F_i = \tau_i = G_i \gamma_i = \frac{G_i}{h_i} x_i \qquad (5.3)$$

where
$$\tau_i = \text{shear stress in soil layer } i$$
$$\gamma_i = \text{shear strain in soil layer } i$$
$$G_i = \text{shear modulus of soil layer } i$$

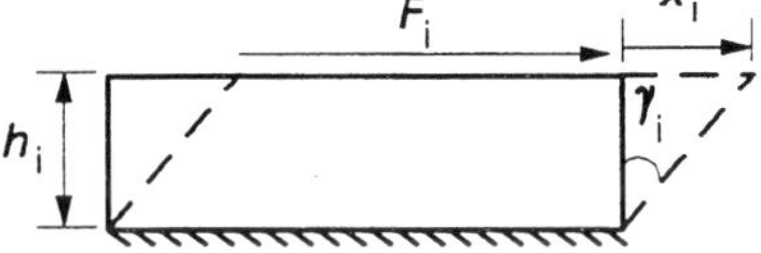

Figure 5.3 Shear deformation of a soil layer.

The stiffness constant k_i is then:

$$k_i = \frac{G_i}{h_i} \qquad (5.4)$$

The equations governing the motion of all soil layers can be written in matrix form by considering the dynamic equilibrium (Newton's second law) of each layer.

$$[m](\ddot{x}) + [c](\dot{x}) + [k](x) = -(m)\ddot{x}_r \qquad (5.5)$$

where $[m]$ = global mass matrix

$$[m] = \begin{bmatrix} m_1 & 0 & 0 & \cdots & 0 \\ 0 & m_2 & 0 & \cdots & 0 \\ 0 & 0 & m_3 & \cdots & 0 \\ \cdots & \cdots & \cdots & \cdots & \cdots \\ 0 & 0 & 0 & 0 & m_N \end{bmatrix}_{N \times N} \qquad (5.6)$$

$[c]$ = global damping matrix

$$[c] = \begin{bmatrix} c_1 & -c_1 & 0 & \cdots & & 0 \\ -c_1 & C_1+c_2 & -c_2 & \cdots & & 0 \\ 0 & -c_2 & c_2+c_3 & \cdots & & 0 \\ \cdots & \cdots & \cdots & \cdots & & \cdots \\ 0 & 0 & 0 & 0 & & -c_N \\ 0 & 0 & 0 & -c_N & & C_{N-1}+c_N \end{bmatrix}_{N \times N} \qquad (5.7)$$

$[k]$ = global stiffness matrix

$$[k] = \begin{bmatrix} k_1 & -k_1 & 0 & \cdots & & 0 \\ -k_1 & k_1+k_2 & -k_2 & \cdots & & 0 \\ 0 & -k_2 & k_2+k_3 & \cdots & & 0 \\ \cdots & \cdots & \cdots & \cdots & & \cdots \\ 0 & 0 & 0 & 0 & & -k_{N-1} \\ 0 & 0 & 0 & -k_{N-1} & & k_{N-1}+k_N \end{bmatrix}_{N [x] N} \qquad (5.8)$$

(x), $(\dot{x})$, $(\ddot{x})$ = relative displacement, velocity, and horizontal acceleration vectors for each layer with respect to the bedrock

$$(x)^T = \left(x_1, \, x_2, \, \ldots, \, x_N\right)_{1 \times N} \qquad (5.9)$$

(m) = vector containing the mass of each soil layer

$$(m)^T = \left(m_1, \, m_2, \, \ldots, \, m_N\right)_{1 \times N} \qquad (5.10)$$

$\ddot{x}_r$ = accelerogram at the bedrock level

We can see that matrices $[m]$, $[c]$, and $[k]$ are of order N (number of layers in the model). The mass matrix $[m]$ is diagonal and symmetrical, but the stiffness and damping matrices $[k]$ and $[c]$ are tridiagonal and symmetrical. The equations of motion (equation 5.5) have to be solved in order to obtain the dynamic response of the soil deposit for a specific accelerogram at the bedrock level. To do so, the dynamic analysis methods presented in chapter 4 can be used. For seismic analysis of a soil deposit, the focus is on the absolute acceleration at the surface of the soil deposit, $\ddot{x}_i(t) + \ddot{x}_r(t)$, called the free-field accelerogram.

5.2.2 Accuracy of Numerical Results

The accuracy of the solution obtained with the lumped mass analysis is a function of the number of soil layers considered in the model. To pick a reasonable number of layers, N, Idriss and Seed (1968) prepared the graph shown in figure 5.4.

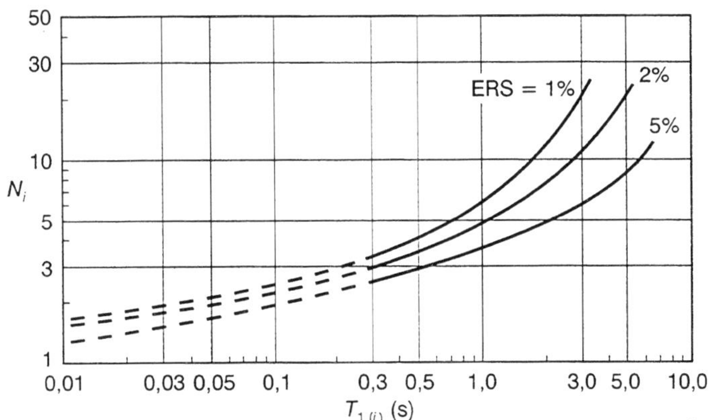

Figure 5.4   Influence of the number of soil layers on the accuracy of a lumped-mass analysis (from Idriss and Seed, 1968).

In this graph, the term ERS represents the error (in %) of the lumped-mass analysis with respect to the exact analysis, which requires solving a system of partial differential equations for the soil deposit. On the horizontal axis of this graph, the period T_1 represents the fundamental period of vibration of each soil type, i, made of a homogeneous material. Writing the differential equation of motion for this soil layer i (Das, 1983) will yield its fundamental period of vibration $T_{1\,(i)}$.

$$T_{1(i)} = \frac{4\,h_i}{\sqrt{\dfrac{G_i}{\rho_i}}} \tag{5.11}$$

where h_i = thickness of layer i
       G_i = shear modulus of layer i
       ρ_i = density (mass per unit of volume) of layer i.

It is important to understand that the graph in figure 5.4 must be used for each type of soil made of a homogeneous material. The graph shows the number of sub-layers, N_i, required to model correctly layer i. The total number of layers required for the analysis, N, is the sum of all of these sub-layers.

$$N = \sum_{1}^{N_m} N_i \tag{5.12}$$

where N_m represents the number of different materials found in the soil deposit.

5.3   DETERMINATION OF PHYSICAL SOIL PROPERTIES
5.3.1 Real Cyclic Behaviour of a Soil Deposit

When a soil layer is subjected to violent vibrations during an earthquake, the shear stress-strain diagram of the layer takes the shape shown in figure 5.5. Note that the response is neither linear nor elastic, as assumed in previous sections. On the contrary, the shear stress-strain diagram is nonlinear with hysteretic energy dissipation in each cycle.

In a dynamic analysis of a lumped mass system, the soil behaviour can be modelled in two ways:

1. First, a linear analysis with an effective shear modulus is performed to model the effective soil stiffness. Then, an equivalent viscous damping is introduced to model the energy dissipation in the ground.
2. A nonlinear time-step dynamic analysis can be performed; the increment of the shear modulus (or stiffness) of each soil layer varies with each time-step to take into account the stress-strain diagram shown in figure 5.5.

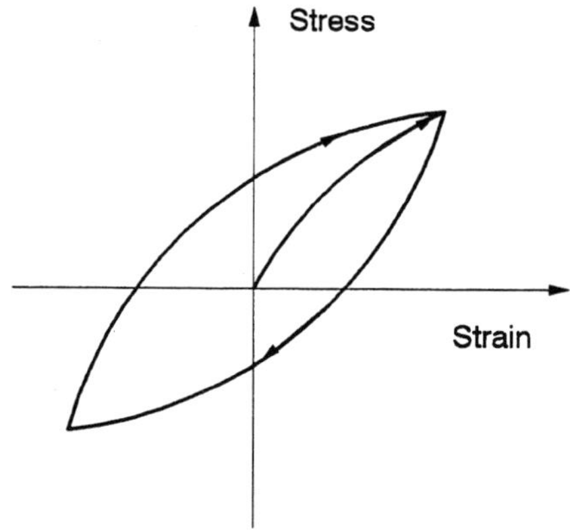

Figure 5.5   Shear stress-strain diagram of a soil layer.

In practice, mainly because of computation time, we often use the linear analysis with effective stiffness and equivalent viscous damping. The sections below will focus on this method for determining the proper shear modulus and viscous damping coefficient to be used for each soil layer. Note that the effective shear modulus (represented by a secant in figure 5.5) of a soil layer, G, reduces with the amplitude of the shear strain (fig. 5.6). Based on the results of numerous experiments, Hardin and Drnevich (1972) concluded that the backbone curve of the shear stress-strain diagram $(\tau - \gamma)$ can be approximated, for all types of soil, by a hyperbolic relation.

$$\tau = \frac{\gamma}{\dfrac{1}{G_{max}} + \dfrac{\gamma}{\tau_{ult}}}$$

(5.13)

where G_{max} = maximum shear modulus of the soil layer corresponding to a very small shear strain

τ_{ult} = ultimate shear stress of the soil layer

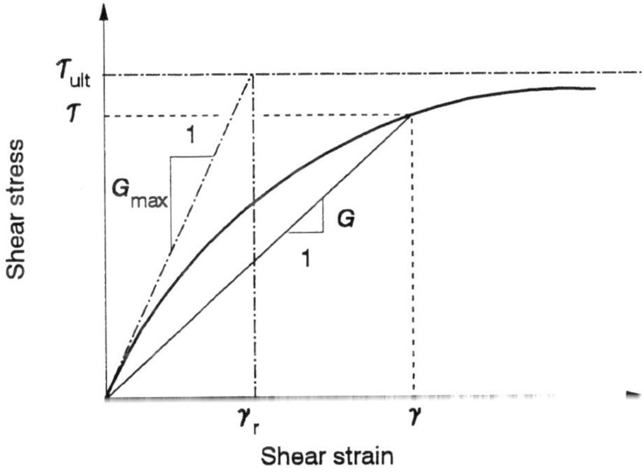

Figure 5.6 Definition of the shear modulus of a soil layer.

By dividing equation 5.13 by γ, the effective shear modulus of the soil, G, can be defined as a function of the shear strain, γ.

$$\frac{G}{G_{max}} = \frac{1}{1 + \dfrac{\gamma}{\gamma_r}}$$

(5.14)

where

$$\gamma_r = \frac{\tau_{ult}}{G_{max}} = \text{reference strain}$$

(5.15)

Figure 5.7 shows the variation of the effective shear modulus with the shear strain for this hyperbolic model. In reality, soils do not behave exactly according to the hyperbolic law given by equation 5.14. To provide a better representation of the experimental results, equation 5.14 is written as a function of an hyperbolic shear strain, γ_h.

$$G = \frac{G_{max}}{1 + \gamma_h}$$

$$\gamma_h = \frac{\gamma}{\gamma_r}\left[1 + ae^{-b\left(\frac{\gamma}{\gamma_r}\right)}\right] \tag{5.16}$$

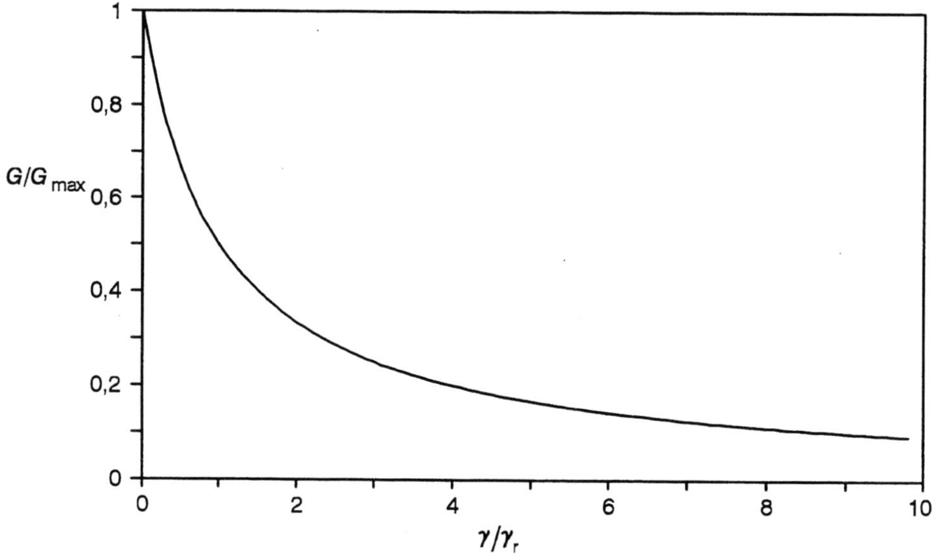

Figure 5.7   Variation of the effective shear modulus of a soil layer with the shear strain (hyperbolic model).

In this equation, the empirical constants a and b are obtained experimentally for different types of soils. Numerical values for these constants are shown in table 5.1.

Table 5.1   Hyperbolic shear strain constants of equation 5.16 (from Hardin and Drnevich, 1972).

Type of soil	Value of a	Value of b
Dry, clean sand	$- 0,5$	0,16
Clean, saturated sand	$- 0,2 \log_{10} N$	0,16
Cohesive, saturated soils	$1 + 0,25 (\log_{10} N)$	1,30
N is equal to the number of loading cycles		

In practice, the linear analysis is repeated in order to adjust the shear modulus as a function of the resulting shear strain level. This iterative procedure will be discussed in section 5.4.

5.3.2 Shear Modulus Estimation by In Situ Testing

The maximum value of the shear modulus of a soil layer, G_{max}, can be determined by measuring the S-wave velocity across the layer. Field (in situ) testing is performed using blast loading. Two methods are often used (fig. 5.8): a) cross-hole method and b) down-hole method.

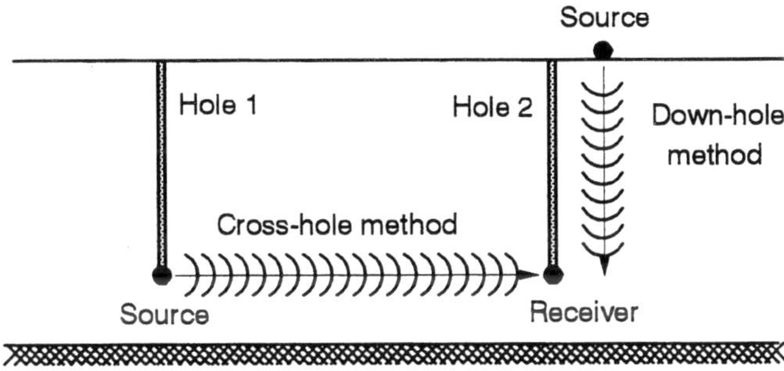

Figure 5.8   In situ methods for measuring G_{max} of a soil layer.

In both cases, the time, t, required for the first shear wave (S-wave) to cross a given distance, d, between a source and a receiver is recorded. The S-wave velocity, v_s, can then be estimated by the well-known relation between distance, velocity, and time.

$$v_s = \frac{d}{t} \tag{5.17}$$

Equation 2.57 is used to estimate the maximum shear modulus, G_{max}.

$$G_{max} = \rho v_s^2 \tag{5.18}$$

where ρ = density (mass per unit of volume) of the soil layer.

5.3.3 Shear Modulus Estimation by Laboratory Testing

Based on numerous laboratory experiments, many researchers have proposed empirical equations to calculate the maximum shear modulus, G_{max}, of a soil layer. Here are a few of the best-known equations. The development of these equations will not be discussed in this section, but the reader can consult the references at the end of the chapter.

- Seed and Idriss's (1970) equation for sand

$$G_{max} = 21{,}7\left(0{,}62\,D_r + 15\right)P_a\sqrt{\frac{\sigma'_m}{P_a}} \tag{5.19}$$

where G_{max} = maximum shear modulus in the same units as P_a
$\quad\quad\ \ D_r$ = relative density of sand
$\quad\quad\ \ \sigma'_m$ = average effective principal stress
$\quad\quad\ \ P_a$ = atmospheric pressure

For gravel, Seed and Idriss (1970) recommended to multiply the value of G_{max} obtained from equation 5.19 by 1,5.

- Hardin and Drnevich's (1972) equation for sand and clay

$$G_{max} = 314\left[\frac{(2{,}973-e)^2}{(1+e)}\right](OCR)^k\,P_a\left(\frac{\sigma'_m}{P_a}\right)^{\frac{1}{2}} \tag{5.20}$$

where G_{max} = maximum shear modulus in the same units as P_a
$\quad\quad\ \ \sigma'_m$ = average effective principal stress
$\quad\quad\ \ P_a$ = atmospheric pressure
$\quad\quad\ \ e$ = void ratio of the soil
$\quad\quad\ \ OCR$ = over-consolidation ratio
$\quad\quad\ \ k$ = exponent depending on the soil plasticity index (table 5.2)

Table 5.2   Exponent k of equation 5.20 (from Hardin and Drnevich, 1972).

Soil plasticity index	Value of k
0	0,00
20	0,18
40	0,30
60	0,41
80	0,48
≥ 100	0,50

5.3.4 Ultimate Shear Strength Estimation

The ultimate shear strength, τ_{ult}, must be known in order to estimate the reference strain, γ_r. Figure 5.9 shows a soil element subjected simultaneously to an effective vertical stress, σ'_o, and to an effective horizontal stress, $K_o\sigma'_o$, where K_o represents the earth pressure coefficient of the soil at rest. The ultimate shear strength of this soil element τ_{ult} is found by the intersection of Mohr's circle, corresponding to the relative stress state, and the line representing Mohr-Coulomb's failure criterion, as shown in figure 5.9. By geometry, we can deduce an expression for the ultimate shear strength, τ_{ult}, of soil.

$$\tau_{ult} = \left(\left[\frac{1}{2}\left(1 + K_o\right)\sigma'_o \, \sin\phi + c\cos\phi \right]^2 - \left[\frac{1}{2}\left(1 - K_o\right)\sigma'_o \right]^2 \right)^{\frac{1}{2}} \tag{5.21}$$

where c = cohesion of the soil
ϕ = friction angle of the soil

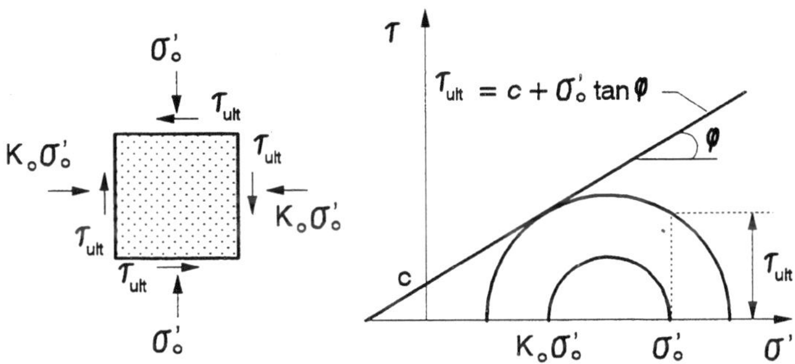

Figure 5.9 Definition of the ultimate shear strength, τ_{ult}, of soil.

5.3.5 Equivalent Viscous Damping Estimation

In the linear analysis of a soil deposit with a lumped-mass model, a third parameter must be taken into account: the viscous damping coefficient for each soil layer. Figure 5.10 shows a shear hysteresis loop for a layer of soil. It can be shown (Clough and Penzien, 1993) that the equivalent critical damping ratio, ζ, is obtained by:

$$\zeta = \frac{(\text{total area under loop})}{4\pi(\text{area of oag triangle})} \tag{5.22}$$

Hardin and Drnevich (1972) presented the following equation relating the critical damping ratio and the shear modulus.

$$\zeta = \zeta_{max}\left(1 - \frac{G}{G_{max}} \right) \tag{5.23}$$

where ζ_{max} = maximum critical damping ratio occurring when $G = 0$

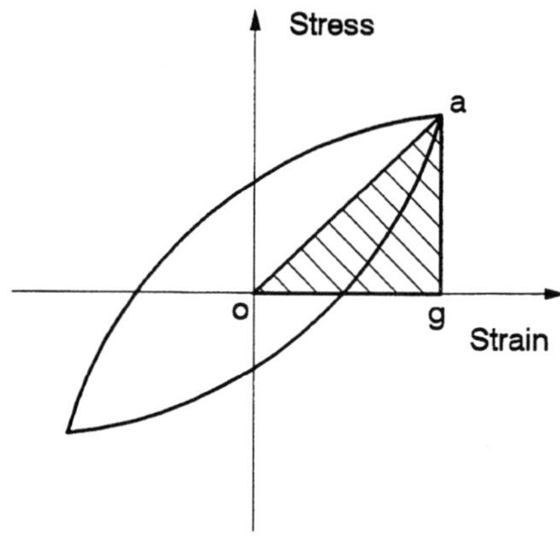

Figure 5.10   Definition of the critical damping ratio, ζ, for a soil layer.

Typical values of ζ_{max} are shown in table 5.3.

Table 5.3   Typical values of ζ_{max} (from Hardin and Drnevich, 1972).

Type of soil	ζ_{max}
Dry, clean sand	$0,01\ (33 - 1,5\ (\log_{10} N))$
Saturated, clean sand	$0,01\ (28 - 1,5\ (\log_{10} N))$
Saturated silt	$0,01\ (26 - 4\sigma_m'^{0,5} + 0,7 f^{0,5} - 1,5\ (\log_{10} N))$
Saturated clay	$0,01\ (31 - (3 + 0,3f)\ \sigma_m'^{0,5} + 1,5f^{0,5} - 1,5\ (\log_{10} N))$
N = number of loading cycles σ_m' = average effective principal stresses (kg/cm^2) f = average loading frequency (Hz)	

Combining equations 5.23 and 5.15, the variation of the viscous damping ratio can be directly written with respect to the shear strain.

$$\frac{\zeta}{\zeta_{max}} = \frac{\dfrac{\gamma}{\gamma_r}}{\left(1 + \dfrac{\gamma}{\gamma_r}\right)} \tag{5.24}$$

Figure 5.11 illustrates this relation. Once again, to better illustrate the real behaviour of soils, the variation of the equivalent critical damping ratio is written with respect to the hyperbolic shear strain, γ_h.

$$\frac{\zeta}{\zeta_{max}} = \frac{\gamma_h}{\left(1 + \gamma_h\right)}$$

$$\gamma_h = \frac{\gamma}{\gamma_r}\left[1 + a_1 e^{-b_1\left(\frac{\gamma}{\gamma_r}\right)}\right]$$

$$(5.25)$$

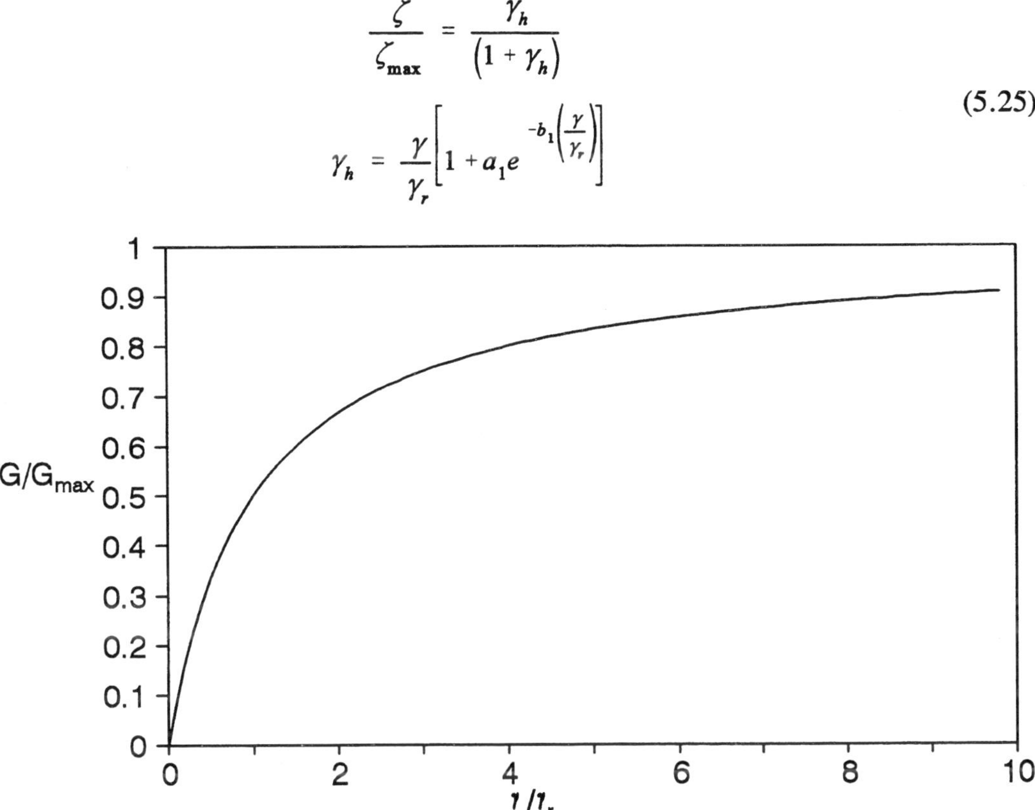

Figure 5.11   Variation of the equivalent critical damping ratio of a soil layer with the shear strain (hyperbolic model).

In this equation, the empirical constants a_1 and b_1 are obtained for different types of soil. Table 5.4 shows typical numerical values for these constants.

Table 5.4   Hyperbolic shear strain constants of equation 5.25 (from Hardin and Drnevich, 1972).

Type of soil	Value of a_1	Value of b_1
Dry, clean sand	$0{,}6\,(N^{-1/6}) - 1$	$1 - N^{-1/2}$
Saturated, clean sand	$0{,}54\,(N^{-1/6}) - 0{,}9$	$0{,}65 - 0{,}65 N^{-1/2}$
Saturated, cohesive soil	$1 + 0{,}2\,(f^{1/2})$	$0{,}2 f\,(e^{-\sigma'_m}) + 2{,}25\,\sigma'_m + 0{,}3\,(\log_{10} N)$
N = number of load cycles		
σ'_m = average effective principal stress (kg/cm^2)		
f = average loading frequency (Hz)		

As for the shear modulus, the equivalent viscous damping ratio is evaluated using an iterative method. Instead of calculating the coefficients c_i in equation 5.7, a new value of ζ is used at each step (iteration) for all modes of vibration of the dynamic system. This new value is based on the corresponding value of G at the centre of the soil deposit. When the values of G_i and ζ are obtained, they are used in an equivalent linear dynamic (seismic) analysis. As the bedrock motion (accelerogram) is known, the displacements, velocities, accelerations, and shear stresses can be calculated for each soil layer.

In particular, two results of this analysis must be noted:

1. The accelerogram at the surface of the soil deposit (free-field accelerogram) and its response spectrum will be used to determine the seismic loads on a structure resting on the soil deposit.
2. The maximum shear stresses and strains in each soil layer will be used to evaluate the liquefaction potential and the settlement resulting from an earthquake.

5.4   SUMMARY OF LUMPED-MASS DYNAMIC ANALYSIS

The lumped-mass analysis is an iterative tool for estimating the seismic response of a soil deposit, with horizontal geometry, subjected to an arbitrary accelerogram at the bedrock level. Here are the important steps for this type of analysis.

Step 1: Estimation of the physical soil properties

For each homogeneous soil layer i, the following physical parameters are estimated:

- maximum shear modulus: $G_{\text{max}\,(i)}$;
- maximum critical damping ratio: $\zeta_{\text{max}\,(i)}$;
- ultimate shear strength: $\tau_{\text{ult}\,(i)}$;
- density (mass per unit of volume): ρ_i;
- correction parameters a, b, a_1 and b_1 of the hyperbolic shear strain.

The reference shear strain, $\gamma_{r(i)}$, takes the following form:

$$\gamma_{r(i)} = \frac{\tau_{\text{ult}(i)}}{G_{\text{max}(i)}} \tag{5.26}$$

Step 2: Selection of the number of soil layers for the analysis

The initial fundamental period, $T_{1\,(i)}$, is calculated for each type of homogeneous soil, using $G_{\text{max}\,(i)}$ as modulus.

$$T_{1(i)} = \frac{4\,h_i}{\sqrt{\dfrac{G_{\text{max}(i)}}{\rho_i}}} \tag{5.27}$$

The required number of layers (fig. 5.4) is chosen to attain an acceptable value of ERS. The effective shear modulus will decrease with the level of shear strain and, therefore, the effective period of the soil layer will be longer. As a result, the value of ERS will be higher. To account for this potential problem, a slightly higher number of layers than the number predicted in figure 5.4 can be used. The total number of layers to be used in the analysis, N, is the sum of all the sub-layers necessary to model the various types of soils present in the soil deposit, N_m.

$$N = \sum_{1}^{N_m} N_i \qquad (5.28)$$

Step 3: Calculation of the mass, stiffness, and damping matrices

$$m_1 = \frac{\rho_1 h_1}{2} \qquad (5.29)$$

$$m_i = \frac{\left(\rho_{i-1} h_{i-1} + \rho_i h_i\right)}{2} \qquad i = 2, 3, ..., N \qquad (5.30)$$

$$k_i = \frac{G_i}{h_i} \qquad (5.31)$$

$$[m] = \begin{bmatrix} m_1 & 0 & 0 & ... & 0 \\ 0 & m_2 & 0 & ... & 0 \\ 0 & 0 & m_3 & ... & 0 \\ ... & ... & ... & ... & ... \\ 0 & 0 & 0 & 0 & m_N \end{bmatrix}_{N \times N} \qquad (5.32)$$

$$[k] \;=\; \begin{bmatrix} k_1 & -k_1 & 0 & \ldots & & 0 \\ -k_1 & k_1+k_2 & -k_2 & \ldots & & 0 \\ 0 & -k_2 & k_2+k_3 & \ldots & & 0 \\ \ldots & \ldots & \ldots & \ldots & & \ldots \\ 0 & 0 & 0 & 0 & & -k_{N-1} \\ 0 & 0 & 0 & -k_{N-1} & k_{N-1}+k_N \end{bmatrix}_{N \times N} \qquad (5.33)$$

- A critical damping ratio is assigned for each mode of vibration.

Step 4: Dynamic response calculation

To solve the equations of motion, one of the dynamic analysis methods discussed in chapter 4 (e.g., modal analysis) is used.

$$[m](\ddot{x}) + [c](\dot{x}) + [k](x) = -(m)\ddot{x}_r \qquad (5.34)$$

Step 5: Calculation of maximum shear strains

For each layer i, the maximum shear strain, $\gamma_{\max (i)}$, is calculated from the maximum inter-layer displacement.

$$\gamma_{\max (i)} = \frac{\left| x_i(t) - x_{i+1}(t) \right|_{\max}}{h_i} \qquad (5.35)$$

Step 6: Update of the effective stiffness

From the maximum shear strains calculated in step 5, the effective shear modulus can be updated for each soil layer. As maximum stress levels are reached only once during the dynamic response, the new effective shear modulus is based on a fraction of the maximum values. Often, 2/3 of the maximum value is used to represent a shear strain level that reflects the global response of each soil layer.

$$G_i = \frac{G_{\max (i)}}{1 + \gamma_{h\,(i)}}$$

$$\gamma_{h\,(i)} = \frac{\frac{2}{3}\gamma_{\max (i)}}{\gamma_{r\,(i)}} \left[1 + ae^{-b\left(\frac{\frac{2}{3}\gamma_{\max (i)}}{\gamma_r} \right)} \right] \qquad (5.36)$$

Step 7: Update of the equivalent critical damping ratio
From the maximum shear strains calculated in step 5, the effective damping ratio can be updated for each soil layer. A fraction of the maximum values is used (e.g., 2/3).

$$\frac{\zeta_i}{\zeta_{max(i)}} = \frac{\gamma_{h\,(i)}}{\left(1 + \gamma_{h\,(i)}\right)}$$

$$\gamma_{h\,(i)} = \frac{\dfrac{2}{3}\gamma_{max(i)}}{\gamma_{r\,(i)}}\left[1 + a_1 e^{-b_1\left(\dfrac{\frac{2}{3}\gamma_{max(i)}}{\gamma_{r\,(i)}}\right)}\right]$$

$$(5.37)$$

We can compute the mean value of the ζ_i across all layers and apply it to each mode in the next iteration. Alternatively, a value of ζ corresponding to a layer at the centre of the soil deposit can be used.

Step 8
Steps 3 to 7 are repeated until the stiffness and damping converge. Usually, only a few iterations are required to attain convergence.

5.5 EXAMPLE

To illustrate the equivalent linear dynamic analysis, let us consider a clay deposit, 16 m thick, resting on bedrock (fig. 5.12). We compute the dynamic response of the soil, when the bedrock level is subjected to the S00E component of the May 18, 1940, El Centro earthquake. We can assume that the soil's behaviour is perfectly modelled by the hyperbolic law, and consequently, $\gamma_h = \gamma/\gamma_r$.

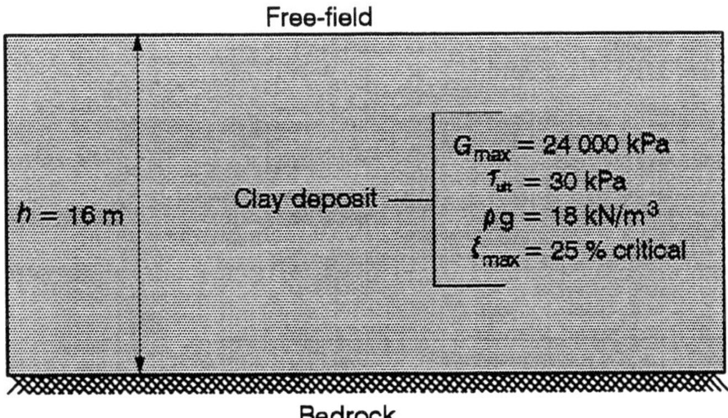

Figure 5.12 Clay deposit resting on bedrock.

Step 1
Using the physical parameters of the soil, the reference shear strain, γ_r, can be calculated.

$$\gamma_r = \frac{30\,\text{kPa}}{24\,000\,\text{kPa}} = 0,00125 \tag{5.38}$$

Step 2
The initial fundamental period of the soil deposit, T_1, is calculated.

$$T_1 = \frac{4\,(16\,\text{m})}{\sqrt{\dfrac{(24\,000\,\text{kPa})(9,81\,\text{m/s}^2)}{18\,\text{kN/m}^3}}} = 0,56 \ \text{s} \tag{5.39}$$

Referring to figure 5.4, if the computation accuracy is to be kept at a maximum of 5% (ERS = 5%), the soil deposit must be split into three layers. Figure 5.13 shows the model with the resulting lumped masses.

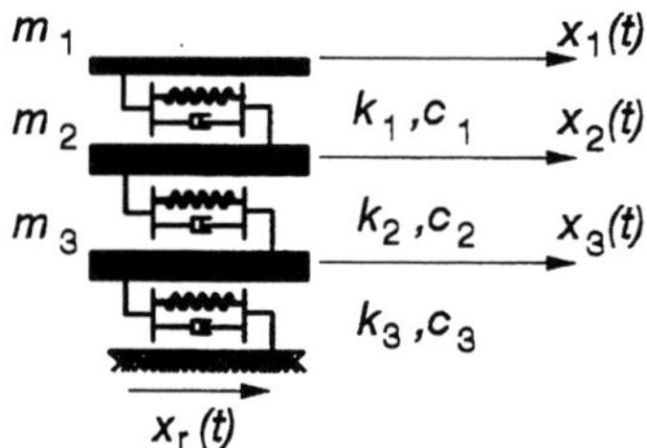

Figure 5.13 Lumped-mass model for clay deposit in figure 5.12.

Step 3
The mass matrix is found.

$$m_1 = \frac{(18\,\text{kN/m}^3)(5,33\,\text{m})}{2\,(9,81\,\text{m/s}^2)} = 4,89\,\frac{\text{kN}-\text{s}^2}{\text{m}^3}$$

$$m_2 = m_3 = 2\,m_1 = 9,78\,\frac{\text{kN}-\text{s}^2}{\text{m}^3} \tag{5.40}$$

$$[m] = \begin{bmatrix} 4,89 & 0 & 0 \\ 0 & 9,78 & 0 \\ 0 & 0 & 9,78 \end{bmatrix} \frac{\text{kN}-\text{s}^2}{\text{m}^3}$$

The stiffness matrix can be written as a function of the effective shear modulus of each layer, G_i.

$$k_i = \frac{G_i(\text{kPa})}{5{,}33\,\text{m}}$$

$$[k] = \frac{1}{5{,}33}\begin{bmatrix} G_1 & -G_1 & 0 \\ -G_1 & G_1+G_2 & -G_2 \\ 0 & -G_2 & G_2+G_3 \end{bmatrix}\frac{\text{kN}}{\text{m}^3} \tag{5.41}$$

Steps 4 to 7

The modal analysis (section 4.7.8) is used iteratively to calculate the dynamic response of the soil deposit. After each iteration, the shear modulus for each layer, G_i, and the critical damping ratio for the three modes of vibration, ζ, are updated using equations 5.36 and 5.37. Table 5.5 summarizes this iterative procedure. Note that the first iteration relates to the maximum shear modulus, G_{max}, and to a zero damping ratio. Convergence is achieved after the sixth iteration. This convergence would be even faster if a more realistic value were used for the critical damping ratio in the first iteration.

Table 5.5   Summary of the equivalent linear dynamic analysis.

Iteration no.	G_1 (kPa)	G_2 (kPa)	G_3 (kPa)	τ^* (%)	$\gamma_{1\,max}$	$\gamma_{2\,max}$	$\gamma_{3\,max}$
1	24 000	24 000	24 000	0	0,0057	0,0129	0,0168
2	5 938	3 038	2 402	21	0,0014	0,0069	0,0113
3	14 282	5 170	3 414	20	0,0006	0,0041	0,0089
4	18 747	7 520	4 208	17	0,0005	0,0036	0,0099
5	19 045	8 244	3 810	16	0,0005	0,0030	0,0105
6	19 478	9 217	3 645	15	0,0005	0,0029	0,0110
7	19 244	9 524	3 506	15			
*Based on $\gamma_{2\,max}$							

From the dynamic analysis results, the maximum shear stress for each layer is calculated by equation 5.13.

$$\tau_{1\,max} = \frac{0{,}0005}{\dfrac{1}{24\,000\,kPa} + \dfrac{0{,}0005}{30\,kPa}} = 9\,kPa$$

$$\tau_{2\,max} = \frac{0{,}0029}{\dfrac{1}{24\,000\,kPa} + \dfrac{0{,}0029}{30\,kPa}} = 21\,kPa \tag{5.42}$$

$$\tau_{3\,max} = \frac{0{,}0110}{\dfrac{1}{24\,000\,kPa} + \dfrac{0{,}0110}{30\,kPa}} = 27\,kPa$$

Figure 5.14 shows the first ten seconds of the bedrock absolute acceleration time-history, $\ddot{x}_r(t)$, and the first ten seconds of the free-field absolute acceleration, $\ddot{x}_r\,(t) + \ddot{x}_1\,(t)$. The frequency content of the free-field accelerogram is very different from the frequency content of the accelerogram at the bedrock. To give a better understanding of the effect of the soil deposit, figure 5.15 compares the absolute acceleration response spectra, at 5% damping, for the two accelerograms. The peak spectral response of the free-field accelerogram is found at 1,3 s, which corresponds to the fundamental period of the soil deposit calculated with the effective shear modulus.

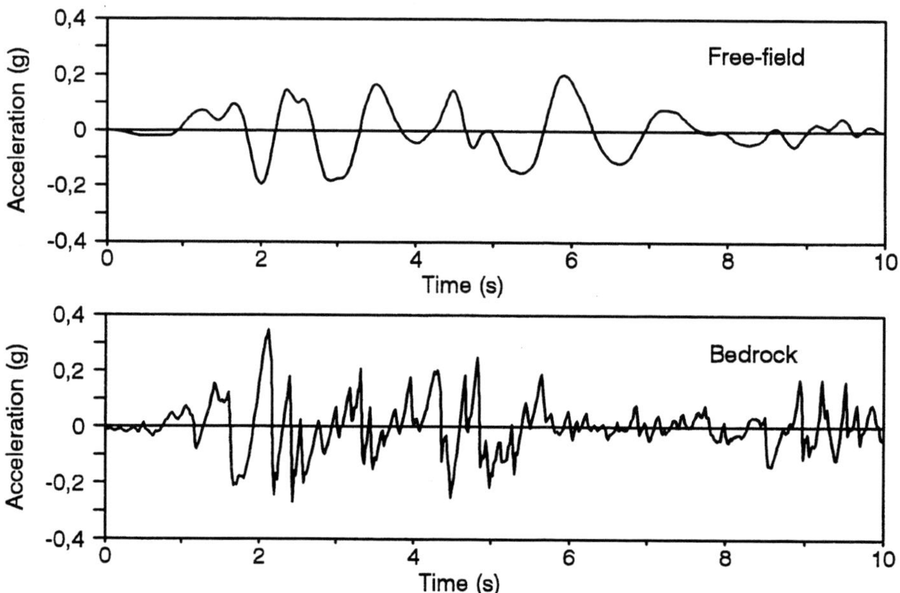

Figure 5.14   Bedrock accelerogram and free-field accelerogram for the soil deposit of figure 5.12.

As we can see, the soil deposit influences the free-field accelerogram in two ways:

1. The soil deposit changes the frequency content of the accelerogram; the free-field accelerogram contains much more energy at the soil deposit's fundamental frequency and attenuates the high frequencies (low periods).
2. The soil deposit may amplify the accelerogram; however, this amplification is not linear with respect to the earthquake intensity; the higher the acceleration at the rock level, the more nonlinearly the soil deposit behaves, which increases the effective damping and decreases the amplification (fig. 5.16).

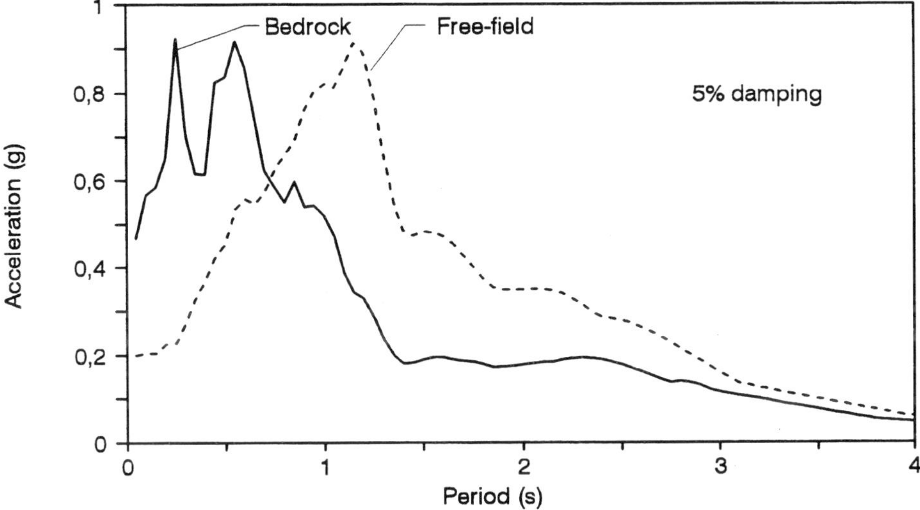

Figure 5.15   Bedrock and free-field absolute acceleration response spectra for the soil deposit of figure 5.12.

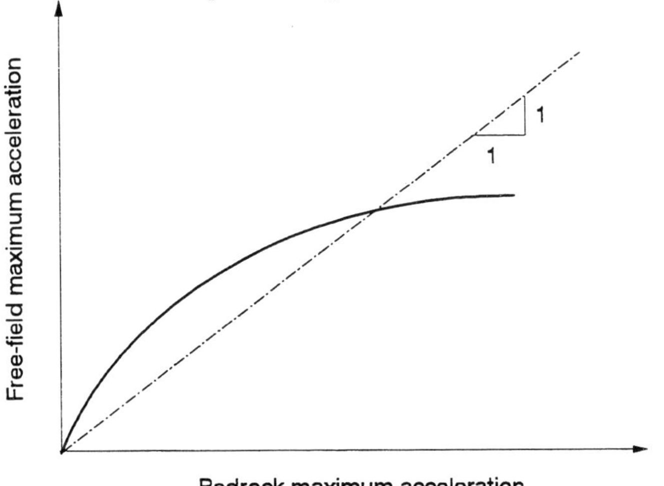

Figure 5.16   Amplification in a soil deposit.

5.6 STRENGTH LOSS AND SOIL LIQUEFACTION

5.6.1 Description of the Phenomenon

The cyclic stresses and strains imposed on a soil deposit during an earthquake may cause total or partial loss of its load-bearing capacity. When a soil layer completely loses its load-bearing capacity, it is said to be liquefied. The types of soil most susceptible to liquefaction are sands and saturated non-compacted silts. Liquefaction of saturated sands has been the major cause of damage to buildings, earthen dams, and retaining walls during a great number of earthquakes. The earthquakes that occurred in Nikata, Japan, in 1964 and in Mexico City in 1985 are good examples of how liquefaction of sands caused considerable damage to civil-engineering structures.

The effect of an earthquake on a soil layer can be simulated in the laboratory by applying a cyclic load on a soil sample (fig. 5.17). Over the last 30 years, a great number of such experiments have taken place. As a result, there is now a better understanding of the liquefaction phenomenon.

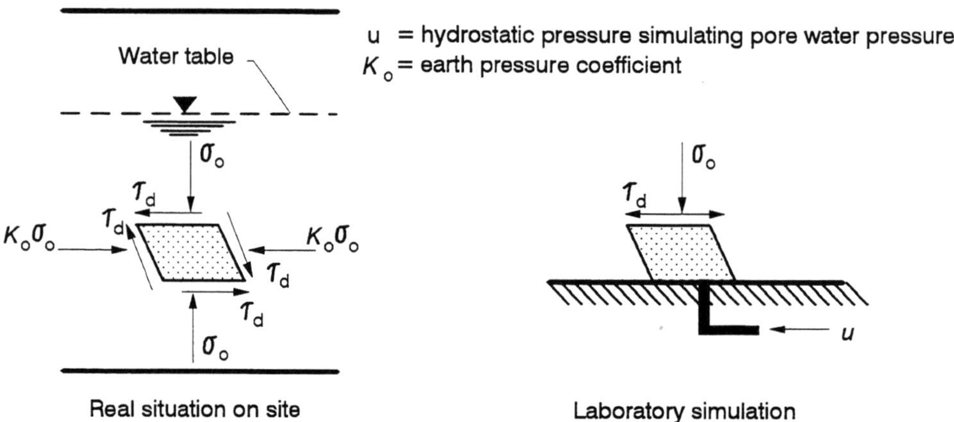

Figure 5.17   Laboratory simulation of the effect of an earthquake on a soil layer.

5.6.2 Physical Explanation of the Liquefaction Phenomenon

The results of cyclic tests on dry sands and on saturated sands for which drainage is possible show a decrease in the sample's volume with a pore water pressure, u, equal to zero. If the sand is saturated and drainage cannot occur, then the volume of the soil cannot decrease, as the water is incompressible. Instead of decreasing in volume, the soil transfers its vertical load to the water, causing an increase in the pore pressure. The shear strength, τ_{ult}, of sand is obtained by the following relation.

$$\tau_{ult} = \left(\sigma_o - u\right)\tan\phi \tag{5.43}$$

where σ_o =   total vertical stress
ϕ =   internal friction angle of the soil

If the pore water pressure increases in such a way that it becomes equal to the total stress $(\sigma_o = u)$, then the shear strength drops to zero and the soil suddenly behaves as a fluid, losing its load-bearing capacity and liquefying. It is the soil's tendency to decrease in volume (to settle) during a cyclic load that creates the problem. If the sand can be drained, the vibrations cause settling without any loss of load-bearing capacity or liquefaction. If, however, the sand does not have time to drain during the vibrations, as is often the case, a loss of bearing capacity (or liquefaction) occurs. Clean gravel has great permeability and will have time to drain during an earthquake, so gravel will not tend to liquefy. If, however, gravel is trapped between two soil layers with very low permeabilities, the gravel will then liquefy. Clays, surprisingly, are not as susceptible to liquefaction as sands are. The reason is that fine layers of water surround the clay particles (adsorbed water). These layers become viscous and reduce soil settlement during a cyclic load; the behaviour of clay is said to be more plastic.

5.6.3 Estimation of Liquefaction Potential by Laboratory Testing

In theory, the liquefaction potential of a soil layer supporting a foundation can be determined by the following two steps:

1. Perform a dynamic analysis for a design earthquake (as discussed above) to determine the maximum dynamic shear stresses the soil layer must sustain.
2. Obtain an undisturbed sample of the soil layer, and simultaneously apply to it a constant vertical stress and the dynamic shear stress calculated in step 1. Then observe whether or not the sample liquefies.

The second step is practically impossible because of the difficulty in acquiring an undisturbed soil sample. Reproducing the calculated dynamic load is also a problem. Sometimes, however, it is possible to measure in the laboratory what is called the liquefaction resistance of sands, which is defined by the number of cycles required to cause liquefaction when the shear stress level is constant. Such tests have been useful in identifying the ratio of the amplitude of the static, vertical, effective shear stresses, σ'_o, to the amplitude of the dynamic, lateral shear stresses, τ_d, as an important parameter controlling the liquefaction of soils.

5.6.4 Estimation of Liquefaction Potential by In Situ Testing

As undisturbed sand samples are very hard to obtain, the liquefaction potential of a particular area is often estimated by field testing *(in situ testing)*. Two types of tests are used: a) the seismic cone test and b) the standard penetration test (or SPT). The SPT is the more accepted and the more widely used. This chapter will focus on this method.

To perform the SPT, a probe is hammered into the ground. The number of blows required for the probe to penetrate 0,3 m into the ground is called the standard penetration value, N.

For soil with uniform density, the value of N increases with depth, as the confining pressure around the probe increases with depth, making it harder for the probe to penetrate into the ground. To take this phenomenon into account, the penetration value is corrected for a normalized confining stress of 1 atmosphere. This corrected penetration value is written N_1. Moreover, the net energy produced by the hammer during the test has to be considered. This net energy is less than the total potential energy because of friction losses. The net transferred energy varies for different types of hammer, as shown in table 5.6. Seed et al. (1984) suggested adjusting the value of N_1 for a net energy equal to 60% of the total potential energy. The final corrected value is written $(N_1)_{60}$.

$$(N_1)_{60} = \frac{C_N E_r N}{60} \tag{5.44}$$

where C_N = correction factor for the confining pressure (fig. 5.18)
$\quad\quad\quad E_r$ = percentage of the energy transmitted to the penetration probe; typical values of E_r (table 5.6)

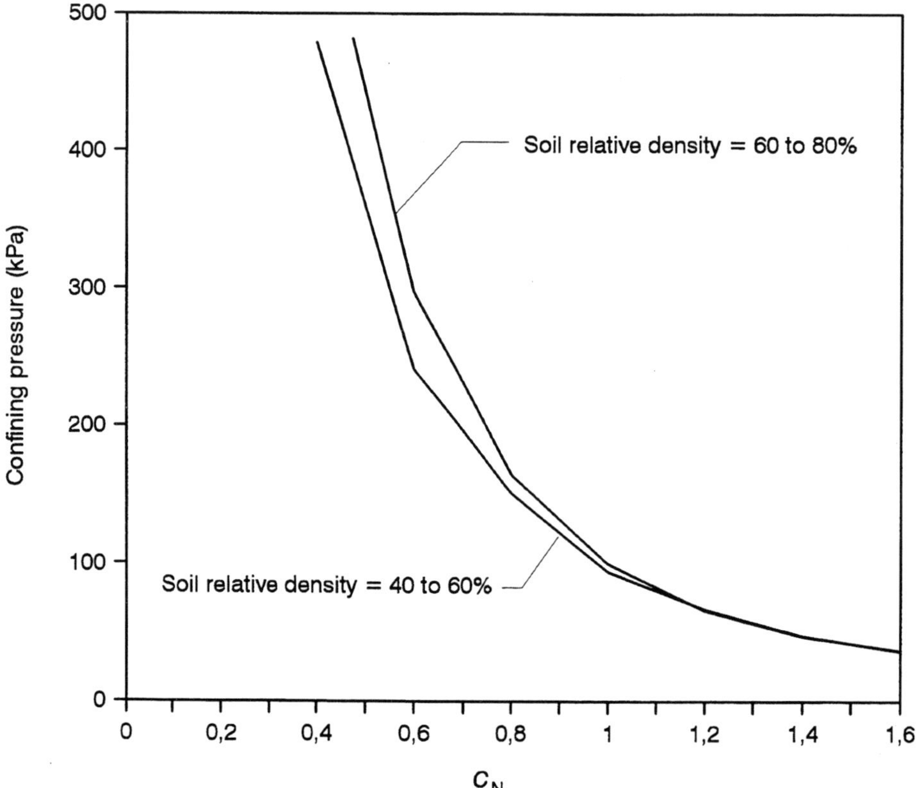

Figure 5.18 Correction factor C_N to take into account the confining pressure in the SPT test (from Seed et al., 1984).

An empirical liquefaction resistance curve is used to evaluate the liquefaction potential of a particular site. Figure 5.19 shows the liquefaction resistance curve developed by Seed et al. (1984). This curve was elaborated by examining a large number of sites, in which liquefaction was either present or absent. These experimental data are also shown in figure 5.19. For each site, the corrected penetration value, $(N_1)_{60}$, and the level of dynamic shear stress were known. The curve of the liquefaction resistance relates the ratio of the maximum shear stress to the vertical confining effective stress, with the value of $(N_1)_{60}$. If a site is located to the right of the boundary curve, liquefaction should not occur. However, if a site is located to the left of the boundary curve, liquefaction occurs. This liquefaction resistance curve was normalized for earthquakes of magnitude $M_L = 7,5$. If the earthquake has a different magnitude, the stress ratio (or the vertical axis of the resistance curve) has to be corrected by the factor shown in table 5.7.

Incidentally, the stress ratio obtained by the dynamic analysis can be divided by the correction factor of table 5.7 and the liquefaction resistance curve can be used as such. This correction is necessary as liquefaction depends not only on the shear stress level, but also on the number of violent vibration cycles, which, in turn, depends on the earthquake's magnitude. Figure 5.19 shows that clean, saturated sand with a $(N_1)_{60}$ value of less than 10 will most likely liquefy even during a moderate earthquake.

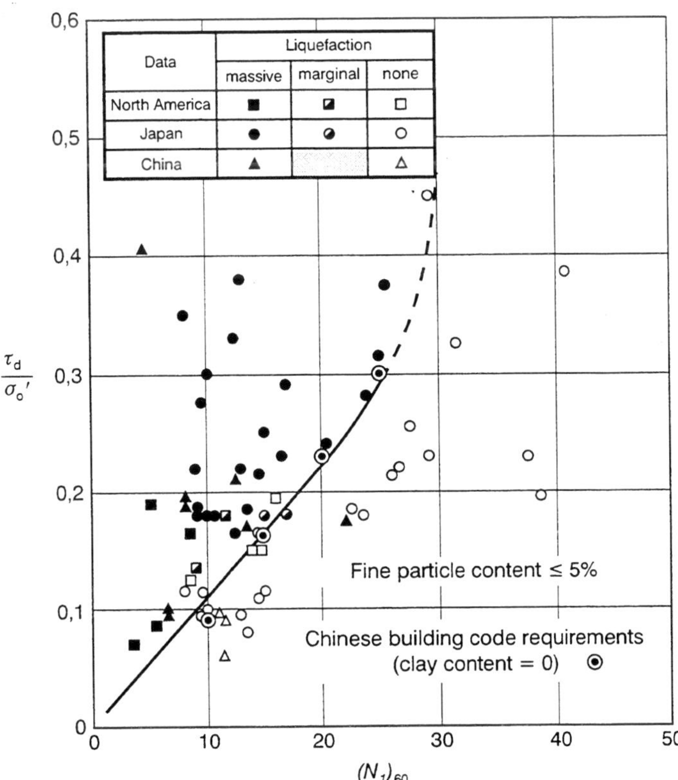

Figure 5.19 Liquefaction resistance of sands during an earthquake of magnitude $M_L = 7,5$ (from Seed et al., 1984).

Table 5.6 Typical values of the percentage of energy transmitted to the probe during an SPT (from Seed et al., 1984).

Country	Type of hammer	Hammer release mechanism	Transmitted energy, E_r (%)
Japan	Donut	Free fall	78
	Donut	Cable and pulley	67
USA	Safety	Cable and pulley	60
	Donut	Cable and pulley	45
Argentina	Donut	Cable and pulley	45
China	Donut	Free fall	60
	Donut	Cable and pulley	50

Table 5.7 Correction of liquefaction resistance for various magnitudes (from Seed et al., 1984).

Magnitude M_L	Number of typical load cycles	Correction factor for liquefaction resistance
8,50	26	0,89
7,50	15	1,00
6,75	10	1,13
6,00	5–6	1,32
5,25	2–3	1,50

For the same value of $(N_1)_{60}$, experiments have shown that silty sands are less likely to liquefy than are clean sands. Seed (1986) suggested that figure 5.19 be used to evaluate the liquefaction potential with a corrected value of $(N_1)_{60}$ according to the following equation.

$$\left(N_1\right)_{60\ \text{corrected}} = \left(N_1\right)_{60\ \text{measured}} + \Delta N_1 \tag{5.45}$$

In equation 5.45, ΔN_1 is a function of the fine-particle content (% of the particles filtered through a #200 sieve), as shown in table 5.8.

Table 5.8 Correction factor ΔN_1 for silty sands (from Seed, 1986).

Fine-particle content (%)	ΔN_1
≤ 5	0
15	3
35	5
50	7

5.6.5 Safety Factor against Liquefaction

A safety factor against liquefaction (SFL) for a soil layer at a given site can be defined by comparing the stress ratio causing liquefaction (liquefaction resistance) with the stress ratio caused by the design earthquake.

$$SFL = \frac{\left(\dfrac{\tau_d}{\sigma_o'}\right)_{\text{resistance}}}{\left(\dfrac{\tau_d}{\sigma_o'}\right)_{\text{earthquake}}} \tag{5.46}$$

It is still unclear what SFL value should be used in practice. The National Academy of Science in the United States (NAS, 1985) proposed a minimum SFL value of 1,33 to avoid liquefaction if the design earthquake at the site is based on a reasonable probability level.

5.6.6 Estimation of Dynamic Shear Stresses

The dynamic shear stresses and strains can be obtained by a dynamic analysis of the soil deposit, as seen in section 5.4. However, use of the maximum shear stress, $(\tau_d)_{\max}$, would be too conservative to estimate the liquefaction potential of a soil layer, as this stress level is attained only once during an earthquake. In practice, it is more realistic to use the equivalent shear stress, τ_{eq}, a stress level more representative of the global response of the soil layer.

$$\tau_{eq} = 0,65\left(\tau_d\right)_{\max} \tag{5.47}$$

To avoid dynamic analysis, which has to be performed for an ensemble of typical earthquakes for a given site, Seed et al. (1984) proposed an empirical equation to estimate the stress ratio caused by an earthquake.

$$\frac{\tau_{eq}}{\sigma_o'} = 0,65\left(\frac{a_{\max}}{g}\right)\left(\frac{\sigma_o}{\sigma_o'}\right)r_d \tag{5.48}$$

where a_{max} = peak horizontal acceleration on the surface of the soil deposit in g
    σ_{o} = total vertical stress
    σ_{o}' = vertical effective stress
    r_{d} = stress-reduction factor to take depth into account (fig. 5.20)

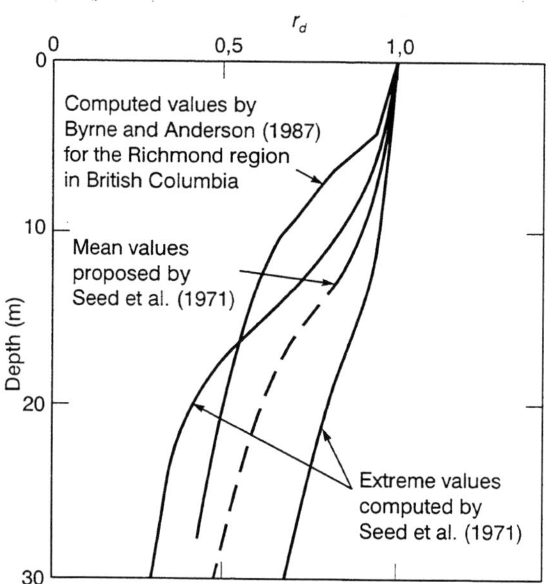

Figure 5.20 Stress-reduction factor for depth.

5.6.7 Procedures to Mitigate Liquefaction

Four different types of interventions are possible to ensure adequate operation of structures located on liquefiable sites. These interventions were defined by Silver (1985), and are presented in table 5.9.

Functional changes. Functional changes are decisions made by the owner or the head of the project in order to minimize the impact caused by liquefaction. The easiest decision is to accept the risk associated with liquefaction of the supporting soil and not take any special measures against it. In many cases, however, simple functional measures can greatly reduce the risk. These measures are: restrict access to vulnerable zones, set up rapid-intervention strategies in collaboration with local authorities; in the case of a reservoir or a dam, define moderate water levels to minimize potential flooding.

Table 5.9 Types of interventions for a liquefiable site (adapted from Silver, 1985).

Type	Description	Possible interventions
1	Functional changes	. acceptance of risk . warning of users . project relocation
2	Treatment of liquefiable soil	. excavation and replacement of liquefiable soil . densification . alteration of liquefiable soil . chemical stabilization
3	Structural changes	. function of the type of structures and foundations
4	Control of pore water pressure	. stone columns . pumping systems . air injection . draining systems . control of the water table

Treatment of liquefiable soil. The first solution when treating a potentially liquefiable soil is to remove and replace the soil. There are different possibilities:
- excavate, backfill, and compact existing soil;
- excavate, backfill, and compact the existing soil with additives;
- excavate existing soil and backfill with new soil.

The second type of treatment is densification of the existing soil to increase the effective static stresses and avoid a loss of volume during vibrations caused by an earthquake. Different techniques are used to densify existing soil:
- drive compaction piles;
- insert vibrating probes;
- use the vibro-flotation technique, which consists of inserting a probe and compacting the soil by vibrations and by injection of water under pressure (Brown, 1977);
- inject chemicals (generally resins) to increase the shear strength of the soil;
- use the dynamic compaction technique, which consists of letting a certain weight free-fall from the top of a crane on a predetermined grid pattern.

The alteration of liquefiable soil consists of mixing the existing soil with chemical additives. This chemical stabilization is done by pressurized injection of additives, such as resins, in the voids of the material.

Structural changes. Different types of modifications – stabilizing or replacing inadequate foundations – can be performed on the structure itself to minimize the effects of liquefaction of the supporting soil. For example, a mat footing can be replaced by piles or box foundations.

Control of pore water pressure. Control of pore water pressure to limit the risk of liquefaction requires extensive knowledge of the nature and extent of the liquefiable soil surrounding the structure. Stone columns (or gravel drains) can be inserted. These columns will be used as drains so that pore water in strategic locations is evacuated during seismic vibrations. Figure 5.21 shows a typical stone-column set-up.

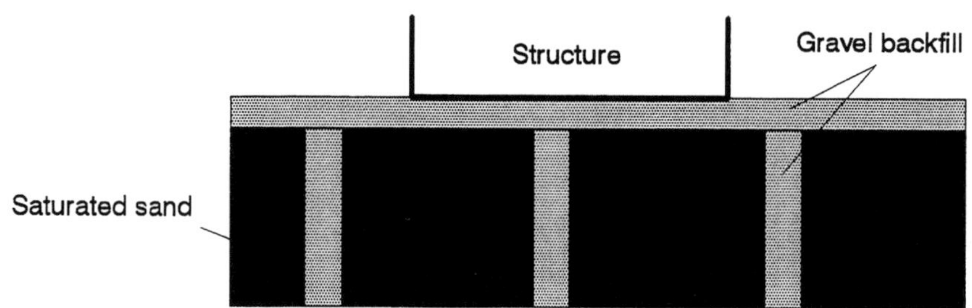

Figure 5.21   Stone columns (or gravel drains).

Reliable pumping systems must be available. These systems must be equipped with gauges such as piezometers. As well, it must be ensured that the system does not pump fine soil particles for the entire duration of the project.

Recently, injection of air into the voids of a liquefiable soil has been proposed as a measure to control pore water pressure. As voids in the soil are filled with air, soil compaction during vibrations does not cause a sizable increase in pore water pressure.

However, conventional drainage systems can be used in new construction. These systems may take different forms depending on the type of structure being built. Finally, the water table level can be controlled by various well-known techniques.

5.6.8 Residual Strength of Sands after Liquefaction

Although sands may liquefy during an earthquake, they may regain some of their strength (resistance) when the vibrations have ceased; this resistance is called the residual strength. Considerable controversy exists regarding the strength after liquefaction has been triggered. Castro (1976), for example, states that sands regain their strength almost totally after vibrations have ceased. However, Seed (1986) believes that the residual strength of sands is related to the value of the standard penetration, $(N_1)_{60}$. Comprehensive testing carried out by Chern (1984) demonstrated that Castro's claim is valid. The undrained strength of Ottawa sand is presented in figure 5.22 as a function of $(N_1)_{60}$. We notice that the residual strength is much higher than the residual strength proposed by Seed (1986).

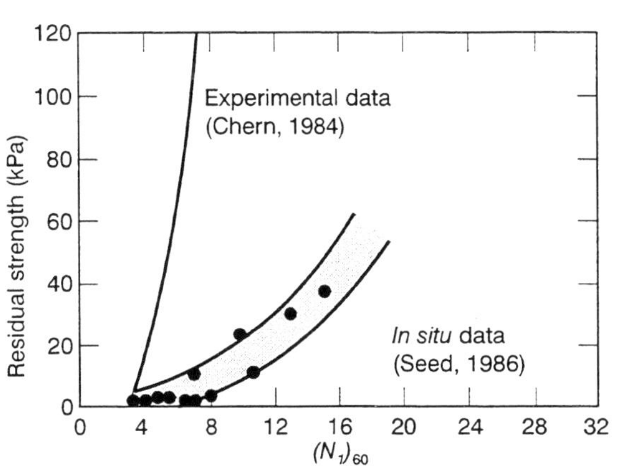

Figure 5.22 Residual strength of sands after liquefaction.

5.7 INTRODUCTION TO SOIL-STRUCTURE INTERACTION

5.7.1 Description of Soil-Structure Interaction

During dynamic analysis of a structure located on a soil deposit, a free-field accelerogram can be used as a motion at the base of a structure while considering this base rigid (fig. 5.23).

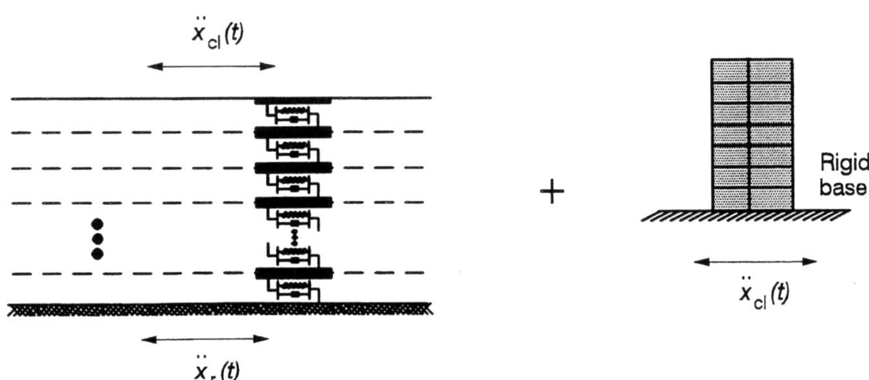

Figure 5.23 Dynamic analysis of a structure resting on a soil deposit, neglecting the soil-structure interaction.

This approach, although frequently used in practice, is controversial, especially if the structure is resting on a very soft soil deposit. The structure can modify the ground motion at its base because the structure and the soil can interact to create a new combined dynamic system. This is known as the soil-structure interaction.

There are several reasons that the soil-structure interaction must be considered:

1. to compute mode shapes that include deformations at the base of the structure: rocking of the foundation, overturning of the footing, etc.;
2. to obtain a better approximation of the period of vibration of the coupled system, which will be longer than the system with a rigid base;
3. to assess the seismic behaviour of critical facilities such as nuclear power plants.

It is important to realize that a direct extension of the lumped-mass dynamic analysis of a soil layer (section 5.2) to consider the presence of a structure at the surface is not valid. The major assumption of this procedure is to consider the soil deposit a semi-infinite continuous medium, while the soil-structure interaction is only a local phenomenon (fig. 5.24).

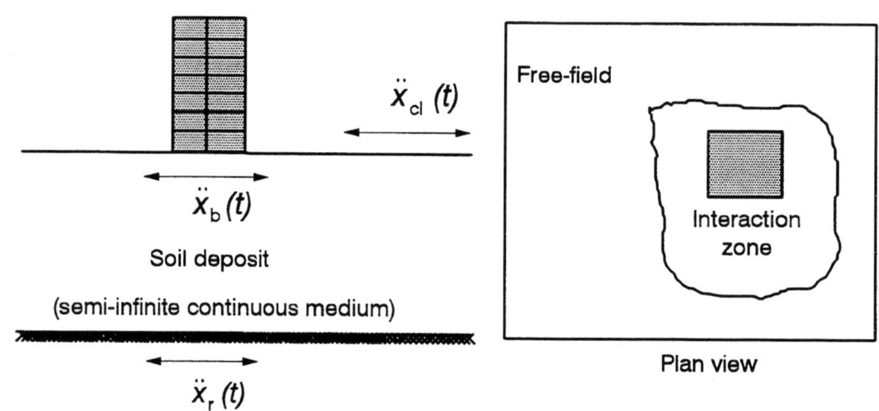

Figure 5.24   Local effect of the soil-structure interaction.

Special analysis techniques must be used to analyze correctly the soil-structure interaction. These techniques are discussed very briefly in the next section.

5.7.2 Soil-Structure Interaction Analysis Techniques

Two classic methods are available to take account of the soil-structure interaction in a dynamic analysis: a) the finite elements method and b) the sub-structure method.

The finite elements method is more direct but often more complex, especially because of its computing time. The soil deposit is discretized into finite elements. The base rock accelerogram is used directly to excite the system. The free-field accelerogram, $\ddot{x}_{cl}(t)$, the accelerogram at the base of the structure, $\ddot{x}_b(t)$, and the complete dynamic coupled response of the soil deposit and of the structure are obtained directly.

The sub-structure method, on the other hand, separates the soil-structure interaction problem into two parts (fig. 5.25).

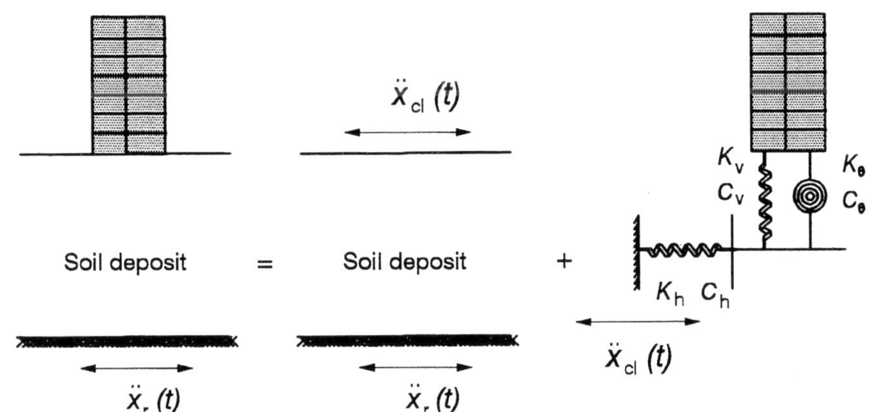

Figure 5.25   The sub-structure method for analyzing a soil-structure interaction problem.

In the first part, the free-field accelerogram is computed without considering the presence of the structure, as discussed above. In the second part, this free-field accelerogram is used to analyze the structure having its base coupled with discrete spring and dashpot elements. These elements are introduced to model the soil behaviour near the foundation. The theory of elasticity can be used to compute the various constants (k_h, k_v, and k_θ) for the spring elements, as shown in table 5.10.

Table 5.10   Spring constants for a rigid circular mat foundation (from Whitman and Richart, 1967).

Movement	Spring constant
Vertical	$k_v = \dfrac{4Gr_0}{(1-\mu)}$
Horizontal sliding	$k_h = \dfrac{32(1-\mu)Gr_0^3}{(7-8\mu)}$
Rocking	$k_\theta = \dfrac{8Gr_0^3}{3\,(1-\mu)}$
Torsion	$k_\alpha = \dfrac{16Gr_0^3}{3(1-\mu)}$
G = effective shear modulus of the soil μ = Poisson's soil ratio r_o = radius of the circular mat foundation	

5.7.3 Final Remarks on Soil-Structure Interaction

One must be extremely careful when performing a soil-structure interaction analysis. The results, even for a very sophisticated model, are very sensitive to the parameters used to model the soil, which cannot be estimated very accurately. Experience and judgment must be called upon to interpret correctly the results of a soil-structure interaction dynamic analysis.

5.8   PROBLEMS

5.1   A soil deposit is made of a 8 m thick sand layer and a 6 m thick clay layer overlying on bedrock. Given:

Layer	Specific weight (kN/m³)	Effective shear modulus (kN/m²)
Sand	16,8	28 000
Clay	18,2	36 000

Determine the number of sub-layers for the sand and the clay such that the maximum error from a lumped-mass dynamic analysis remains under 2%.

5.2   The results of standard penetration tests (SPTs) for a saturated sand deposit in the Richmond area, near Vancouver, are given below.

Depth (m)	$(N_1)_{60}$ value
3	8
5	14
10	16
15	17
20	17
25	17
30	17

The design earthquake for Richmond is based on a local magnitude of 7 and a peak ground horizontal acceleration of 0,20 g. The stress ratios, τ_{eq}/σ'_0, induced in the sand deposit by the design earthquake are given below.

Depth (m)	τ_{eq}/σ'_0
4	0,28
6	0,23
8	0,21
10	0,19
15	0,16
20	0,14
30	0,12

Assuming that the water table is at the surface, draw a profile of the safety factor against liquefaction (SFL) with depth and estimate the liquefaction potential of the region. Use the liquefaction-resistance curve proposed by Seed et al. (1984) shown in figure 5.19.

5.3   The central concrete core of the 21-storey office building is shown in figure 5.26 is founded on a circular mat footing of radius R.

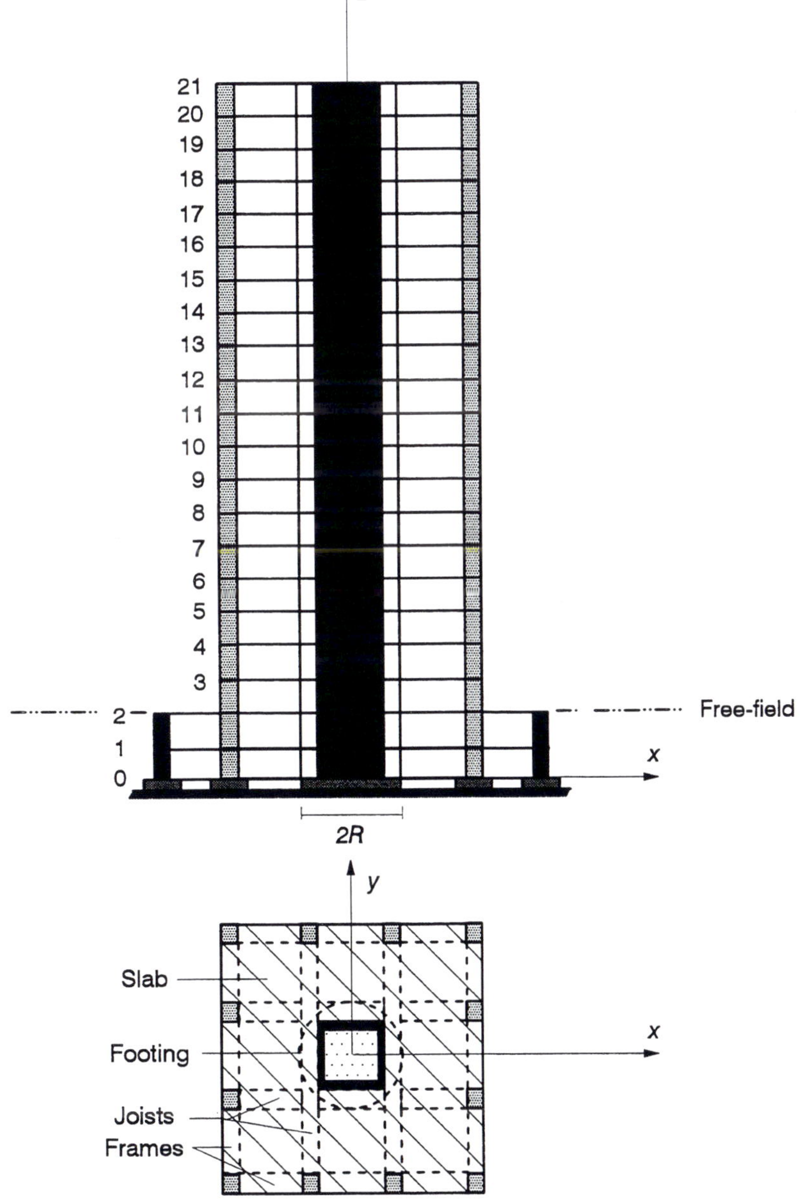

Figure 5.26   Twenty-one-storey office building.

A two-dimensional dynamic analysis model of the building, including the soil-structure interaction effect, is shown in figure 5.27. The peripheral frames are much more flexible than is the concrete core, and can be neglected in the model. The influence of the soil near the foundation is modelled by a group of N linear springs with stiffness coefficients k_i $(i = 1 ... N)$.

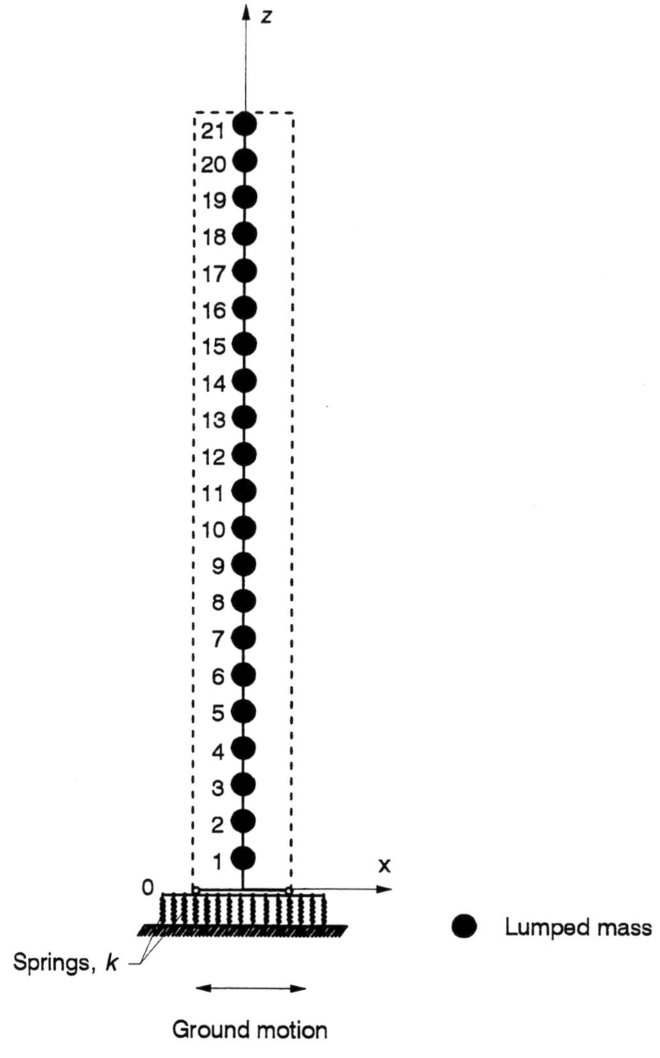

Figure 5.27   Two-dimensional model of a 21-storey office building.

a)   Compute a constant vertical stiffness per unit area of soil corresponding to the equivalent rocking stiffness of the foundation (see table 5.10).

b)   For the same assumption as a), derive an expression for the stiffness coefficient k_i for each spring. Give your answer as a function of the spacing between the springs, s, the distance between the spring and the centre of the footing, d_i, the radius of the footing, R, the shear modulus of the soil, G, and Poisson's soil ratio, μ.

5.4   Figure 5.28 shows a sand deposit, 20 m thick, founded on bedrock. The water table is located at a depth of 10 m below the surface. The properties of the sand are as follows:

Condition	Specific weight (kN/m³)	Shear modulus (kPa)
Dry	16,8	28 000
Saturated	19,2	28 000

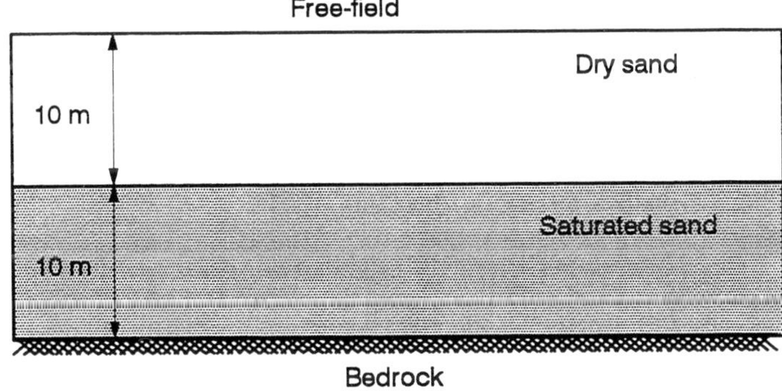

Figure 5.28   Sand deposit founded on bedrock.

The results of standard penetration tests (SPTs) for this sand deposit are:

Depth (m)	$(N_1)_{60}$ value
4	8
8	12
11	15
14	17
18	17

The design earthquake on bedrock for the region is based on a local magnitude of 7,5, a peak ground horizontal acceleration of 0,2 g, and a peak ground horizontal velocity of 0,1 m/s. It can be assumed that the design response spectrum, for 5% damping at the bedrock, corresponds to the NBCC response spectrum (fig. 5.29).

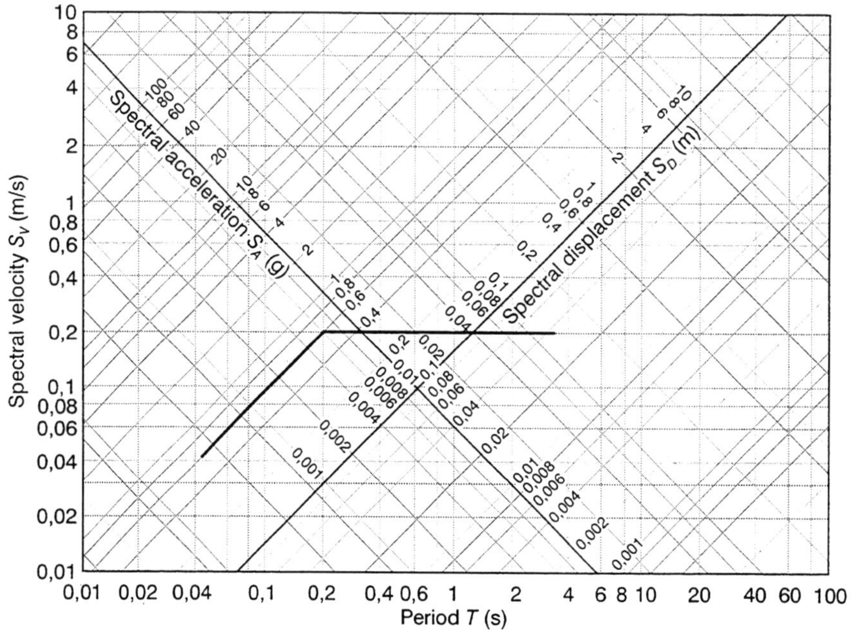

Figure 5.29　NBCC response spectrum for a seismic zone in Canada with a peak ground horizontal acceleration of 0,2 g and a peak ground horizontal velocity of 0,1 m/s.

a)　Divide the sand deposit into two layers, and perform a spectral analysis to estimate the maximum probable dynamic shear stress above and below the water table caused by the design earthquake at the bedrock. As a first approximation, the sand is considered linear and elastic with an equivalent viscous damping ratio of 5% in each mode of vibration.

b)　Using the liquefaction resistance curve proposed by Seed (1984) (fig. 5.19), estimate the liquefaction potential of the sand deposit.

5.9　REFERENCES

Byrne, P.M., and Anderson, D.L. (1987). *Liquefaction Potential for Richmond, B.C.* Soil Mechanics Report Series. Department of Civil Engineering, University of British Columbia, Vancouver, BC.

Brown, R.E. (1979). "Vibroflotation Compaction of Cohesionless Soils." ASCE, *Journal of the Geotechnical Division* 103(GT12): 1437–51.

Castro, G. (1976). "Comments on Seismic Stability Evaluation of Embankment Dams." *Proceedings of the Conference on the Evaluation of Dam Safety,* Pacific Grove, CA, pp. 377–90.

Chern, J.C. (1984). *Undrained Response of Saturated Sands with Emphasis on Liquefaction and Cyclic Mobility.* Ph.D. diss., Department of Civil Engineering, University of British Columbia, Vancouver, BC.

Clough, R.W., and Penzien, J. (1993). *Dynamics of Structures.* 2d ed. New York: McGraw-Hill.

Das, B.M. (1983). *Fundamentals of Soil Dynamics.* New York: Elsevier.

Hardin, B.O., and Drnevich, V.P. (1972). "Shear modulus and damping in soils: Design equations and curves." ASCE, *Journal of the Soil Mechanics and Foundations Division,* 98(SM7): 667–92.

Idriss, I.M., and Seed, H.B. (1968). "Seismic Response of Horizontal Soil Layers." ASCE, *Journal of the Soil Mechanics and Foundations Division,* 94(SM4): 1003–31.

National Academy of Science. Committee on Earthquake Engineering. (1985). *Liquefaction of Soils During Earthquakes.* Washington, DC: National Academy Press.

Seed, H.B. (1986). *Design Problems in Soil Liquefaction.* Report No. UCB/EERC-86/02. Earthquake Engineering Research Center, University of California, Berkeley, CA.

Seed, H.B., and Idriss, I.M. (1970). *Soil Moduli and Damping Factors for Dynamic Response Analysis.* Report No. EERC 70-10. Earthquake Engineering Research Center, University of California, Berkeley, CA.

— (1971). "Simplified Procedure for Evaluating Soil Liquefaction Potential." ASCE, *Journal of the Soil Mechanics and Foundations Division,* 97(SM9): 1249–73.

Seed, H.B., et al. (1984). *The Influence of SPT Procedures in Soil Liquefaction Resistance Evaluations.* Report No. UBC/EERC-84/15. Earthquake Engineering Research Center, University of California, Berkeley, CA.

Silver, M.L. (1985). *Remedial Measures to Improve the Seismic Strength of Embankment Dams.* Report No. 85-10. Department of Civil Engineering, University of Illinois, Chicago, IL.

Whitman, R.V., and Richart, Jr., F.E. (1967). "Design Procedure for Dynamically Loaded Foundations." ASCE, *Journal of the Soil Mechanics and Foundations Division,* 93(SM6): 169–93.

Earthquake-Resistant Design for Buildings in Canada

6.1 CONCEPT OF DUCTILITY

The concept of ductility is the key element of earthquake-resistant design of structures. To illustrate this concept, consider the simple portal frame shown in figure 6.1. It is assumed that the bending stiffness of the girder is much greater than the bending stiffness of the columns; in other words, the columns have no rotation at their ends.

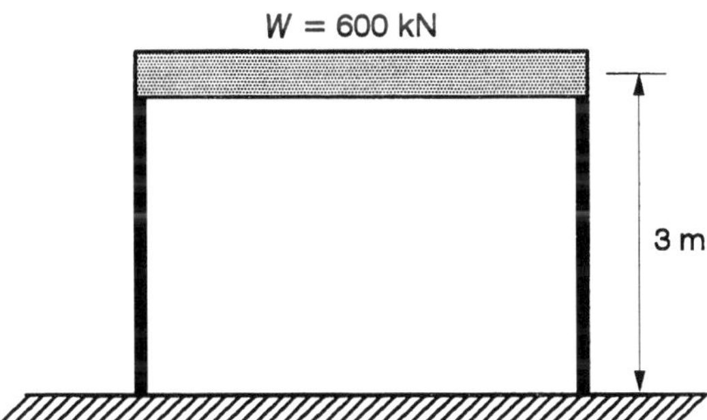

Figure 6.1 Simple portal frame.

For earthquake-resistant design, the National Building Code of Canada (NBCC, 1995) requires that the structure be able to sustain a base shear, V, proportional to the weight of the floor (this approach will be studied in detail below). If we consider a seismic zone in Canada for which $a_{max} = 0{,}3$ g and $v_{max} = 0{,}3$ m/s, the base shear will be equal to:

$$V = CW \approx 0{,}067\,W \tag{6.1}$$

For our case:

$$V = 40\,\text{kN} \tag{6.2}$$

Each column bends without any rotation at its ends, causing an inflection point to occur at mid-height, as shown in figure 6.2.

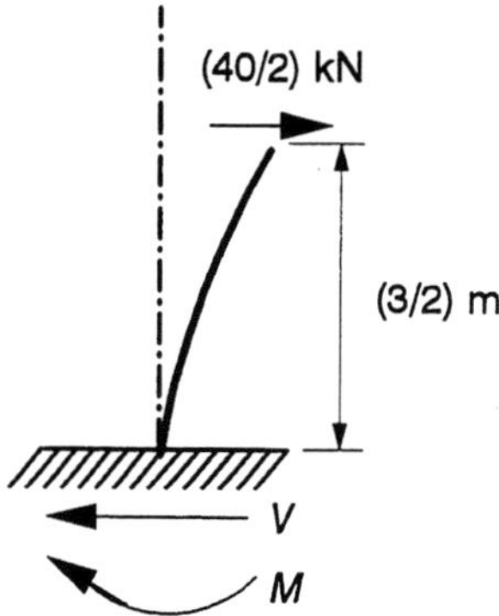

Figure 6.2   Free-body diagram for a column fixed at both ends.

The required bending moment for each column is then:

$$M_{required} = \left(\frac{40\,\text{kN}}{2}\right)\left(\frac{3\,\text{m}}{2}\right) = 30\,\text{kN-m} \tag{6.3}$$

Using a safety factor of 2, the required yield moment M_y is:

$$M_y = 60\,\text{kN-m} \tag{6.4}$$

An adequate steel column section has a depth $d = 125$ mm with flanges having each an area of 1 600 mm^2.

$$S_{required} = \frac{I}{C} = \frac{M_y}{\sigma_y} = \frac{2AC^2}{C} \tag{6.5}$$

$$(2AC)_{required} = \frac{60 \times 10^6\,\text{N-mm}}{300\ \text{MPa}} = 2 \times 10^5\,\text{mm}^3$$

The properties of the portal frame designed with these columns are shown in table 6.1.

Table 6.1   Properties of the portal frame designed according to the NBCC (1995).

Young's modulus, E	200 000 MPa
Yield strength, σ_y	300 MPa
Moment of inertia, I (for each column)	$12,5 \times 10^{-6}$ m^4
Lateral stiffness coefficient, k $k = 2(12EI/L^3)$	2 222 kN/m
Fundamental period, T $T = 2\pi(W/gk)^{1/2}$	1,04 s
Yield moment, M_y $M_y = \sigma_y I/C$ (for each column)	60 kN-m
Base shear corresponding to M_y, V_y $V_y = 2(2M_y/3)$	80 kN
Lateral deflection corresponding to V_y, δ_y $\delta_y = V_y/k$	36 mm

Figure 6.3 illustrates the idealized force-lateral displacement relationship of the frame. Let us now subject this structure to the S00W component of the El Centro earthquake (1940-05-18). Figure 2.12 illustrates the accelerogram for this component. This record shows: $a_{max} = 0,34$ g and $v_{max} = 0,33$ m/s; these values are very close to the seismic parameters used to design the portal frame. A nonlinear dynamic analysis is performed to obtain a realistic response from the structure subjected to this accelerogram.

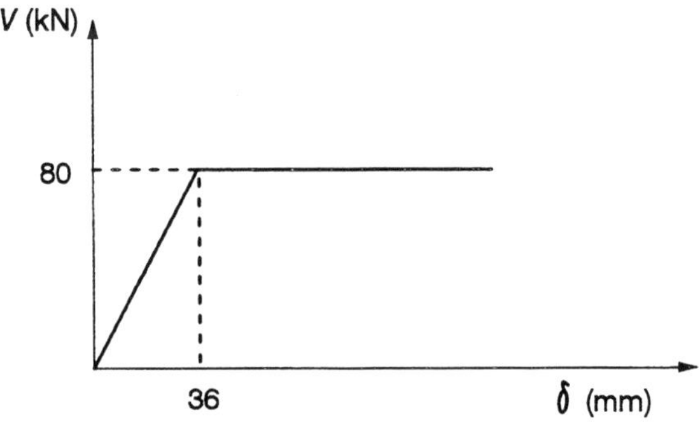

Figure 6.3   Idealized force-lateral displacement relation of the portal frame designed according to the NBCC (1995).

As seen in section 4.9, this analysis takes yielding of the structure into account and performs a time-step integration of the motion equation. To perform this nonlinear analysis, the damping ratio is set at 5% critical, the behaviour of the structure is idealized as elasto-plastic, and the integration time-step used is 0,01 s.

Figure 6.4 illustrates the results of this analysis in terms of the first 10 seconds of the lateral displacement time-history and of the base shear-lateral displacement hysteresis loops. The maximum lateral displacement is 94 mm, which is by far greater than the yield displacement of the columns. Even when the seismic provisions of the code are applied, the columns of the frame will yield during an earthquake! If the nonlinear analysis is repeated for different lateral strengths, V_y, the results shown in figure 6.5 are obtained. Although the details of the response vary, the maximum lateral displacement is not greatly influenced by the lateral strength. This phenomenon, first observed by Newmark and Veletsos (1960), is called the equal displacement principle, or Newmark's assumption. It forms the basis for seismic provisions in the majority of modern building codes, including the National Building Code of Canada. Figure 6.6 illustrates this ideal behaviour.

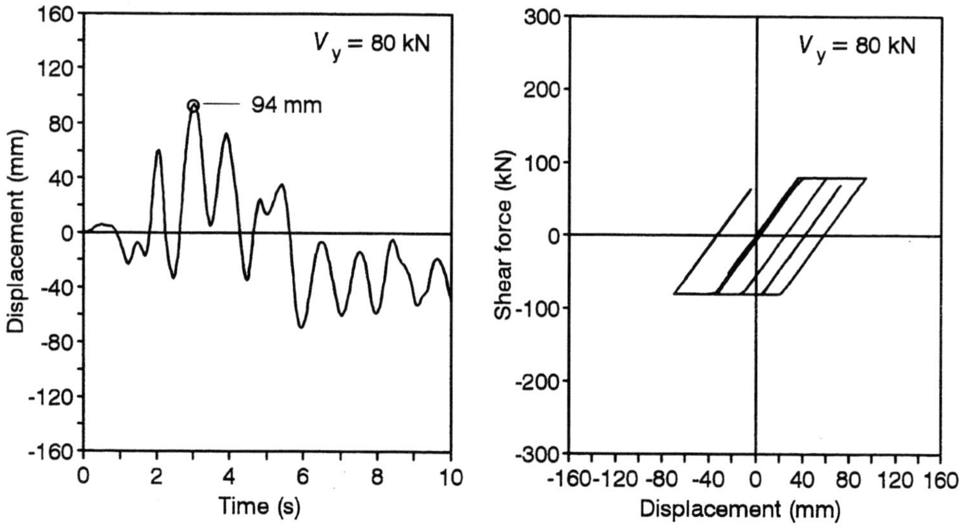

Figure 6.4   Results of the nonlinear dynamic analysis on a portal frame.

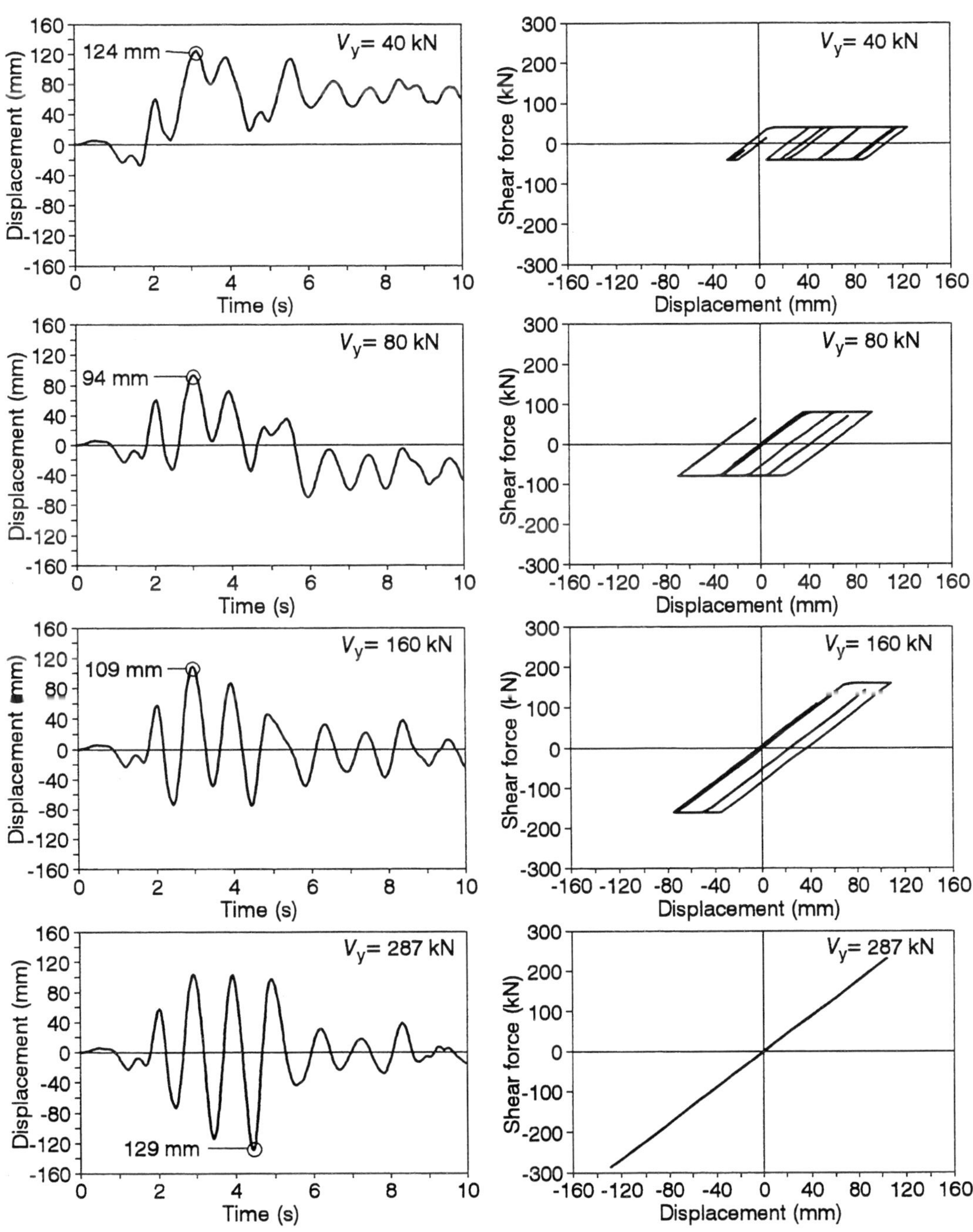

Figure 6.5 Dynamic response of a portal frame with various lateral strengths.

The length of the yield plateau (fig. 6.6) is defined as the ductility ratio, μ.

$$\mu = \frac{\delta_{max}}{\delta_y} = \frac{V_{max}}{V_y} \geq 1 \tag{6.6}$$

where δ_{max} = maximum displacement
δ_y = displacement corresponding to the structure's first yield

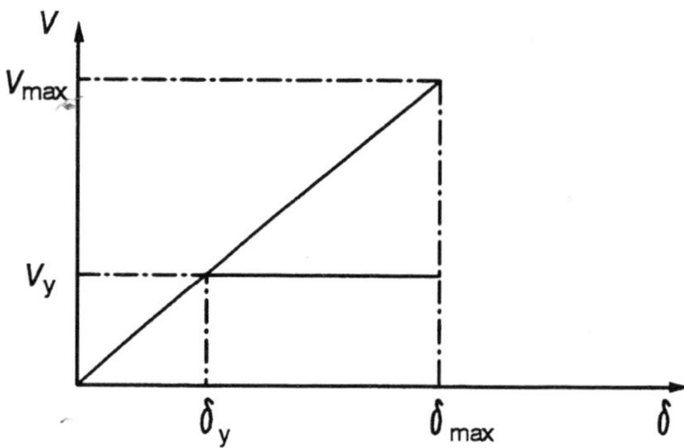

Figure 6.6 Idealized behaviour of a structure according to the
equal-displacement principle.

It is clear that if the structure has no ductility, then $\mu = 1$. As we can see, seismic-design provisions are based not on the load level that a structure can sustain, but on the displacement level that it can withstand without collapsing. Because of economic factors, yielding of certain members is tolerated as long as the inelastic behaviour is ductile and the structure's integrity remains. In other words, the structure has to demonstrate a certain level of ductility in order to survive during an earthquake. Ductility is achieved by appropriate structural details (connections, etc.).

It must be understood that the seismic design loads $(V = CW)$ specified by the NBCC (1995) are not real loads that the structure must sustain in the elastic range of the material during a major earthquake, but rational design criteria aiming to obtain a structure that is reasonable in terms of its ductility capacity.

6.2   CONCEPT OF INELASTIC RESPONSE SPECTRA

The fact that maximum lateral displacement of a nonlinear system is almost equal to maximum displacement of the corresponding linear system allows us to define inelastic seismic response spectra from the elastic seismic response spectra discussed in section 4.3. Here again, the equal displacement principle shown in figure 6.6 is used. For a given earthquake, the two systems have the same maximum lateral deflection. Using similar triangles, it yields the following equation:

$$\frac{V_{max}}{V_y} = \frac{\delta_{max}}{\delta_y} = \mu \tag{6.7}$$

or

$$V_y = \frac{V_{max}}{\mu} \tag{6.8}$$

Equation 6.8 can be written as a function of the seismic coefficients (section 4.5).

$$C_{(inelastic)} W = C_{(elastic)} \frac{W}{\mu} \tag{6.9}$$

or

$$S_{A_{(inelastic)}} \frac{W}{g} = S_{A_{(elastic)}} \frac{W}{\mu g} \tag{6.10}$$

The absolute inelastic acceleration response spectrum, S_A (inelastic), can then be defined as a function of the elastic acceleration response spectrum and the ductility ratio.

$$S_{A_{(inelastic)}} = \frac{S_{A_{(elastic)}}}{\mu} \tag{6.11}$$

The relative inelastic displacement spectrum, S_D (inelastic), can also be defined as the relative displacement of the structure when first yield is reached.

$$\delta_y = S_{D_{(inelastic)}} = \frac{\delta_{max}}{\mu} = \frac{S_{D_{(elastic)}}}{\mu} \tag{6.12}$$

Therefore, a tripartite graph can be used to construct an inelastic response spectrum for a certain level of ductility from an elastic response spectrum (fig. 6.7). On the top elastic curve, we can read:

- maximum absolute acceleration of the elastic system;
- maximum relative velocity of the elastic system;
- maximum relative displacement for the elastic and inelastic systems.

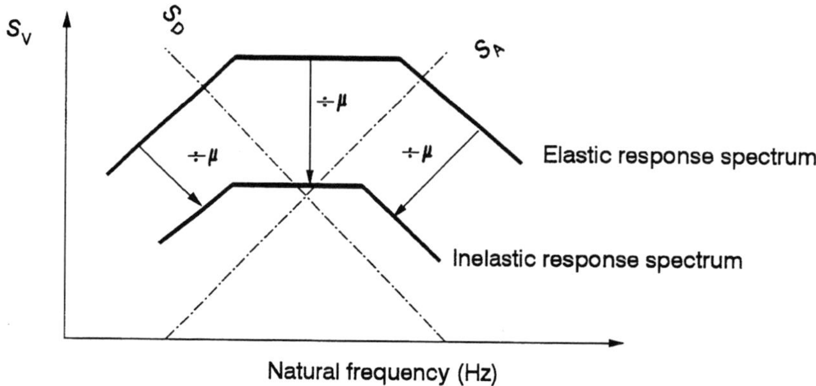

Figure 6.7 Construction of an inelastic response spectrum from an elastic response spectrum.

On the bottom inelastic curve, we can read:

- maximum absolute acceleration of the inelastic system only.

The bottom curve is really an absolute inelastic acceleration response spectrum. This spectrum can be used to determine the yield base shear, V_y, required to design the structure based on a given available ductility factor.

6.3   ENERGY CRITERION FOR SHORT-PERIOD STRUCTURES

If a system is very stiff and has a much shorter natural period than the predominant period of the accelerogram, then the equal-displacement principle cannot be used. In fact, for this specific case, the nonlinear dynamic analysis shows that, generally, the inelastic system produces more deformations than does the corresponding elastic system (fig. 6.8). This phenomenon can be explained by the fact that when a system with a short initial period yields, its period becomes longer and shifts toward the predominant period of the accelerogram.

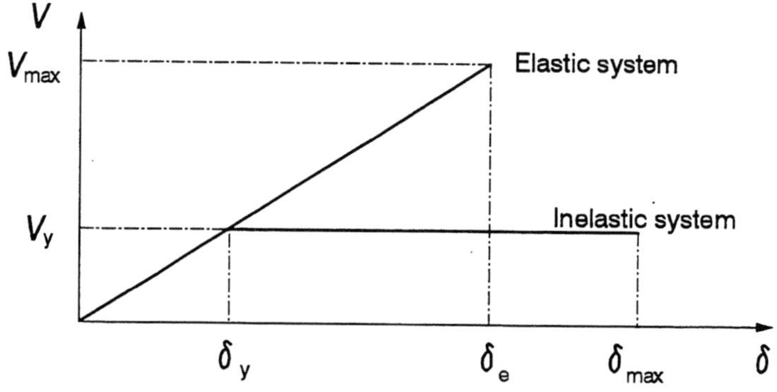

Figure 6.8   Energy criterion of a short-period inelastic system.

It is, in fact, a quasi-resonance phenomenon. A typical example of this phenomenon was observed during the 1985 Mexico City earthquake, in which movement at the ground surface was virtually sinusoidal with a period of vibration of 2 s. As a result, a great number of 10- to 15-storey buildings collapsed during the earthquake because of increased resonance after yielding. For systems with short initial periods of vibration, an energy criterion seemed to represent their behaviour better. This criterion states that the strain energy of an inelastic system and the strain energy of a corresponding elastic system are equal. From figure 6.8, the surfaces under the two curves are assumed to be equal.

$$\frac{V_{max}\,\delta_e}{2} = \frac{V_y\,\delta_y}{2} + V_y\left(\delta_{max} - \delta_y\right) \tag{6.13}$$

Using similar triangles, the following relation is obtained:

$$\frac{V_{max}}{\delta_e} = \frac{V_y}{\delta_y} \tag{6.14}$$

or

$$\delta_e = \frac{V_{max}\,\delta_y}{V_y} \tag{6.15}$$

Equation 6.15 is substituted into equation 6.13.

$$\frac{\left(V_{max}\right)^2 \delta_y}{2\,V_y} = V_y\left(\frac{\delta_y}{2} + \left(\delta_{max} - \delta_y\right)\right) \tag{6.16}$$

Simplifying, the following expression is obtained:

$$\left(V_{max}\right)^2 = \left(V_y\right)^2\left(2\frac{\delta_{max}}{\delta_y} - 1\right) = \left(V_y\right)^2\left(2\mu - 1\right) \tag{6.17}$$

The maximum base shear of the inelastic system can then be obtained from the base shear corresponding to the elastic system and from the available ductility ratio.

$$V_y = \frac{V_{max}}{\sqrt{2\mu - 1}} \tag{6.18}$$

In general, the elastic acceleration spectrum is reduced as follows:

$$S_{A_{(inelastic)}} = \frac{S_{A_{(elastic)}}}{\mu} \quad \text{for } T \geq 0{,}5 \text{ s}$$

$$S_{A_{(inelastic)}} = \frac{S_{A_{(elastic)}}}{\sqrt{2\mu - 1}} \quad \text{for } T < 0{,}5 \text{ s}$$

(6.19)

6.4 PRINCIPLES AND OBJECTIVES OF EARTHQUAKE-RESISTANT DESIGN FOR BUILDINGS IN CANADA

The seismic provisions of the NBCC (1995) are minimum requirements providing an acceptable degree of seismic safety. Structures designed according to these requirements must have a certain level of stiffness, strength, and ductility. In low-intensity earthquakes, the stiffness allows for protection against architectural damage to a building. In moderate-intensity earthquakes, the strength limits important damage to the main structure of a building. In major earthquakes, the ductility of the structure must allow for large inelastic displacements without collapse. In this context, a building is said to collapse when evacuation of its occupants is prevented. Structures could be designed to sustain the greatest earthquakes without any damage, but, in most cases, such designs are deemed neither economical nor justified because of the low probability of such seismic events occurring in Canada. The purpose of the building code's provisions is to reduce the probability of major damage to buildings when moderate earthquakes occur and to prevent the collapse of the main structure during severe earthquakes.

6.5 SEISMIC PARAMETERS INCLUDED IN THE NATIONAL BUILDING CODE OF CANADA (NBCC, 1995)

Earthquake design must be considered for all structures built in Canada, even though seismic activity varies considerably from region to region. To take into account this great variation in seismicity across the country, the 1995 edition of the NBCC includes seismic zoning maps for maximum horizontal accelerations and maximum horizontal velocities. These seismic zoning maps (fig. 3.7) are based on a statistical analysis of earthquakes that have occurred in Canada and neighbouring regions since 1534, as seen in section 3.4. A probability of excedance of 10% over 50 years (mathematically equivalent to an annual probability of excedance of 0,0021, or a return period of 475 years) are considered adequate for development of seismic zoning maps. Table 6.2 shows, for each zone, the ranges of the ratio of peak ground horizontal acceleration to acceleration of gravity, a_{max}/g, and of peak ground horizontal velocity ratio, v_{max}, in units of m/s. Table 6.3 shows typical values of acceleration and velocity ratios for different probabilities of excedance in some Canadian cities. Finally, two important points must be recognized. The first refers to identification of zones 6 on the seismic zoning maps. These zones correspond to western Canada, the Arctic, the St. Lawrence Valley, and the Grand Banks region in Newfoundland. The second refers to the accelerations and velocities of zone 6. We note that they are not identical (table 6.3), but are minimal values that structural engineers must consider in their design.

Table 6.2 Characteristics of seismic zones in Canada (adapted from NBCC, 1995).

Acceleration or velocity zones Z_a, Z_v	Peak ground horizontal acceleration, g, or peak ground horizontal velocity, m/s, for a probability of excedance of 10% in 50 years	Acceleration ratio of the zone, (a_{max}/g) Velocity ratio of the zone (v_{max})
0	$0{,}00 \leq g < 0{,}04$	0,00
1	$0{,}04 \leq g < 0{,}08$	0,05
2	$0{,}08 \leq g < 0{,}11$	0,10
3	$0{,}11 \leq g < 0{,}16$	0,15
4	$0{,}16 \leq g < 0{,}23$	0,20
5	$0{,}23 \leq g < 0{,}32$	0,30
6	$0{,}32 \leq g$	0,40

It is important to note that before the 1985 edition, the code considered only one seismic parameter-peak ground horizontal acceleration, a_{max}, to specify the design earthquake. The 1985 edition introduced a second seismic parameter: the peak ground horizontal velocity, v_{max}. The use of a_{max} alone would be adequate if damage to a structure during an earthquake always showed a good correlation with a_{max}. However, experience shows that this is not the case for all structures. To illustrate this phenomenon, let us consider a specific site influenced by two different seismic sources (fig. 6.9). Let us assume that source 1, which is close to the site, causes moderate earthquakes, S1, and that source 2, which is far from the site, causes major earthquakes, S2, so that a_{max} at the site is the same for both sources. The accelerograms from both sources, measured at the site, are shown in figure 6.10. The accelerogram caused by S1 has a strong component in P-waves with short periods (high frequencies). The accelerogram caused by S2, however, contains mostly S-waves with longer periods (low frequencies).

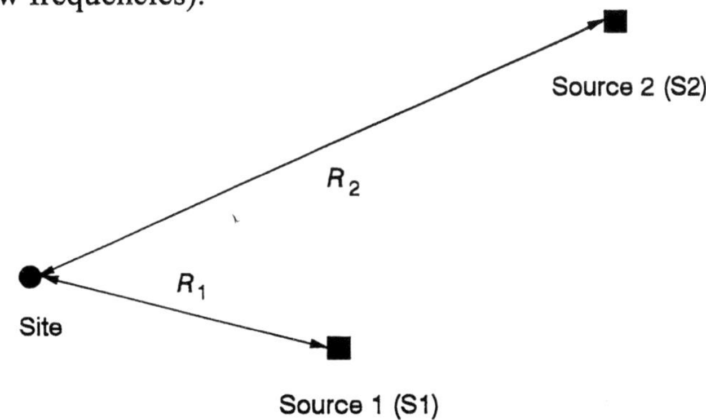

Figure 6.9 Specific site influenced by two seismic sources.

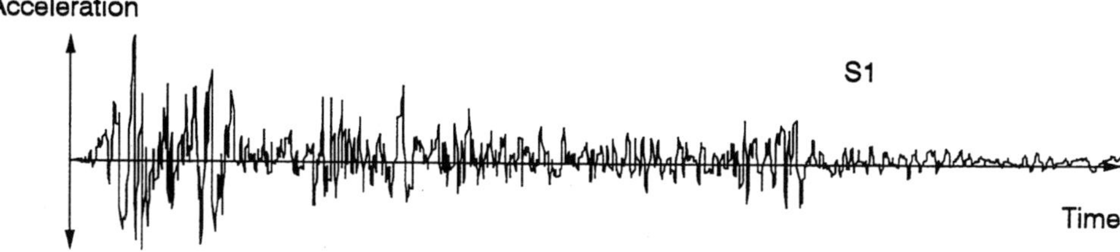

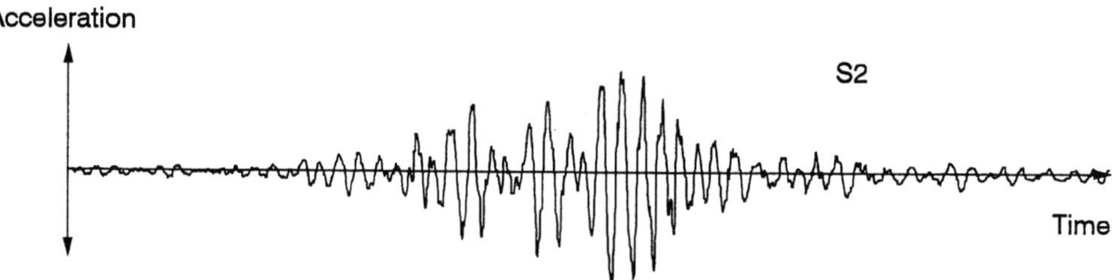

Figure 6.10   Measured accelerograms at a site influenced by two seismic sources.

Table 6.3   Seismic parameters for different localities in Canada (adapted from NBCC, 1995).

Locality and geographic coordinates (%N%W) used in calculations	Annual probability of excedance					
	0,01		0,005		0,0021	
	a_{max}	v_{max}	a_{max}	v_{max}	a_{max}	v_{max}
Inuvik (68,30; 133,48)	0,033	0,052	0,043	0,066	0,060	0,083
Prince Rupert (54,30; 130,43)	0,074	0,13	0,096	0,20	0,13	0,27
Victoria (48,65; 123,43)	0,12	0,088	0,18	0,15	0,28	0,26
Vancouver (49,18; 123,17)	0,089	0,077	0,13	0,12	0,21	0,21
Calgary (51,10; 114,02)	0,011	0,026	0,014	0,032	0,019	0,040
Toronto (43,67; 79,63)	0,029	0,014	0,039	0,023	0,056	0,038
Ottawa (45,32; 75,67)	0,084	0,031	0,12	0,054	0,20	0,098
Montreal (45,47; 73,75)	0,078	0,031	0,12	0,054	0,20	0,098
Quebec City (46,80; 71,38)	0,075	0,035	0,11	0,066	0,19	0,14
Fredericton (45,87; 66,53)	0,046	0,020	0,066	0,036	0,096	0,066
Halifax (44,88; 63,52)	0,027	0,016	0,038	0,030	0,056	0,056
St. John's (47,61; 52,75)	0,022	0,013	0,033	0,026	0,054	0,052

Let us now consider two adjacent buildings located at the site:

- a low-rise building with a short natural period;
- a high-rise building with a longer natural period.

Considering only a_{max}, the two buildings should experience similar damage, as a_{max} is the same for both earthquakes. But this is not the case because of the resonance phenomenon. The low-rise building will be more damaged by the S1 earthquake and the high-rise building will be more damaged by the S2 earthquake, as shown in figure 6.11. This result proves the weakness of a_{max} being used as the sole seismic parameter in building design. To improve its seismic provisions, the NBCC introduced v_{max} as the second seismic parameter in 1985. To demonstrate the influence of v_{max}, let us consider the time-histories of the ground velocity at the site caused by earthquakes S1 and S2. As a first approximation, let us assume that the ground motion at the site is almost harmonic.

$$v_{max} \approx \left(\frac{T_s}{2\pi}\right) a_{max} \tag{6.20}$$

where $\qquad T_s$ = predominant period of the ground motion

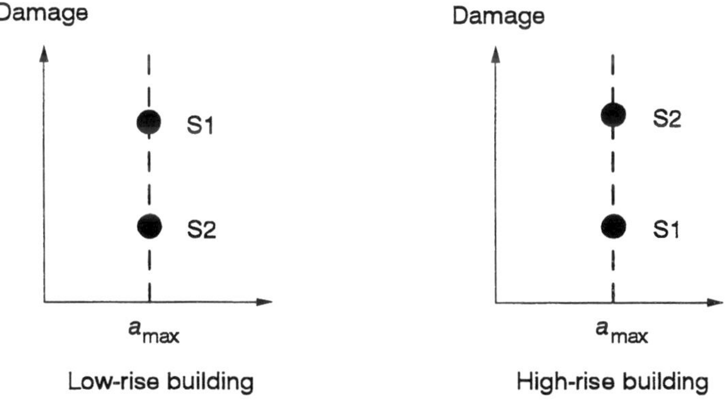

Figure 6.11   Damage caused to buildings located at the site.

As a_{max} is the same for both earthquakes, we note that v_{max} is proportional to T_s. We then note that:

- S1 = short period earthquake = v_{max} low
- S2 = long period earthquake = v_{max} high

A positive correlation can then be found between damage to tall buildings (long period) and v_{max} (fig. 6.12). In general, we will obtain:

- Moderate earthquake produced by a source close to the site.
  = a_{max} high   = v_{max} low
- Large earthquake produced by a source far from the site.
  = a_{max} low   = v_{max} high

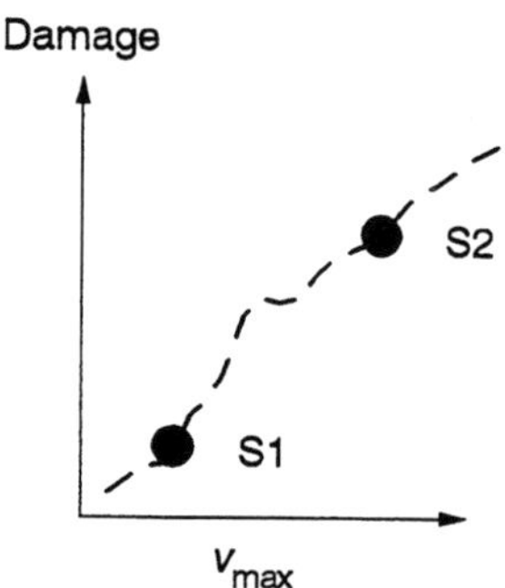

Figure 6.12   Correlation of damage to tall buildings with v_{max}.

A structure with a short natural period will sustain more damage under a moderate earthquake produced by a source close to the site; we say that damage is proportional to the acceleration. A structure with a long natural period will sustain more damage under a severe earthquake produced by a source located far from the site; we say that damage is proportional to the velocity.

In other words, if we measure:

- a_{max} high and v_{max} low

we conclude that the accelerogram will have short periods; if we measure:

- a_{max} low and v_{max} high

we conclude that the accelerogram will have long periods.

6.6   STATIC METHOD OF THE NATIONAL BUILDING CODE OF CANADA (NBCC, 1995)

For regular structures, the NBCC allows engineers to use a static design method. It is important to note that this method cannot be used for all types of structures. Unusual structures, irregularly shaped buildings, and industrial buildings for specific applications such as nuclear reactors, power stations, and industrial chimneys must be treated as special cases requiring specific design criteria and perhaps even a complete dynamic analysis.

6.6.1 Base Shear Formula

The NBCC states that regular structures must be designed to sustain a base shear, V, in each of their principal directions given by the following formula:

$$V = \left(\frac{V_e}{R} \right) U \tag{6.21}$$

where V_e =   base shear corresponding to an elastic response ($\mu = 1$) of the structure
$U = 0,60$ =   calibration factor so that seismic forces calculated with the 1995 edition of the Code be comparable to values calculated with prior editions of the Code
R =   force modification factor ≥ 1

The elastic shear V_e is given by:

$$V_e = v \cdot S \cdot I \cdot F \cdot W \tag{6.22}$$

The factors v, S, R, I, F, and W are defined as follows:
- Peak ground horizontal velocity, v

The peak ground horizontal velocity, v, in m/s is obtained by the seismic zoning maps presented in figure 3.7. These values can also be found in Appendix C of the NBCC (1995).

- Seismic response factor, S

The seismic response factor S represents the value of an elastic response spectrum and is a function of the fundamental period of the structure. Table 6.4 shows the values of S stipulated by the NBCC.

Table 6.4   Seismic response factor, S (NBCC, 1995).

T	Z_a/Z_v	S
	$> 1,0$	4,2
$\leq 0,25$	$1,0$	3,0
	$< 1,0$	2,1
	$> 1,0$	4,2 - 8,4 $(T$-$0,25)$
$0,25 < T > 0,50$	$1,0$	3,0 - 3,6 $(T$-$0,25)$
	$< 1,0$	2,1
$\geq 0,50$	All values	$1,5/\sqrt{T}$

The fundamental period of the structure, T, can be estimated by dynamic procedures or by Rayleigh's formula, demonstrated in section 4.7.7 (Rayleigh's method).

$$T \approx 2\pi \sqrt{\frac{\sum\limits_{i=1}^{N} W_i \delta_i^2}{g \sum\limits_{i=1}^{N} W_i \delta_i}} \tag{6.23}$$

where W_i =   weight of floor i
N =   number of floors
δ_i =   static lateral displacement at floor i produced by all forces W_i $(i = 1...N)$

However, the Code allows for the use of other analytical procedures to estimate the fundamental period of the structure, as long as the value of V_e is not less than 80% of the value of V_e obtained using the fundamental period of vibration based on the empirical formulas presented in table 6.5.

Table 6.5 Empirical formulas of the NBCC (1995) to estimate the fundamental period of a building.

Formula	Lateral-load resisting system
$0{,}1\,N$	All moment-resisting frames[1]
$0{,}085\,h_n^{3/4}$	Steel moment-resisting frames[1]
$0{,}075\,h_n^{3/4}$	Reinforced-concrete moment-resisting frames[1]
$(0{,}09\,h_n)/(D_s^{1/2})$	All other lateral-load-resisting systems

[1]   If the frame resists 100% of the required lateral loads and if the frame is not enclosed or adjoined by more rigid elements that would tend to prevent it from resisting lateral loads.

N = number of floors
h_n = total height of the building with respect to the base (m)
D_s = plan dimension of the building parallel to the direction of excitation (m)

The objective of these empirical formulas is to limit the force reduction obtained when using a procedure other than the static method and give coherence to this method, whatever the fundamental period used. These formulas may be very important for short-period buildings for which the S factor varies very rapidly.

- Importance factor, I

The importance factor is set at:

1,5   for buildings that must remain operational immediately after an earthquake: hospitals, police stations, fire stations, radio stations and towers, power stations, water pumping stations, etc.;

1,3   for schools;

1,0   for other types of buildings.

- Foundation factor, F

The foundation factor takes into consideration the influence of soil on ground motion and response spectra. Table 6.6 shows different types of soil and the corresponding value of F required by the NBCC. The product of $F \times S$ must not exceed 3 if $Z_a \le Z_v$ and must not exceed 4,2 if $Z_a > Z_v$.

Table 6.6 Foundation factor, F (NBCC, 1995).

Category	Type and depth of soil measured from the foundation or pile cap level	F
1	Rock, very dense coarse-grained soils; very stiff and hard fine-grained soils; compact coarse-grained soils and firm and stiff fine-grained soils from 0 to 15 m	1,0
2	Compact coarse-grained soils, firm and stiff fine-grained soils with a depth greater than 15 m; very loose and loose coarse-grained soils and very soft and soft fine-grained soils from 0 to 15 m deep	1,3
3	Very loose and loose coarse-grained soils with depth greater than 15 m	1,5
4	Very soft and soft fine-grained with depth greater than 15 m	2,0

- Seismic weight, W

The seismic weight of a structure is made of the total dead load plus the following loads: 25% of the snow load (specified by the Code), 60% of the live load in storage areas and 100% of the live load for tanks.

- Force modification factor, R

The factor, R, given to different types of structural systems arises from the experience acquired in terms of design and construction and also from the study of building behaviour during earthquakes. Table 6.7 shows values of R varying from 1,0 to 4,0 that take into account the following criteria:

1. The seismic energy absorption capability of the structure (ductility).
2. The redundancy of the lateral-load-resisting system. It is preferable to have a statically indeterminate system in order to reduce the risk of collapse.
3. The stiffness of the lateral-load-resisting system. Rigid systems attract higher base shear forces than do flexible systems. This is why a lower value of R is assigned to stiff systems.

The base shear V is the algebraic sum of the inertia (or seismic) forces reacting on the masses of the structure as a result of a horizontal movement at the base. If the structure reacts mostly in its first mode of vibration, these inertial forces are distributed in a triangular shape with its base at the top of the structure (fig. 4.56). As the inertia forces, F_i, at level i are proportional to the weights, W_i, the distribution of the seismic forces is obtained by:

$$F_i = \left(\frac{W_i\, h_i}{\sum\limits_{j=1}^{N} W_j\, h_j} \right) V \tag{6.24}$$

where h_i = height at level i over the base
     N = number of floors

Table 6.7 Force modification factors, R (NBCC, 1995).

Category	Type of lateral-load-resisting system	R
	Steel structure designed and detailed according to CAN/CSA-S16.1-M	
1	Ductile moment-resisting frame	4,0
2	Ductile eccentrically braced frame	4,0
3	Ductile steel plate shear wall	4,0
4	Ductile braced frame	3,0
5	Moment-resisting frame with nominal ductility	3,0
6	Nominally ductile steel plate shear wall	3,0
7	Braced frame with nominal ductility	2,0
8	Ordinary steel plate shear wall	2,0
9	Other lateral-force-resisting systems not defined in cases 1 to 8	1,5
	Reinforced-concrete structures designed and detailed according to CSA-A23.3	
10	Ductile moment-resisting frame	4,0
11	Ductile coupled wall	4,0
12	Other ductile wall systems	3,5
13	Moment-resisting frame with nominal ductility	2,0
14	Wall with nominal ductility	2,0
15	Other lateral-force-resisting systems not defined in cases 10 to 14	1,5
	Timber structures designed and detailed according to CSA-O86.1	
16	Nailed shear panel with plywood, waferboard or OSB	3,0
17	Concentrically braced heavy timber frame with ductile connections	2,0
18	Moment-resisting wood frame with ductile connections	2,0
19	Other systems not included in cases 16 to 18	1,5
	Masonry structures designed and detailed according to CSA S304.1	
20	Reinforced masonry wall with nominal ductility	2,0
21	Reinforced masonry	1,5
22	Unreinforced masonry	1,0
23	Other lateral-force-resisting systems not defined in cases 1 to 22	1,0

6.6.2 Distribution of Base Shear

However, the Code also recommends that a portion F_t of the total lateral force be applied at the top of the structure to take into account the possible influence of higher modes of vibration:

$$F_t = 0,07\,TV \le 0,25\,V \qquad (6.25)$$

If the fundamental period of the structure is less than 0,7 s, the value of F_t is neglected. If F_t is not equal to zero, then the remaining total force is distributed as follows:

$$F_i = \left(\frac{W_i h_i}{\sum\limits_{j=1}^{N} W_j h_j} \right) (V - F_t) \tag{6.26}$$

6.6.3 Overturning Moments

The design seismic loads take the effect of higher modes into account by defining a force F_t at the top of the building. Although this force is a reasonable representation of the effects of higher modes on inter-storey shear, it overestimates the effects of the higher modes on the overturning moment at each floor. In order to take this phenomenon into account, the code suggests that the overturning moment of a given level x of the building be multiplied by a force modification factor, J_x, given by:

$$J_x = J + (1 - J)\left(\frac{h_x}{h_n} \right)^3 \tag{6.27}$$

where h_x = height of level x
h_n = total height of the building

The reduction factor for the overturning moment at the base of the structure, J, has the form:

- $J = 1$ for $T \le 0,5$ s
- $J = (1,1 - 0,2\ T)$ for $0,5$ s $< T \le 1,5$ s
- $J = 0,8$ for $T > 1,5$ s

6.6.4 Limit State Design

When designing buildings and their structural elements, the most unfavourable effect of the factored loads must be considered according to the limit state design method. The effect of the factored loads is obtained by combining the effects caused by the dead load, D; live loads from the normal use of the building, including rain, ice, and snow loads, L; wind loads, W; seismic loads, E; and additional loads caused by the effects of temperature variation, Q, multiplied by the load factors, α, the simultaneity factor, ψ, and the risk factor, γ. The following formulas represent the simultaneity of the factored loads.

$$\alpha_D D + \gamma \psi \left[\alpha_L L + \alpha_W W + \alpha_T T \right] \quad \text{and}$$

$$\alpha_D D + \gamma \left[\alpha_L L + \alpha_E E \right] \tag{6.28}$$

The earthquake load factor $\alpha_E = 1{,}0$ for the seismic loads. The live load factor $\alpha_L = 1{,}5$ except when it is added to the seismic loads. In this case, $\alpha_L = 1{,}0$ or $0{,}0$ when live loads arise from storage or meeting areas and $0{,}5$ or $0{,}0$, for all other live loads, including snow. The dead load factor $\alpha_D = 1{,}25$ except when the seismic loads are used; then in this case, $\alpha_D = 1{,}0$.

Note that in the 1995 edition of the NBCC, for the first time, the seismic loads, E, were separated from the wind loads, W. In previous editions, the seismic and wind loads were combined. To consider the seismic loads just another lateral load is a mistake, as the effects of earthquakes are dependent on the level of displacement and the ductility of the lateral-load-resisting system, as seen above. It is thus reasonable to differentiate the two loads by labelling them with two different symbols.

6.6.5 Example

To illustrate the NBCC's static method, let us consider the ductile reinforced-concrete moment-resisting frame acting as the lateral-load-resisting system in a principal direction of a 10-storey office building (figure 6.13). For this frame, we want to calculate the seismic-load distribution according to the NBCC's (1995) static method. The building is located on a rock site in Montreal, and the seismic weight at each level of the frame is 262 kN.

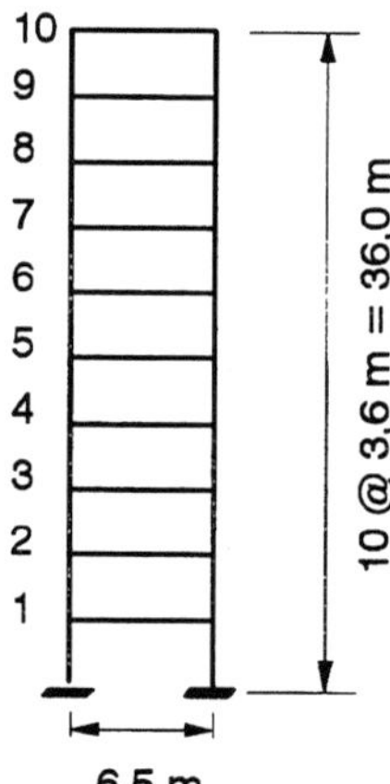

Figure 6.13   Moment-resisting frame part of a 10-storey building located in Montreal.

Seismic response factor: $S = 1{,}0$; $Z_a/Z_v = 2 > 1$; $S = 1{,}5/\sqrt{1{,}1} = 1{,}43$
Seismic zones for Montreal: $Z_a = 4$; $Z_v = 2$
Peak ground horizontal velocity for Montreal $v = 0{,}10$ m/s
Estimate of the fundamental period: $T = 0{,}075\, h_n^{3/4} = 1{,}1$ s
Importance factor: $I = 1{,}0$ (office building)
Foundation factor: $F = 1{,}0$ (rock)
Seismic weight: $W = 10(262\ \text{kN}) = 2\ 620$ kN

The elastic base shear V_e is as follows:

$$V_e = (0,10)(1,43)(1,0)(1,0)(2\ 620\,\text{kN}) = 375\,\text{kN} \tag{6.29}$$

Force modification factor: $R = 4,0$ (ductile moment-resisting frame)
The design base shear V is then:

$$V = \left(\frac{375\,\text{kN}}{4}\right)0,6 = 56,25\,\text{kN} \tag{6.30}$$

As a result:

$$F_t = 0,07\,(1,1)\,56,25\,\text{kN} = 4,33\,\text{kN} \tag{6.31}$$

According to equation 6.26, it yields:

$$F_i = \left(\frac{W_i\,h_i}{\displaystyle\sum_{j=1}^{10} W_j\,h_j}\right)(V - F_t) \tag{6.32}$$

As all floors bear the same weight (262 kN) and are the same height (3,6 m), it yields:

$$F_i = \frac{(56,25 - 4,33)(262)(3,6)(i)}{(262)(3,6)\displaystyle\sum_{j=1}^{10} j} = 0,944\,i\ \text{kN} \tag{6.33}$$

It is then easy to evaluate the distribution of seismic loads:

$$
\begin{aligned}
F_{10} &= 4,33\ +\ 9,44\ =\ 13,77\,\text{kN}\\
F_9 &= 8,50\,\text{kN}\\
F_8 &= 7,55\,\text{kN}\\
F_7 &= 6,61\,\text{kN}\\
F_6 &= 5,66\,\text{kN}\\
F_5 &= 4,72\,\text{kN}\\
F_4 &= 3,78\,\text{kN}\\
F_3 &= 2,83\,\text{kN}\\
F_2 &= 1,89\,\text{kN}\\
F_1 &= 0,94\,\text{kN}
\end{aligned}
\tag{6.34}
$$

6.6.6 Special Requirements

The Code specifies the following special requirements for seismic design.

1. For Z_a or $Z_v \geq 2$ and buildings exceeding three storeys, a dedicated lateral-load-resisting system must be designed (categories 1 to 8, 10 to 14, 16 to 18, 20 and 21 in table 6.7).
2. For buildings with $R = 2$ or 1,5 and height > 60 m, in seismic zones with $Z_v \geq 4$, V must be increased by 50% (Bagotville, La Malbaie, Montmagny, Cacouna, Tadoussac, Rivière-du-Loup, for example).
3. Elevated tanks not supported by a building must be designed for $R = 1$ and $1,5 \leq SI \leq 3,0$.
4. For seismic zones with Z_a or $Z_v \geq 2$ where buildings have discontinuous elements, the rest of the structure must be able to sustain the additional loads caused by failure at the points of discontinuity.
5. For seismic zones with Z_a or $Z_v \geq 2$, reinforced masonry must be used in buildings with:
   a) load-bearing masonry resisting lateral loads;
   b) elevator-shaft masonry, staircase masonry, or masonry used as siding;
   c) masonry partition walls except for partition walls not exceeding 200 kg/m^2 or not exceeding 3 m in height and laterally supported at the top.

6.6.7 Calculation of Lateral Deflections

The NBCC requires that the inter-storey drifts, Δ, be computed according to the following formula:

$$\Delta = (R) \times (\text{elastic displacements}) \tag{6.35}$$

Elastic displacements are calculated from the static loads $(V_e U/R)$ of the NBCC. If the P-Δ effects are taken into account, non-factored gravity loads $(\alpha_D = 1,0)$ must be used when calculating elastic deflections. The NBCC limits inter-storey drifts to 1% of the height of the floors for post-disaster buildings that must remain in operation immediately after an earthquake and to 2% of the height of the floors for other types of buildings. Nonstructural elements, considered isolated elements, must be able to withstand the predicted motion caused by the horizontal deflection. However, if these elements affect the internal load distribution of the structure, their influence on the structural behaviour must be taken into account in the elastic, yielding, and failure phases.

An adequate distance between adjacent buildings is needed to avoid pounding during an earthquake. The NBCC requires that this spacing be equal to the sum of the displacements of each adjacent building (fig. 6.14). If this specific distance cannot be respected, the NBCC requires that the adjacent buildings be structurally linked together.

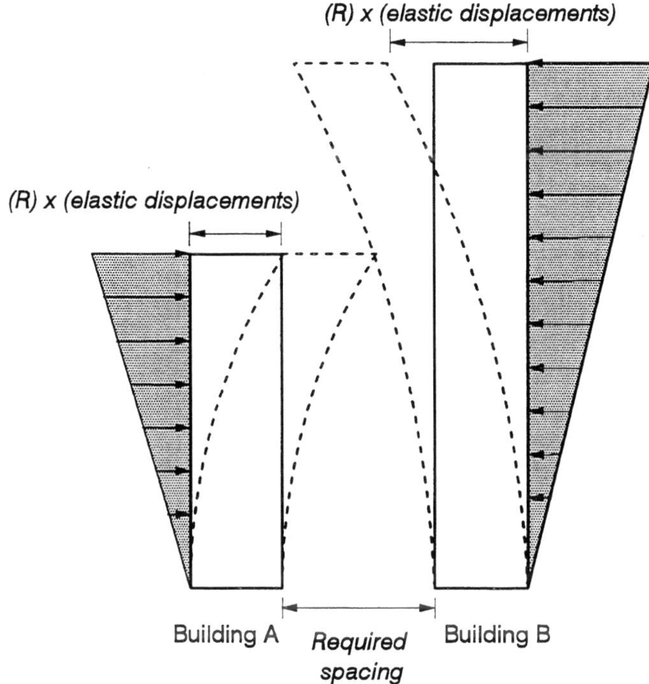

Figure 6.14   Spacing required by the NBCC (1995) between two adjacent buildings to avoid seismic pounding.

Recent studies (Filiatrault et al., 1994; Filiatrault and Cervantes, 1995) show that the spacing required by the NBCC is much too large. These requirements do not take into account the phases between the relative dynamic movements of adjacent buildings. The same studies also show that the approximate spectral difference method, proposed by Jeng, Kasai, and Maison (1992), produces spacing values that are much more realistic with respect to the results of nonlinear dynamic analyses. According to this method, the necessary distance, s_{AB}, between two adjacent buildings is obtained by:

$$s_{AB} = \sqrt{u_{A\max}^2 + u_{B\max}^2 - 2\gamma u_{A\max} u_{B\max}} \tag{6.36}$$

where $u_{A\max}$ and $u_{B\max}$ correspond to the maximum lateral displacements (in absolute values) of buildings A and B at the pounding level. These maximum displacements can be calculated using the spectral analysis and by considering only the first mode of each building.

As seen in section 4.8, these displacements are determined by:

$$u_{A\,max} = A_{pA}^{(1)} \alpha_{1A} S_D^{1A}$$

$$u_{B\,max} = A_{pB}^{(1)} \alpha_{1B} S_D^{1B} \tag{6.37}$$

where $A_{PA}^{(1)}, A_{PB}^{(1)}$ = components of the first mode for each building at the anticipated pounding level

α_{1A}, α_{1B} = first mode participation factors for each building

S_D^{1A}, S_D^{1B} = spectral displacements for each building based on the NBCC design spectrum (figure 4.35) for a given site

The influence factor, γ, in equation 6.36 is obtained by:

$$\gamma = \frac{8\,\zeta^2 \left(1+\dfrac{T_B}{T_A}\right)\left(\dfrac{T_B}{T_A}\right)^{3/2}}{\left[1-\left(\dfrac{T_B}{T_A}\right)^2\right]^2 + \left[4\,\zeta^2\left(1+\dfrac{T_B}{T_A}\right)^2\left(\dfrac{T_B}{T_A}\right)\right]} \tag{6.38}$$

where T_A, T_B = fundamental periods of the buildings

ζ = viscous damping ratio, considered the same for both buildings

According to this method, if the two adjacent buildings have the same period of vibration, $\gamma = 1$, the necessary spacing becomes equal to the difference between the maximum lateral displacements.

$$s_{AB}(T_A = T_B) = |u_{A\,max} - u_{B\,max}|$$

6.6.8 Design of Foundations

In general, the NBCC recommends that foundations be designed so that yielding occurs first in the superstructure. In other words, the foundations must be stronger than the superstructure. However, a recent dynamic analysis (Filiatrault et al., 1992), conducted on a 21-storey core-type building on a mat footing that could not sustain the flexural capacity of the core, showed that even though the footing was weak, it did not have an adverse influence on the building's global seismic performance.

6.7 TORSIONAL EFFECTS

6.7.1 Cause of Torsion in Asymmetric Structures

The inertia forces produced by an earthquake act through the centre of mass, *CM*, of the structure. If the structure is not uniform, the *CM* and the centre of rigidity, *CR*, do not coincide and torsional moments are produced (figure 6.15).

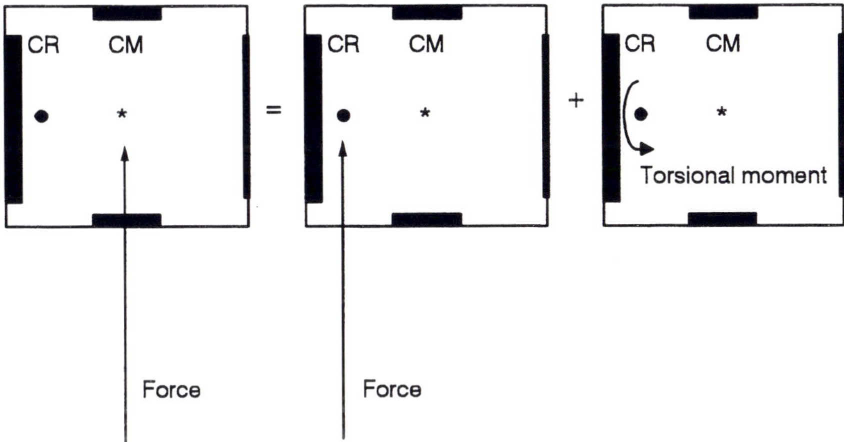

Figure 6.15   Torsional moment in an asymmetric structure.

Distributing shear walls (or lateral-load-resistant elements) far away from the *CR*, as shown in figure 6.16, is always preferable for resistance of torsional moments.

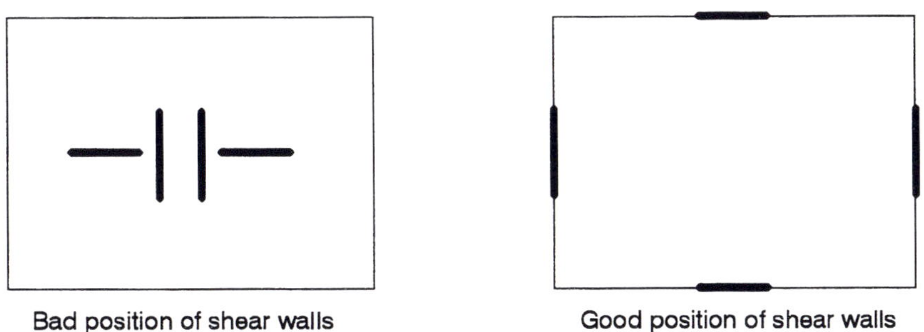

Figure 6.16   Distribution of shear walls for efficient resistance to torsion.

To calculate these torsional moments, the position of the *CM* and the *CR* must be evaluated: the position of the *CM* is easy to evaluate, but the position of the *CR* is very complex to determine.

6.7.2 Position of the Centre of Rigidity *(CR)*

The position of the *CR* is very hard to evaluate for multi-storey buildings; many definitions can be found.

Relative stiffness criterion. According to this definition, proposed by Blume, Newmark, and Corning (1961), the position of the *CR* can be evaluated in a similar fashion as the position of the centre of gravity of a section.

$$X_{CR} = \frac{\sum_{i=1}^{n} K_{yi} X_i}{\sum_{i=1}^{n} K_{yi}}$$

$$Y_{CR} = \frac{\sum_{i=1}^{n} K_{xi} Y_i}{\sum_{i=1}^{n} K_{xi}}$$

$$(6.40)$$

where K_{xi} and K_{yi} correspond to the lateral stiffness of the n lateral-load-resisting elements in the x and y directions, respectively. This technique requires the same type of resisting elements over the entire height. As a result, the *CR* is, in theory, constant over the entire height.

No rotation of a particular floor criterion. According to this definition, proposed by Humar (1984), the position of the *CR* at a given floor is the location where the lateral load will not cause a rotation of that floor, while other floors may rotate. This technique requires a three-dimensional analysis of the structure.

No rotation of all floors criterion. This definition, proposed by Stafford-Smith and Vézina (1985), states that the position of the *CR* at each floor is the location where the lateral loads will prevent rotation of the entire structure. This technique depends on the vertical distribution of the lateral loads.

$$X_{CR,\,r} = \frac{\sum_{i=1}^{n} \left(\left(V_{yi,\,r} - V_{yi,\,r+1} \right) X_i \right)}{F_{y,\,r}}$$

$$Y_{CR,\,r} = \frac{\sum_{i=1}^{n} \left(\left(V_{xi,\,r} - V_{xi,\,r+1} \right) Y_i \right)}{F_{x,\,r}}$$

$$(6.41)$$

where $V_{xi,r}$, $V_{yi,r}$ = shear forces in the x and y directions, respectively, of the lateral load resisting element i at floor r

$F_{x,r}$, $F_{y,r}$ = applied forces in the x and y directions, respectively, at the same level

Shear centre criterion. When the shear force is known for each level, we can evaluate the location of the shear centre, *SC,* where the resultant of all the shear forces is centred. This definition, proposed by Tso (1990), differs from the *CR* and requires calculation of the resultant of the seismic loads applied above the level considered.

Discussion. It is important to consider the facts concerning the different criteria defining the centre of rigidity.

- All definitions of eccentricity are identical for one-storey buildings.
- The extrapolation to multi-storey buildings obviously produces great differences.
- If the *CR* is not approximately located on a vertical line, the NBCC requires a dynamic analysis.

6.7.3 Consideration of Torsional Effects by the National Building Code of Canada (NBCC, 1995)

The 1995 edition of the NBCC proposes two ways of calculating the torsional moments around the vertical axis of a building: the static method described in section 6.6 and a three-dimensional dynamic analysis.

Static method. In a static analysis, the torsional moment, T_x, on a horizontal plan at a given floor x must be calculated by one of the following equations:

$$T_x = F_x\left(1{,}5\,e \pm 0{,}1\,D_{nx}\right)$$
$$\text{or}$$
$$T_x = F_x\left(0{,}5\,e \pm 0{,}1\,D_{nx}\right)$$

$$(6.42)$$

where F_x = seismic load at the floor level x calculated by the static method

D_{nx} = plan dimension of the building at level x perpendicular to the direction of the seismic load

e = distance measured perpendicularly to the direction of the seismic load between the *CM* and the *CR* at the given level

The value of $0{,}1D_{nx}$ represents the accidental torsion and is equal to 10% of the floor plan dimension. This value tries to take into account the rotational motion of the ground, the variations in the calculated stiffness and weight values, the addition of partition walls after the completion of the building, and so on.

Dynamic method. When a three-dimensional analysis is done, the NBCC requires that the effects of accidental torsion be added to each level x based on a torsional moment calculated with the following equation:

$$T_x = \pm 0{,}1\,D_{nx}F_x$$

$$(6.43)$$

This approach is equivalent to reducing the term 1,5 in equation 6.43 to 1,0. However, the NBCC considers that this reduction does not modify significantly the amplitude of the member forces. In practice, the torsional moments are applied by moving the position of the *CM* of $\pm 0,1\, D_{nx}$ in the dynamic analysis.

6.7.4 Torsional Analysis without Determining the Position of the Centre of Rigidity

Goel and Chopra (1993) showed that torsion could be taken into account, according to equation 6.42, without directly calculating the position of the *CR*. To do so, the results of three different three-dimensional static analyses must be combined. This method involves four steps.

- **Step 1**

First, the building is analyzed with seismic lateral loads applied to the *CM* of each floor, but the floors are not allowed to rotate. To do this, zero rotation constraint is imposed on the central node of each floor. The results (deflections and internal forces) for this first analysis are represented by r_1.

- **Step 2**

Then, the same analysis as step 1 is performed, but this time, the floors are allowed to rotate. The results of the second analysis are represented by r_2.

- **Step 3**

Then, the building is analyzed when subjected only to accidental torsional moments, T_x, applied to the CM of each level x and given by the following equation:

$$T_x = 0,1\, D_{nx} F_x \tag{6.44}$$

The results of this third analysis are represented by r_3.

- **Step 4**

Finally, the total effect of the lateral loads and the torsional moments, r_c, are obtained by combining the results of the three previous analyses with one of the two equations that will produce the most unfavourable effect in the members.

$$r_c = -0,5\, r_1 + 1,5\, r_2 + r_3$$
$$\text{or} \tag{6.45}$$
$$r_c = 0,5\, r_1 + 0,5\, r_2 - r_3$$

6.7.5 Example for a Three-Storey Building

Building Description. To illustrate calculation of torsional effects according to the NBCC, let us consider the three-storey building shown in figure 6.17. Tinawi (1991) has already analyzed the torsional effects of this building. The building is located in Montreal and is made of the following elements:

Line A-A: two coupled shear walls (12 m high)
Line B-B: a two-bay moment resisting frame (12 m high)
Line C-C: two cantilevered shear walls (8 m high)
Line D-D: one cantilevered shear wall (12 m high)

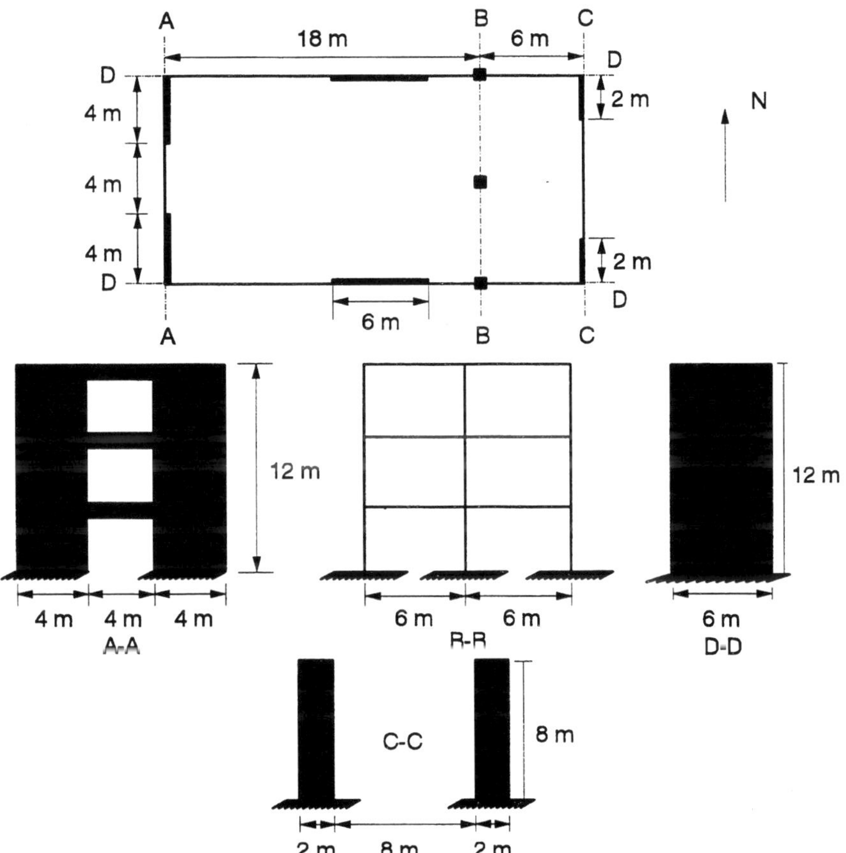

Figure 6.17 Three-storey building located in Montreal (from Tinawi, 1991).

The structure is symmetrical with respect to the E-W axis, and all of the lateral-load-resisting elements are linked together by a rigid floor (0,2 m thick) acting as a horizontal diaphragm. The physical properties of the structure are as follows:

Weight of the structure (self weight: 4 kN/m^2 per floor)
Weight of level 1: $W_1 = 1\,152$ kN
Weight of level 2: $W_2 = 1\,152$ kN
Weight of level 3: $W_3 = 864$ kN

Properties of the structural elements
Line A-A. Coupled shear walls:
for the walls:

$$\begin{aligned} A &= 0{,}4 \text{ m}^2 \\ A_s &= 0{,}4 \text{ m}^2 \\ I &= 0{,}533 \text{ m}^4 \end{aligned}$$

for the link beams:

$$\begin{aligned} A &= 0{,}075 \text{ m}^2 \\ A_s &= 0{,}075 \text{ m}^2 \\ I &= 0{,}00352 \text{ m}^4 \end{aligned}$$

with rigid extensions of 2 m

Line B-B. Moment-resisting frame:
for the columns:

$$\begin{aligned} A &= 0{,}024 \text{ m}^2 \\ A_s &= 0 \\ I &= 0{,}0143 \text{ m}^4 \end{aligned}$$

for the beams:

$$\begin{aligned} A &= 0{,}02 \text{ m}^2 \\ A_s &= 0 \\ I &= 0{,}01 \text{ m}^4 \end{aligned}$$

Line C-C. Cantilevered shear walls:
for each wall:

$$\begin{aligned} A = A_s &= 0{,}4 \text{ m}^2 \\ I &= 0{,}1333 \text{ m}^4 \end{aligned}$$

Line D-D. Cantilevered shear wall:
for this wall:

$$\begin{aligned} A = A_S &= 0{,}6 \text{ m}^2 \\ I &= 1{,}8 \text{ m}^4 \end{aligned}$$

Computation of lateral loads. The design base shear, V, is evaluated according to the NBCC, for an earthquake in the N-S direction.

$$V = \left(\frac{V_e}{R}\right) U \tag{6.46}$$

$$V_e = vSIFW$$

For Montreal, the seismic parameters are the following:

$$\begin{aligned} Z_a &= 4 \\ Z_v &= 2 \\ v &= 0{,}1 \text{ m/s} \end{aligned}$$

We assume that

$$\begin{aligned} I &= 1 \text{ (ordinary building)} \\ F &= 1 \text{ (building built on a rock site).} \end{aligned}$$

We also assume that the structure will have a nominal ductility of $R = 2$.

The natural period of vibration, T, is computed by the empirical formulas of the NBCC (mixed lateral-load-resisting system in the N-S direction).

$$T = \frac{0,09\,h_n}{\sqrt{D_s}} = \frac{(0,09)(12)}{\sqrt{12}} = 0,31\text{ s} \tag{6.47}$$

Then, for $Z_a/Z_v > 1$ and $0,25\text{ s} < T < 0,5\text{ s}$, the seismic response factor will be:

$$\begin{aligned} S &= 4,2 - 8,4\,(T-0,25) \\ &= 4,2 - 8,4\,(0,31-0,25) \\ &= 3,7 \end{aligned} \tag{6.48}$$

then:

$$V_e = (0,1)(3,7)(1)(1)(3\ 168\,\text{kN}) = 1\ 172\,\text{kN}$$

$$V = \left(\frac{1\ 172\,\text{kN}}{2}\right) 0,6 = 352\,\text{kN} \tag{6.49}$$

$F_t = 0$ since $T < 0,7$ s. The vertical distribution of the seismic loads are shown in figure 6.18.

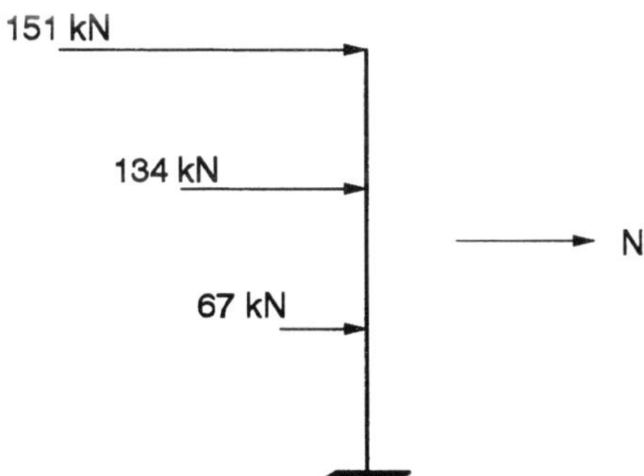

Figure 6.18 Vertical distribution of the lateral loads for a three-storey building.

Analysis with calculation of the position of the *CR*. If the seismic loads in the N-S direction go through the *CR*, then the resisting elements along lines A-A, B-B, and C-C will share the same lateral displacements. To determine the load distribution in these elements, a three-dimensional analysis can be performed by blocking the rotation of each floor around the vertical axis. A two-dimensional analysis can also be performed by linking the different structural elements with rigid bars (fig. 6.19). Shear forces resulting from this analysis for each level and each vertical element are shown in figure 6.20.

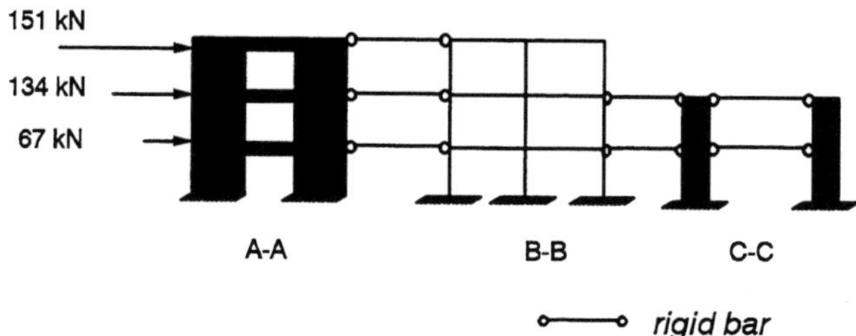

Figure 6.19 Two-dimensional analysis with rigid bars.

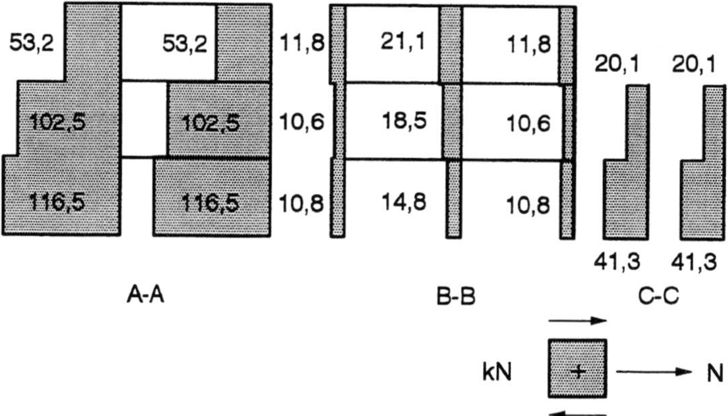

Figure 6.20 Shear force distribution from two-dimensional analysis with rigid bars.

For each level x, the total shear forces are evaluated for all vertical elements along lines A-A, B-B, and C-C. These forces are written V_{Ax}, V_{Bx}, and V_{Cx} in table 6.8. Knowing the shear force over and under a given floor and the applied lateral load on this floor, the *CR* of the floor can be evaluated using equation 6.41. For example, let us consider the moments with respect to line A-A for all levels, the positions of the *CR* can be calculated at each level using the following equation:

$$X_{CR,3} = \frac{(44,66)(18)}{(151)} = 5,32\,\text{m}$$

$$X_{CR,2} = \frac{\left[(40,22-0)(24)+(39,75-44,66)(18)\right]}{(134)} = 6,54\,\text{m} \tag{6.50}$$

$$X_{CR,1} = \frac{\left[(82,66-40,22)(24)+(36,29-39,75)(18)\right]}{(67)} = 14,27\,\text{m}$$

Table 6.8 Calculation of the *CR* and of the theoretical eccentricity of a three-storey building.

Level x	F_x (kN)	V_{Ax} (kN)	V_{Bx} (kN)	V_{Cx} (kN)	Position of the *CR* (m)	Position of the *CM* (m)	e (m)
3	151	106,34	44,66	0	5,32	9	3,68
2	134	205,02	39,75	40,22	6,54	12	5,46
1	67	233,06	36,29	82,66	14,27	12	-2,27

In order to evaluate the torsional moments according to the NBCC, the amplification factors and accidental torsion must be taken into account according to equation 6.42. Table 6.9 shows the value of the torsional moment at each floor. Figure 6.21 illustrates these values according to the right-hand rule.

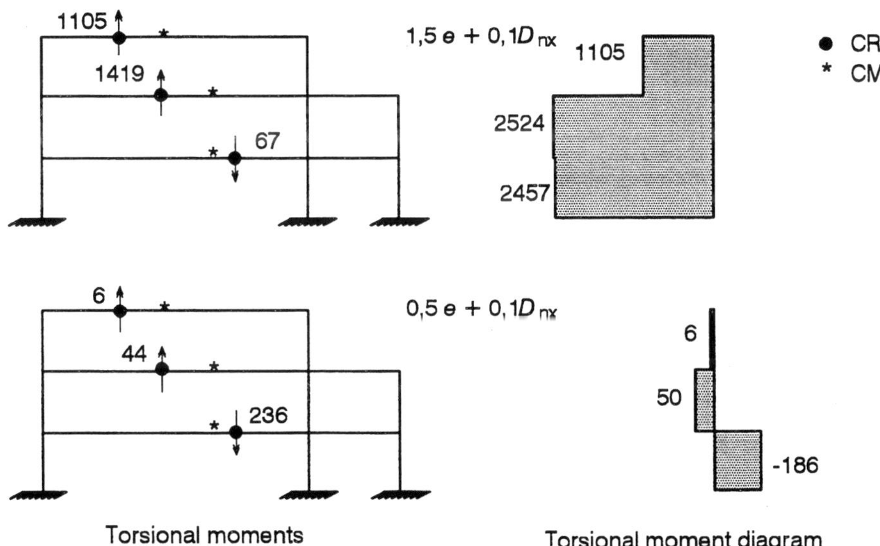

Figure 6.21 Torsion at each floor and torsional moment diagram for a three-storey building.

Table 6.9 Calculation of the torsional moments according to the NBCC's static method for a three-storey building.

Level x	F_x (kN)	e (m)	D_{nx} (m)	$e_1 = 1,5e + 0,1D_{nx}$ (m)	$F_x e_1$ (kN-m)	$e_2 = 0,5e - 0,1D_{nx}$ (m)	$F_x e_2$ (kN-m)
3	151	3,68	18	7,32	1 105	0,04	6
2	134	5,46	24	10,59	1 419	0,33	44
1	67	-2,27	24	-1,01	-67	-3,54	-236

Figure 6.22 shows shear forces induced in each vertical element based on a three-dimensional analysis. The seismic loads and the torsional moments of table 6.9 were applied at the *CR* of each level.

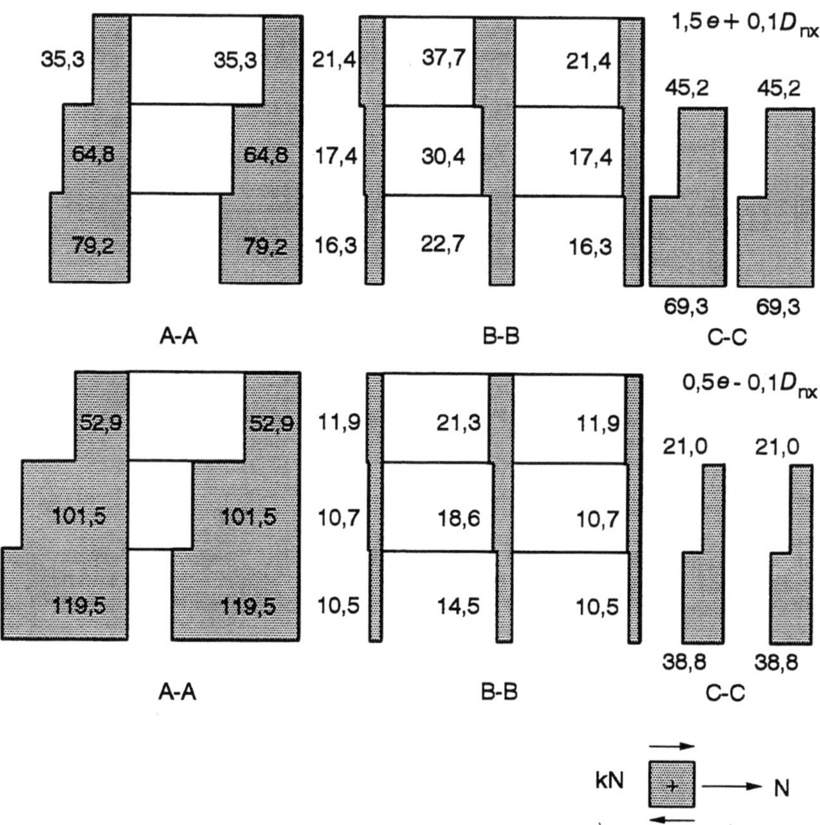

Figure 6.22   Shear force distribution from a three-dimensional analysis: torsional effect and calculation of the position of the *CR*.

Analysis without determining the position of the *CR*. Using the approach proposed by Goel and Chopra (1993), the results of three different three-dimensional analyses must be combined. The first analysis applies the lateral seismic loads to the *CM* while preventing the floors from rotating. This analysis was performed in the previous section; figure 6.20 shows the distribution of the shear forces in each vertical element. The second analysis is identical to the first one but the floors are allowed to rotate. The shear force distribution for each vertical element of this analysis are shown in figure 6.23. The shear forces in the coupled shear walls have decreased with respect to the results obtained in figure 6.20; all levels are subjected to a counterclockwise rotation with respect to the vertical axis. Finally, in the third analysis, the accidental torsional moments are applied to the *CM* of each floor, as shown in table 6.10. The shear force distribution from this analysis is shown in figure 6.24.

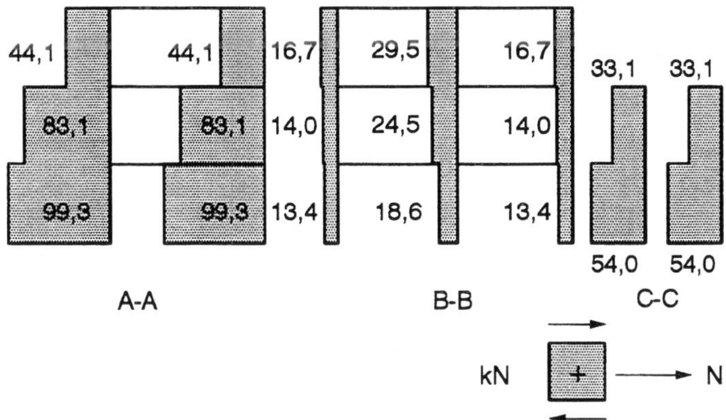

Figure 6.23   Shear force distribution for the seismic loads applied to the *CM* with free rotation of the floors.

Table 6.10   Accidental torsional moments (NBCC, 1995).

Level x	F_x (kN)	D_{nx} (m)	$0{,}1\,D_{nx}\,F_x$ (kN-m)
3	151	18	272
2	134	24	322
1	67	24	161

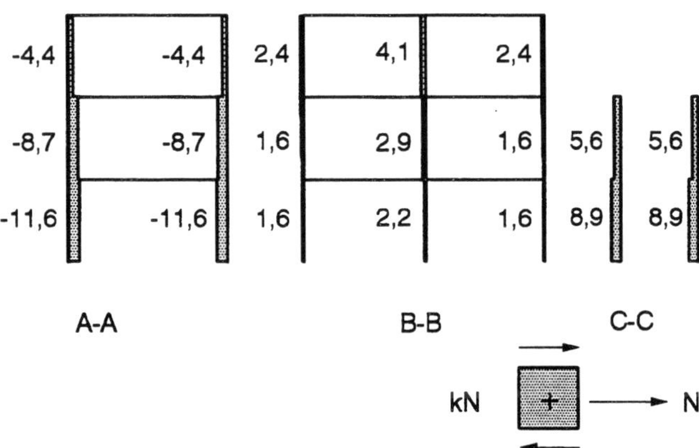

Figure 6.24   Shear force distribution for the torsional moments applied to the *CM* with free rotation of the floors.

To obtain the design shear forces, which include the combined effects of the lateral loads and the torsion, the results of the three analyses must be combined according to equation 6.45. If the shear force distributions shown in figures 6.20, 6.23, and 6.24 are written as r_1, r_2, and r_3, respectively, it yields the distributions shown in figure 6.25, which are identical to the ones in figure 6.22.

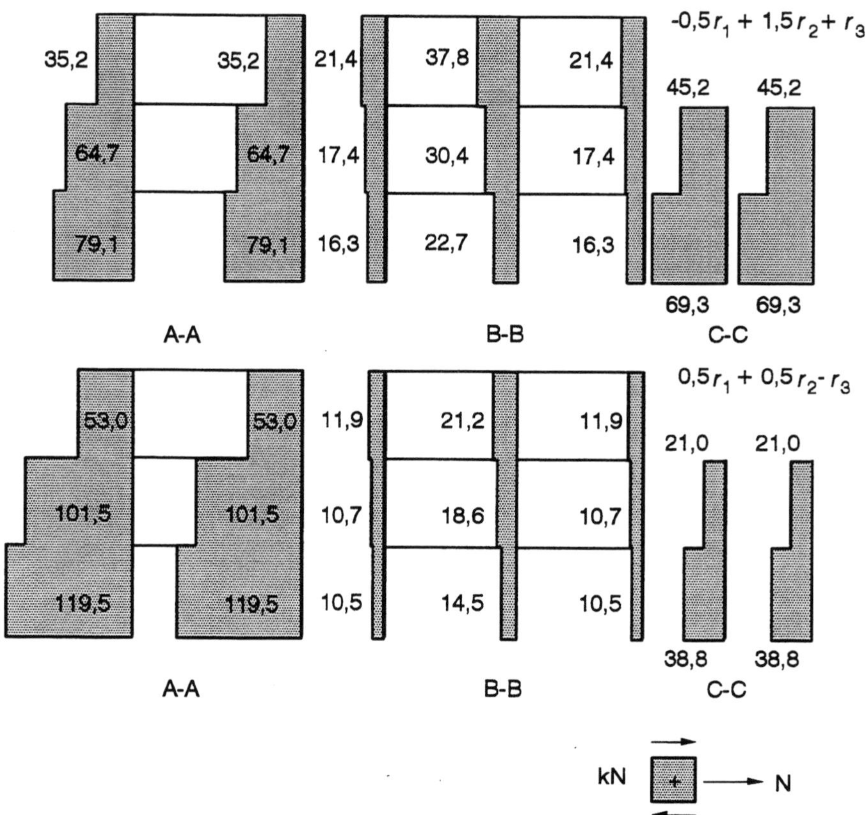

Figure 6.25   Shear force distribution from a three-dimensional analysis: torsional effect without determining the position of the *CR*.

6.7.6 A Final Remark on Torsional Effects

It is important to realize that there is a certain paradox when the NBCC considers the torsional effects. Even though the philosophy of seismic design is based on the concept of ductility, where the structure exhibits an inelastic behaviour (without collapse) during a major earthquake, the torsional shear force distribution in the different members is based on an elastic stiffness of these members with a fixed position of the *CR* at each level. In reality, if certain members yield during the structure's dynamic response, the relative stiffness of the different elements will be modified, causing the *CR* of each level to shift and changing the torsional shear force distribution. Only a nonlinear three-dimensional dynamic analysis can take into account the change in the position of the *CR* over time.

6.8   PRACTICAL ASPECTS OF EARTHQUAKE-RESISTANT BUILDING DESIGN

6.8.1 Objectives of Seismic Analysis

The lateral stiffness of a building is generally determined in the early stages of the design phase. The factors governing the lateral stiffness are:

- the choice of materials;
- the lateral-load-resisting system (moment-resisting frames, braced frames, shear walls, etc.);
- the position of the resisting elements.

When these factors are determined, changes in individual strengths of the structural elements will be of minor influence. The first goal of an earthquake-resistant design is to select an adequate level of lateral strength so that the available ductility of the system is mobilized following implementation of adequate structural detailing. Several methods are used to compute the seismic design forces; the choice of an analysis method depends on the size and the importance of the building.

6.8.2 Equivalent Static Analysis

For regular and uniform buildings up to 20 storeys, the NBCC (1995) static analysis can be used, with torsional effects, as described in sections 6.6 and 6.7.

6.8.3 Simplified Dynamic Spectral Analysis

For buildings with an irregular configuration in plane or in elevation (with large set-backs, for example), the NBCC (1995) states that the lateral load distribution must be determined along the height of the building using a dynamic spectral analysis, which is based on the spectral analysis studied in section 4.8. However, the code requires that the modal responses be scaled so that the computed base shear equals the base shear calculated with the NBCC (1995) static method in order to take into account the anticipated inelastic behaviour of the building. The steps of this simplified spectral analysis procedure are described as follows:

Step 1: Choice of dynamic degrees of freedom and computation of the mass and stiffness matrices

The most important degrees of freedom are selected. In a two-dimensional analysis, one degree of freedom per floor is chosen; in a three-dimensional analysis, three are chosen for each floor. The mass and stiffness matrices are then computed.

Step 2: Computation of natural frequencies and modes of vibration
We solve the eigenvalues problem for each mode i:

$$\left[[k] - \omega_i^2[m]\right](A^{(i)}) = (0) \tag{6.51}$$

Step 3: Computation of generalized masses
The generalized mass, M_i, is calculated for each mode i.

$$M_i = (A^{(i)})^T[m](A^{(i)}) \tag{6.52}$$

Step 4: Computation of modal participation factors
The modal participation factor, α_i, is calculated for each mode i.

$$\alpha_i = \frac{(A^{(i)})^T[m](I)}{M_i} \tag{6.53}$$

Step 5: Computation of spectral responses
The design spectrum of the NBCC (1995), shown in figure 4.35, is multiplied by the maximum ground velocity, v, and the modal response for each mode of vibration is calculated. It is recommended that the number of modes used be equal to the number of building storeys, though not exceeding 5.

Step 6: Computation of the probable base shear
The probable base shear, V_m, is obtained by combining the modal base shears using one of the methods presented in section 4.8.4. For most structures, the SRSS combination is adequate. For structures with closely spaced modes, however, the CQC method is recommended.

Step 7: Scaling of the modal contributions.
The modal forces are multiplied by the following α factor:

$$\alpha = \frac{V}{V_m} \tag{6.54}$$

where V is the base shear calculated by the static method of the NBCC (1995) as shown in equation 6.21. The design shear force at each floor and other response parameters are obtained by combining the scaled modal values for each mode.

6.8.4 Nonlinear Time-Step Dynamic Analysis

For very large or highly irregular structures, a nonlinear time-step analysis must be used (section 4.9). An important stage of this analysis is to determine a set of accelerograms that are representative of a given site. This is not an easy task; several approaches are possible:

- use of historical accelerograms recorded in the same region as the given site;
- use of historical accelerograms recorded in regions far from the given site but with compatible seismic parameters to the ones anticipated at the site;
- use of artificially generated accelerograms that contain compatible seismic parameters to the ones anticipated at the given site.

Filiatrault, D'Aronco, and Tinawi (1994), for example, used a computerized data base of historical seismic records (SMCAT, 1989) in order to gather a set of representative accelerograms for three Canadian cities: Montreal $(Z_a > Z_v)$, Vancouver $(Z_a = Z_v)$, and Prince Rupert $(Z_a > Z_v)$. Only free-field seismic records were chosen for which the maximum horizontal acceleration, a_{max}, and maximum horizontal velocity, v_{max}, had to be simultaneously within the range of values prescribed by the NBCC (1995) for each city. Table 6.11 shows a summary of these selection criteria.

Table 6.11   Selection parameters for historical seismic records.

City	Z_a	Z_v	Selection parameters	
			a_{max} (g)	v_{max} (m/s)
Montreal	4	2	$0{,}16 \leq a_{max} \leq 0{,}23$	$0{,}08 \leq v_{max} \leq 0{,}11$
Vancouver	4	4	$0{,}16 \leq a_{max} \leq 0{,}23$	$0{,}16 \leq v_{max} \leq 0{,}23$
Prince Rupert	3	5	$0{,}11 \leq a_{max} \leq 0{,}16$	$0{,}23 \leq v_{max} \leq 0{,}32$

Table 6.12 shows the results of this selection. Three historical records for the city of Montreal were identified. To put together a more representative set of records for Montreal, artificial accelerograms generated by Atkinson (1992) were added to the historical records. These artificial accelerograms represent design ground motions anticipated for eastern Canada at different magnitudes and distances from the epicentre. The same procedure was used to identify six records from five different earthquakes for Vancouver. Finally, eight records from four different earthquakes were found for Prince Rupert.

Figure 6.26 compares the absolute acceleration response spectrum, for 5% critical damping, corresponding to the average plus one standard deviation (MSD) of the seismic record ensemble with the NBCC (1995) design spectrum for each city. D'Aronco (1993) calculated the response spectrum for each record. Each accelerogram was normalized for the design peak ground horizontal acceleration of each city: 0,18 g for Montreal, 0,12 g for Vancouver, and 0,13 g for Prince Rupert. The MSD spectra show a good correlation with the NBCC (1995) design spectrum for each city.

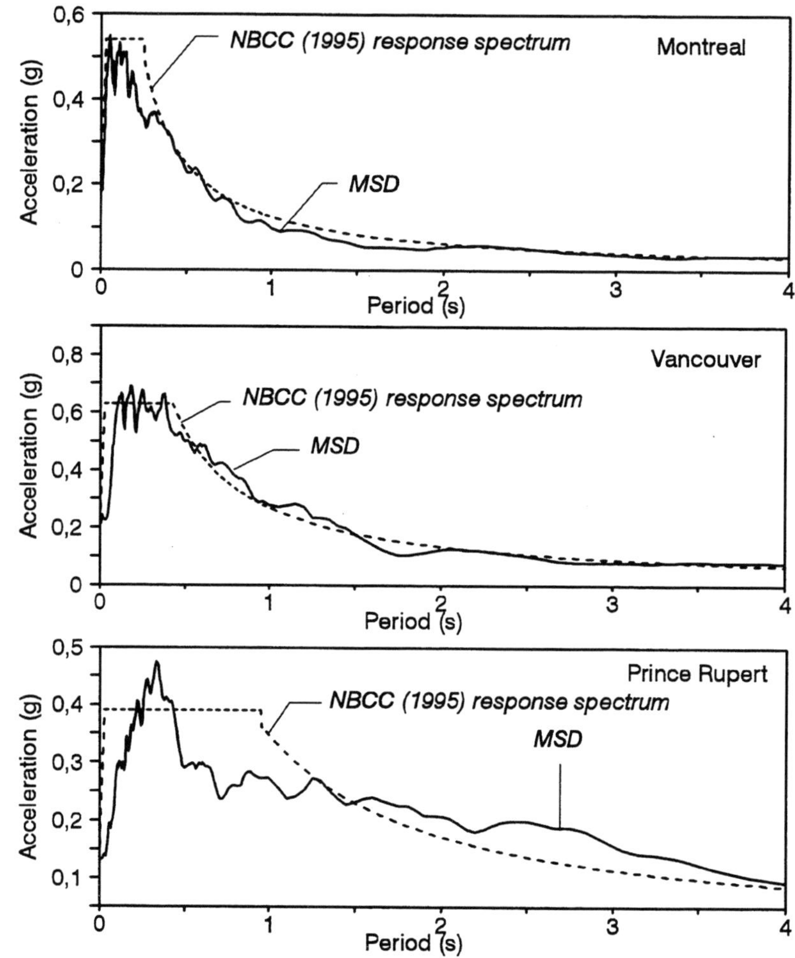

Figure 6.26   Absolute acceleration response spectra of the seismic records and NBCC (1995) design response spectrum.

6.8.5 Specific Details for Dynamic Analysis of Buildings

Young's modulus for reinforced concrete structures. Young's modulus, prescribed by reinforced and pre-stressed concrete codes, is generally based on the 28-day compressive strength of concrete. Concrete grows stronger over time, which may cause a considerable increase in Young's modulus. Experimental studies on reinforced and pre-stressed concrete bridges (Moss and Carr, 1979) have shown that the effective Young's modulus of these structures was higher than the prescribed design value.

Table 6.12 Selection results for the historical earthquake records.

City	Name, station, and component	a_{max} (g)	v_{max} (m/s)
Montreal	Whittier 1987, Hollywood Strg., LA 360	0,20	0,09
	Imperial Valley aftershock, 1979 Anderson RD, El Centro 230	0,17	0,09
	Adak 1971, Naval Base West	0,19	0,08
	Atkinson S32, M_L = 5-50 km	0,02	-
	Atkinson S61, M_L = 6-50 km	0,05	-
	Atkinson S81, M_L = 7-20 km	0,52	-
	Atkinson S92, M_L = 7-50 km	0,13	-
	Atkinson S102, M_L = 7-80 km	0,07	-
Vancouver	Coalinga, aftershock 1983	0,22	0,16
	Whittier 1987, Union Oil Yard, 90	0,22	0,16
	Morgan Hill 1984, San Ysidro Sch., 270	0,22	0,19
	Puget Sound 1949, Hwy Test Lab., N.04W.	0,16	0,21
	San Fernando 1971, Hollywood St., S.00W.	0,17	0,17
	San Fernando 1971, Hollywood St., N.90E.	0,21	0,21
Prince Rupert	Long Beach 1933, Vernon, S.08W.	0,13	0,29
	Borrego Mountain 1968, El Centro, S.00W.	0,13	0,26
	Chili 1971, Santiago, N.10W.	0,16	0,23
	San Fernando 1971, Orion Blvd., LA, S.90W.	0,13	0,24
	San Fernando 1971, Owen St., LA, S.00W.	0,11	0,32
	San Fernando 1971, Water & Power Blvd., LA, N.50W.	0,15	0,23
	San Fernando 1971, Sunset Blvd., LA, S.89W.	0,15	0,23
	San Fernando 1971, Ventura Blvd. LA, N.79W.	0,15	0,24

In this study, in order adequately to represent the natural frequencies and modes of vibration obtained experimentally, Young's modulus was increased in the numerical models by 20% for reinforced concrete and by 40% for pre-stressed concrete.

Sectional properties for reinforced-concrete elements. The sectional properties to be used in a dynamic model of a reinforced-concrete structure must reflect the state of the elements during earthquake response, including the effects of concrete cracking. It should also take into account shear deformations for short and deep beams, such as link beams of coupled shear walls. The dynamic model should also consider the contributing portion of the slabs in the computation of the beams' stiffness. Table 6.13, inspired by the New Zealand reinforced-concrete code, shows the sectional properties to be used in a dynamic model of a reinforced-concrete structure for different levels of seismic intensity.

Table 6.13 Sectional properties to be used in the dynamic model of a reinforced-concrete structure for different levels of seismic intensity (adapted from Carr, 1994).

Element	Condition	Seismic intensity		
		Low; **ductility $\approx$ 1,5**	**Average;** **ductility $\approx$ 3**	**High;** **ductility $\approx$ 6**
Beam	Rectangular	I_g	$0{,}70\,I_g$	$0{,}40\,I_g$
	T or L	I_g	$0{,}60\,I_g$	$0{,}35\,I_g$
Column	$P > 0{,}5 f_c A_g$	I_g	$0{,}90\,I_g$	$0{,}80\,I_g$
	$P = 0{,}2 f_c A_g$	I_g	$0{,}80\,I_g$	$0{,}60\,I_g$
	$P = -0{,}05 f_c A_g$	I_g	$0{,}70\,I_g$	$0{,}40\,I_g$
Wall	$P = 0{,}2 f_c A_g$	$I_g;\ A_g$	$0{,}70\,I_g;\ 0{,}90\,A_g$	$0{,}45\,I_g;\ 0{,}80\,A_g$
	$P = 0$	$I_g;\ A_g$	$0{,}50\,I_g;\ 0{,}75\,A_g$	$0{,}25\,I_g;\ 0{,}50\,A_g$
	$P = -0{,}1 f_c A_g$	$I_g;\ A_g$	$0{,}50\,I_g;\ 0{,}65\,A_g$	$0{,}15\,I_g;\ 0{,}30\,A_g$
Link Beam	Diagonal reinforcement	$\dfrac{I_g}{1{,}7 + 1{,}3(h/L)^2}$	$\dfrac{0{,}70\,I_g}{1{,}7 + 2{,}7(h/L)^2}$	$\dfrac{0{,}40\,I_g}{1{,}7 + 2{,}7(h/L)^2}$
	Horizontal reinforcement	$\dfrac{I_g}{1{,}0 + 5(h/L)^2}$	$\dfrac{0{,}70\,I_g}{1{,}0 + 8(h/L)^2}$	$\dfrac{0{,}40\,I_g}{1{,}0 + 8(h/L)^2}$

I_g = moment of inertia of the gross concrete section
A_g = area of the gross concrete section
P = compression axial load
f_c = compressive strength of concrete
h = depth of link beams
L = free span of link beams

Rigid links. The use of rigid links at the end of the beams, columns, link beams, and walls may significantly influence the structure's stiffness. The natural frequencies may increase by 10% to 20% when rigid links are included in the model. This effect is more noticeable when modelling coupled walls. The computer time associated with modelling rigid links is generally negligible. For reinforced-concrete frames, however, the beam-column assembly generally mobilizes high shear deformations; it is therefore preferable to omit rigid links.

Mass distribution. For most buildings, use of a lumped mass on each floor is generally adequate. In a two-dimensional analysis, the mass generates only one degree of freedom per floor. For a three-dimensional analysis, the hypothesis can be made that this mass reacts in two orthogonal directions and also generates a rotational inertia around a vertical axis going through the floor's centre of mass. This hypothesis requires three degrees of freedom per floor and is valid only if the in-plane stiffness of the floors is high enough that the floors themselves are considered to be rigid. If the displacements on each levels do not correspond to the centre of mass of each floor, coupling will occur and the mass matrix will not be diagonal. Most three-dimensional dynamic analysis software packages will allow specification of an arbitrary point on the floor where the mass is lumped and where all other nodes can be connected to this point by rigid links.

Flexible foundations. Even though soil deposits cannot be considered rigid, most dynamic seismic analyses make the assumption that the foundation is rigid. As shown in section 5.7, the flexibility of bearing soils can be estimated in a dynamic analysis by introducing springs and dashpots at the base of the structure. These elements may cause important changes to the structure's natural periods and modes of vibration.

Table 6.14 shows, for example, the influence of flexible foundations on the first two natural periods of a 21-storey reinforced-concrete core-type building (Filiatrault et al., 1992). The presence of flexible foundations greatly modifies the fundamental period.

Table 6.14 Influence of flexible foundations on the natural periods of a reinforced-concrete core-type building (from Filiatrault et al., 1992).

Model	Natural period(s)	
	First mode	**Second mode**
Rigid foundation	1,72	0,29
Flexible foundation	2,30	0,34

Orientation of base excitation. The designer must consider the base excitation along two horizontal orthogonal directions. If necessary, when slender beams or long cantilevers are present, the vertical base excitation must also be taken into account. However, peak response does not necessarily occur along the geometric axis of the building, especially if the structure is not symmetrical. The in-plane orthogonal excitation axes should represent the minor and major axes of the structure's lateral stiffness.

Simultaneous components of base excitation. The National Building Code of Canada (NBCC, 1995) does not consider the possibility of simultaneous base excitations. The designer must thus run an analysis for each orthogonal direction. However, some standards, such as Norway's standards for offshore structures, require that analysis results be combined along orthogonal axes using the SRSS method. Other standards (SEAOSC, 1977) require that results be combined along an axis with 30% to 40% of the results according to the corresponding orthogonal axis. The effect of simultaneous excitations will be of greater importance in the columns and joints of a building, but will be minimal in the beams unless important torsional loads occur.

6.9   ARCHITECTURAL PRINCIPLES TO ENSURE GOOD SEISMIC BEHAVIOUR

To ensure that a structure behaves well during a major earthquake, certain construction principles must be taken into account:

- building simplicity and symmetry;
- building elevation;
- structure continuity and uniformity;
- choice of building materials;
- control of failure modes.

These construction principles are briefly described below.

6.9.1 Building Simplicity and Symmetry

Experience with past earthquakes has shown that simple structures have a better chance of sustaining major earthquakes. There are three reasons for this:

- It is easier to understand the dynamic behaviour of a simple structure compared to a complex structure. In particular, the torsional effects in irregularly shaped structures are very hard to predict.
- It is easier to predict simple connections than complex ones.
- A simple structural system is easy to build.

Symmetry is desired for essentially the same reasons. An asymmetric structure may be excited in torsional modes that are very hard to predict. These very destructive torsional effects can be avoided by using a concentric (or symmetric) construction, as shown in figure 6.27.

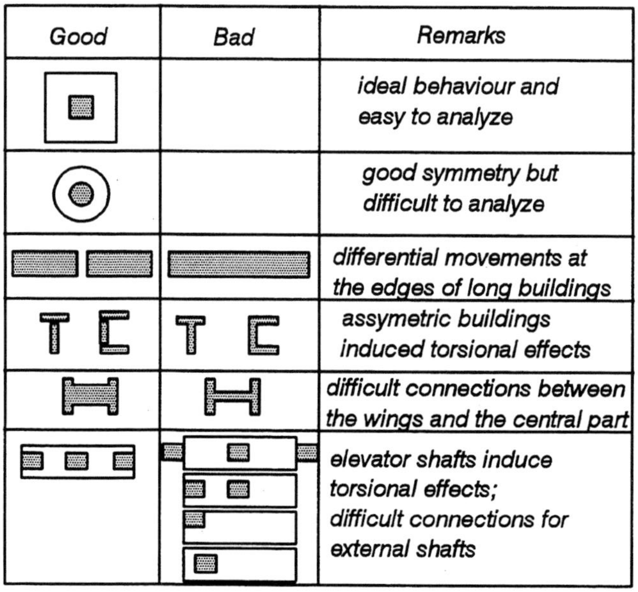

Good	Bad	Remarks
		ideal behaviour and easy to analyze
		good symmetry but difficult to analyze
		differential movements at the edges of long buildings
		assymetric buildings induced torsional effects
		difficult connections between the wings and the central part
		elevator shafts induce torsional effects; difficult connections for external shafts

Figure 6.27 Simple rules for positioning in-plane structural elements.

6.9.2 Building Elevation

As shown in figure 6.28, structures that are very slender or have large setbacks should be avoided in regions of high seismicity. The problems caused by slender structures are:

- the axial loads in the columns may be very high;
- the foundation's stability becomes difficult to ensure;
- higher modes can significantly contribute to the dynamic response.

Good	Bad	Remarks
$h < 4b$	$h > 4b$	very slender buildings can exhibit very large horizontal deflections (P-Delta effects)
		buildings with setbacks are difficult to analyze and induce torsional effects

Figure 6.28   Simple rules concerning building elevation.

6.9.3 Structure Continuity and Uniformity

This concept is closely linked with simplicity and symmetry. Sudden changes in lateral stiffness along the height of the building are not recommended because important loads may develop at the interface. Figure 6.29 illustrates this concept.

Good	Bad	Remarks
		lateral load resisting should be well distributed to increase redundency
Shear wall	Shear wall	sudden change in stiffness between floors causes concentrations of forces at the interface soft storey clearly vulnerable

Figure 6.29   Simple rules concerning the continuity and uniformity of a building.

6.9.4 Choice of Construction Materials

The choice of materials for construction of the main structure, architectural elements, and equipment greatly influences the building's chances of survival during a major earthquake. It is rare that optimum materials are chosen for every element. The main obstacles limiting choices are local availability of materials, local construction expertise, budget constraints, and political decisions.

When focusing only on the seismic behaviour, the construction materials should have the following attributes:

- high ductility;
- high strength-to-weight ratio;
- high homogeneity;
- capacity to produce high-strength connections.

In general, the importance of these factors will increase with the size of the structure. For example, table 6.15 shows the applicability of the most common construction materials for different types of buildings.

Table 6.15   Seismic applicability of construction materials.

	Type of building		
	High-rise	**Medium-rise**	**Low-rise**
Best	(1) steel	(1) steel	(1) wood
	(2) reinforced concrete	(2) reinforced concrete	(2) reinforced concrete
Materials in order of preference		(3) prefabricated concrete	(3) steel
		(4) pre-stressed concrete	(4) pre-stressed concrete
		(5) reinforced masonry	(6) pre-stressed concrete
Worst			(7) non-reinforced masonry

6.9.5 Control of Failure Modes

Good seismic design does not simply maintain an acceptable failure probability, but ensures that undesirable failure modes (nonductile) will not occur before ductile failure modes. These undesirable failure modes are:

- total premature failure of a structure: rupture of columns before rupture of beams;
- sudden failure modes without energy dissipation: shear, buckling, etc.

Figure 6.30 shows, for example, two possible failure modes of a moment-resisting frame. Here, it is clear that a plastic mechanism in the columns will cause a premature collapse of the structure with negligible energy dissipation before failure. A plastic mechanism in the beams, however, will give rise to very important energy dissipation before failure. This concept is called "the weak-beams and strong-columns philosophy" and will be discussed in detail in chapters 7 and 8.

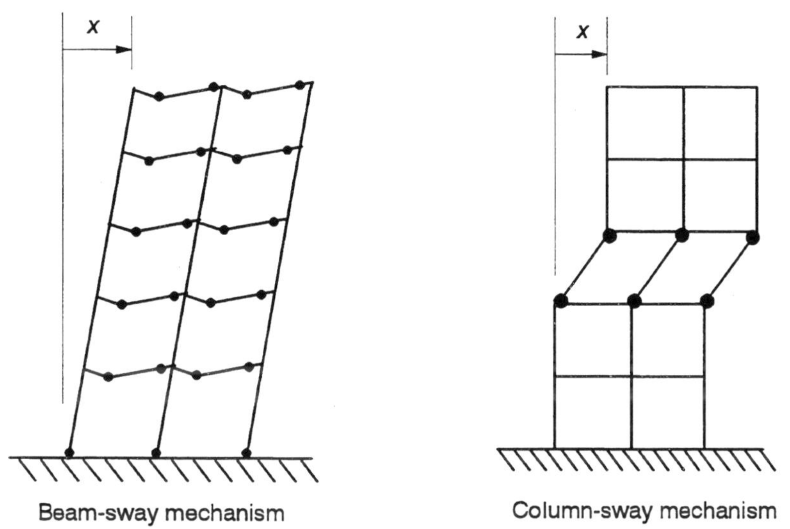

Figure 6.30   Two possible failure modes of a moment-resisting frame.

6.10 DUCTILITY: FINAL REMARKS

From the previous discussion, it is clear that ductility is the primordial factor in the seismic design of buildings. It is also important to note that a certain level of ductility should always be incorporated even though seismic loads may not represent the determining factor in a design. Let us consider, for example, that the lateral seismic loads prescribed by the code cause a base shear of 100 kN, but lateral loads caused by the wind yield a base shear of 150 kN, as shown in figure 6.31.

Inexperienced engineers may think that since wind loads govern the choice of members, ductility should not be an important factor, but this could not be farther from the truth. If the seismic loads depend on a ductility factor equal to 4, a ductility factor of 2,7 ((4)(100)/150) must still be used to ensure the survival of the structure. Even though a peak wind shear of 150 kN will generally not occur during the entire life span of the structure, this peak state will nevertheless be reached during the design earthquake. So, only a ductility factor of 2,7 will ensure the survival of the structure. It would not be necessary to include ductile details if the base shear caused by wind were greater than the elastic seismic base shear, (100)(4)/(0,6) = 667 kN.

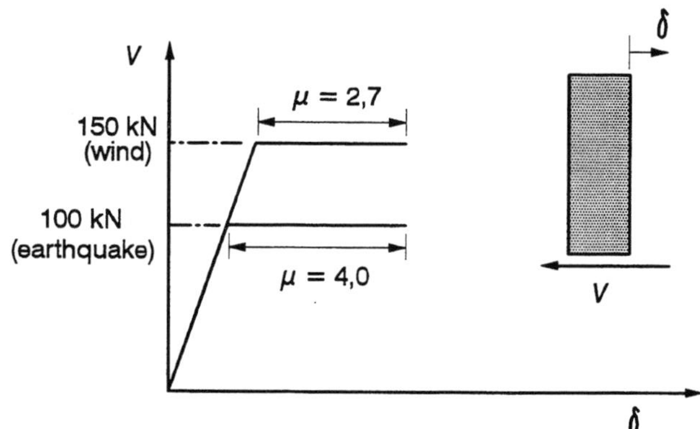

Figure 6.31   Examples of base shears generated by wind loads and by seismic loads.

6.11 PROBLEMS

6.1   Figure 6.32 shows a three-storey, reinforced-concrete, symmetric building located on a rock site in Montreal. The lateral-load-resisting system is made of the following elements:

Lines A-A and B-B: two coupled shear walls 12 m in height.

Lines C-C and D-D: a two-bay moment-resisting frame.

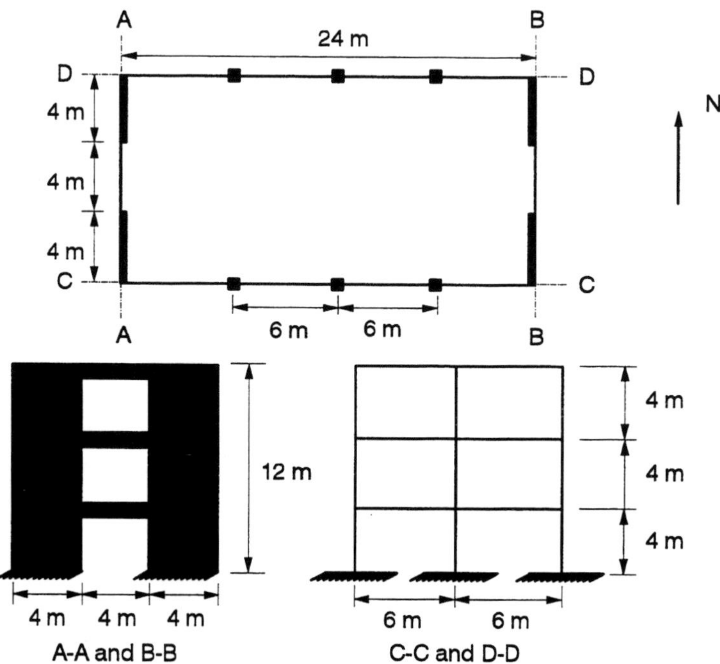

Figure 6.32   Three-storey symmetric building.

The structure is symmetric with respect to the E-W and N-S axes. A rigid floor links all elements of the lateral-load-resisting system. The dead load of each floor is 4 kN/m². The structure incorporates only nominally ductile details.

Evaluate, for each direction, the distribution of the lateral seismic design loads according to the static method of the National Building Code of Canada (NBCC, 1995). Include the effect of accidental eccentricity.

6.2   Briefly describe the problems associated with the seismic behaviour of each of the structure illustrated in figure 6.33.

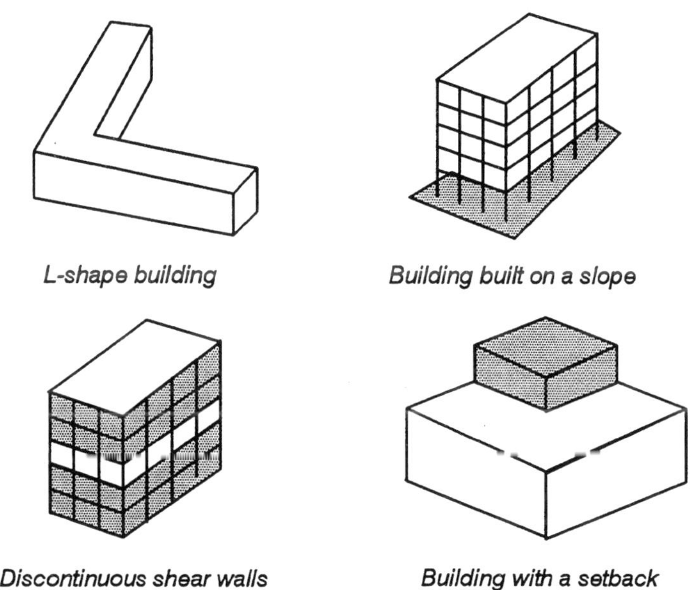

Figure 6.33   Problems associated with the seismic behaviour of structures.

6.12 REFERENCES

Atkinson, G. (1992). Private communication. Department of Geology, Carleton University, Carleton, ON.

Blume, J., Newmark, N.M., and Corning, L.H. (1961). *Design of Multi-Storey Reinforced Concrete Buildings for Earthquake Motions*. Chicago, IL: Portland Cement Association.

Carr, A.J. (1994). "Dynamic Analysis of Structures." *Bulletin of the New Zealand National Society for Earthquake Engineering, 27*(2): 129–46.

National Building Code of Canada and Commentary. 1995. 18th ed. Ottawa: National Research Council of Canada.

D'Aronco, D. (1993). *Évaluation du cisaillement sismique des murs ductiles en béton armé au Canada*. Master's thesis, Department of Civil Engineering, École Polytechnique de Montréal, Montreal.

Filiatrault, A., and Cervantes, M. (1995). "Separation Between Buildings to Avoid Pounding during Earthquakes." *Canadian Journal of Civil Engineering,* 22(1): 164–79.

Filiatrault, A., Anderson, D.L., and DeVall, R.H. (1992). "Effect of Weak Foundation on the Seismic Response of Core Wall Type Buildings." *Canadian Journal of Civil Engineering,* 19(3): 530–39.

Filiatrault, A., D'Aronco, D., and Tinawi, R. (1994). "Seismic Shear Demand of Ductile Cantilever Walls: A Canadian Code Perspective." *Canadian Journal of Civil Engineering,* 21(3): 363–76.

Filiatrault, A., et al. 1994. "Pounding of Buildings During Earthquakes: A Canadian Perspective." *Canadian Journal of Civil Engineering,* 21(2): 251–65.

Goel, R.K., and Chopra, A.K. (1993). "Seismic Code Analysis of Buildings Without Locating Centers of Rigidity." ASCE, *Journal of Structural Engineering,* 119(10): 3039–55.

Humar, J.L., and Awad, A.M. (1984). "Dynamic Response of Buildings to Ground Rotational Motion." *Canadian Journal of Civil Engineering,* 11(1): 48–56.

Jeng, V., Kasai, K., and Maison, B.F. (1992). "A Spectral Difference Method to Estimate Building Separations to Avoid Pounding." *Earthquake Spectra,* 8(2): 201–03.

Moss, P.J., and Carr, A.J. (1979). *Vibration Tests on the Buller River Bridge at Westport.* Research Report No. 79-16. Department of Civil Engineering, University of Canterbury, New Zealand.

Newmark, N.M., and Veletsos, A.S. (1960). "Effect of Behaviour of Simple Systems to Earthquake Motion." *Proceedings of the 2nd World Conference on Earthquake Engineering,* Tokyo, pp. 895–912.

SEAOSC, Electronic Computation Committee (1977). *Seismic Analysis by Computer.* Report. Los Angeles: Structural Engineers Association of Southern California.

SMCAT (1989). *An Earthquake Strong Motion Data Catalog for Personal Computers.* Computerized Data Bank. Boulder, CO: National Geophysical Data Center.

Stafford-Smith, B., and Vézina, S. (1985). "Evaluation of Centres of Resistance in Multi-Storey Building Structures." *Proceedings of the Institution of Civil Engineers,* vol. 79, pt. 2, pp. 623–35.

Tinawi, R. (1991). "Exigences du CNBC 1990 relatives au calcul des charges sismiques." *Séminaires sur la conception parasismique.* École Polytechnique de Montréal, Montreal.

Tso, W.K. 1990. "Static Eccentricity Concept for Torsional Moment Equations." ASCE, *Journal of Structural Engineering,* 116(5): 1199–1212.

CHAPTER 7

Elements of Seismic Detailing for Reinforced-Concrete Structures in Canada

7.1 INTRODUCTION

To ensure the ductile behaviour of a reinforced-concrete structure, flexure must be the weakest link in order to avoid brittle failure modes such as shear. In other words, shear forces must remain in the elastic range of the material when a structural element reaches its peak flexural resistance. A classical example of a brittle shear failure is the collapse of the ambulance shelter at Olive View Hospital in California during the 1971 San Fernando earthquake. This structure was made of a series of simple portal frames, as shown in figure 7.1.

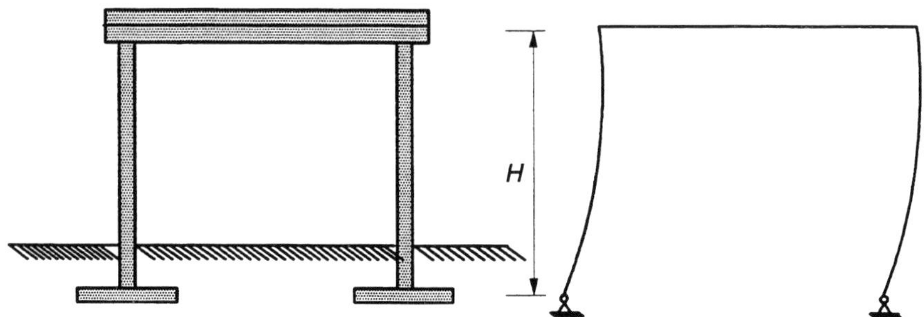

Figure 7.1 Illustration of the ambulance shelter at Olive View Hospital.

The designer assumed that the columns were simply supported at the footing level. He computed the maximum shear force for each column, V_{max}, assuming the formation of plastic hinges in the columns at the girder level.

$$V_{max} = \frac{M_{nc}}{H} \tag{7.1}$$

where M_{nc} represents the nominal (unfactored) flexural resistance of each column.

In reality, however, the concrete slab was restraining the columns and acting as a fixed support (figure 7.2) well above the footings. The actual maximum shear force in each column was thus much higher.

$$V_{max} = \frac{\left[M_{high} + M_{low}\right]}{h} \tag{7.2}$$

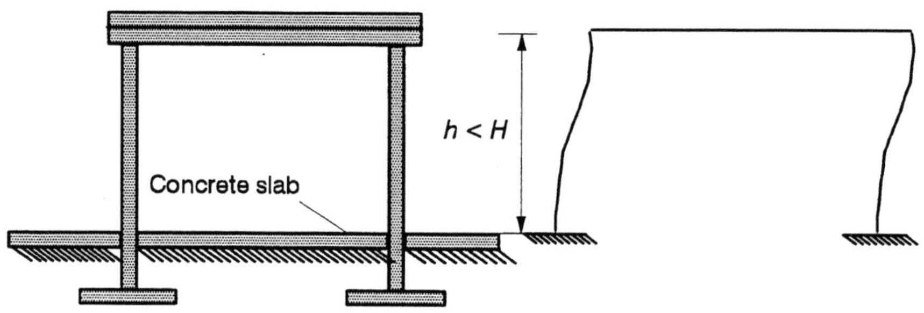

Figure 7.2   Actual behaviour of the ambulance shelter at Olive View Hospital.

The shear force was essentially doubled. As a result, a brittle shear failure occurred in the columns at the girder level.

7.2   SECTIONAL CURVATURE DUCTILITY

The ductility of a reinforced-concrete section is called the curvature ductility, μ_c. It is defined as the ratio of its sectional curvature at ultimate, ϕ_u, to the first yield curvature, ϕ_y (figure 7.3).

$$\mu_c = \frac{\phi_u}{\phi_y} \tag{7.3}$$

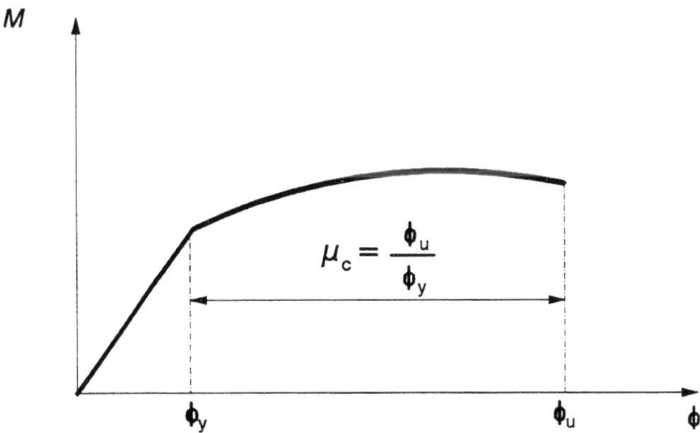

Figure 7.3 Definition of curvature ductility.

The curvature ductility is a function of:

- the geometry of the section;
- the reinforcement arrangement;
- the stress-strain relationships of the steel and concrete.

7.2.1 Curvature Ductility for a Singly Reinforced Concrete Section

For this type of section, yielding of the tensile steel will occur before crushing of the concrete in compression. According to figure 7.4, the first yield curvature of the section, ϕ_y, is easily obtained by strain compatibility.

$$\phi_y = \frac{\epsilon_{sy}}{\left[(1-k)d\right]} = \frac{f_y}{\left[E_s(1-k)d\right]} \tag{7.4}$$

where

$$k = \sqrt{\left((\rho n)^2 + 2\rho n\right)} - \rho n$$

$$\rho = \frac{A_s}{bd} = \text{tensile reinforcement ratio}$$

$$n = \frac{E_s}{E_c} = \text{modulus ratio} \tag{7.5}$$

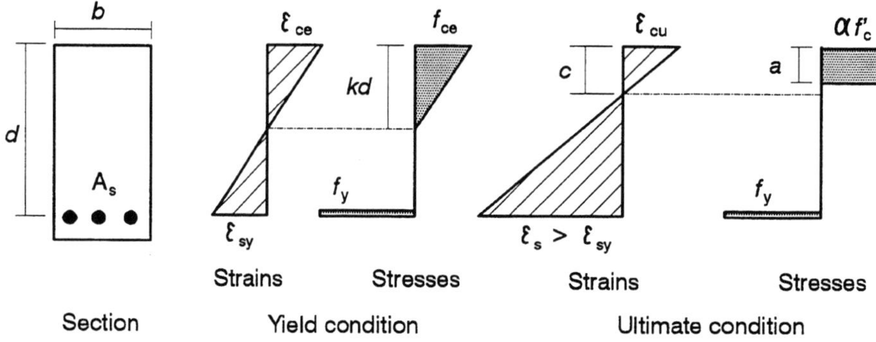

Figure 7.4 Flexural behaviour of a singly reinforced concrete section.

The value of k given by equation 7.5 is valid only then the concrete behaves elastically in compression. However, this equation is reasonably accurate even for higher compressive stresses in concrete.

Referring again to figure 7.4, the ultimate sectional curvature is obtained by strain compatibility when the concrete reaches its ultimate compressive stress.

$$\phi_u = \frac{\epsilon_{cu}}{c} = \frac{\beta_1 \epsilon_{cu}}{a} \tag{7.6}$$

where

$$a = \frac{A_s f_y}{\alpha_1 f_c' b} \tag{7.7}$$

where α_1 and β_1 define the equivalent compressive stress block of concrete. In Canada we have (CSA, 1994):

$$\alpha_1 = 0,85 - 0,0015 f_c' \geq 0,67$$

$$\beta_1 = 0,97 - 0,0025 f_c' \geq 0,67 \tag{7.8}$$

The available curvature ductility can then be calculated for the section.

$$\mu_c = \frac{\phi_u}{\phi_y} = \frac{\epsilon_{cu} d (1 - k) E_s}{c f_y} \tag{7.9}$$

It is important to note that in Canada, the ultimate compressive strain of concrete is taken to be equal to 0,0035 (CSA, 1994).

7.2.2 Curvature Ductility of a Doubly Reinforced Concrete Section

The curvature ductility of a doubly reinforced concrete section (figure 7.5) is calculated in the same way as above.

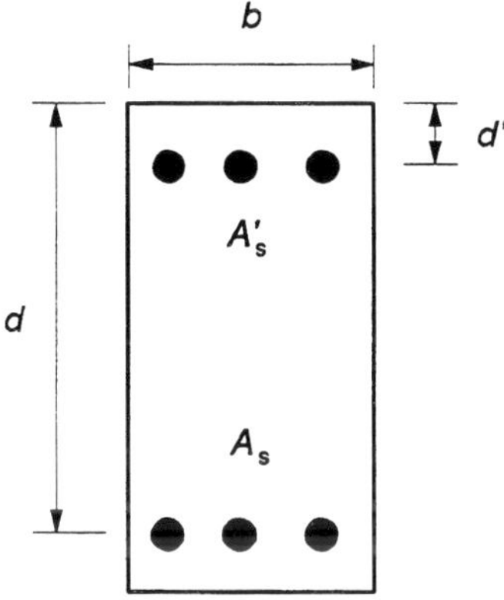

Figure 7.5 Doubly reinforced concrete section.

The equation for the available curvature ductility of a doubly reinforced concrete section is the same as the equation for a singly reinforced section.

$$\mu_c = \frac{\phi_u}{\phi_y} = \frac{\epsilon_{cu} d(1-k) E_s}{c f_y} \tag{7.10}$$

To take into account the compressive steel ratio, ρ', equations for c and k become:

$$c = \frac{a}{\beta_1} = \frac{(\rho A_s - \rho A_s') f_y}{0,85 f_c' b \beta_1} \tag{7.11}$$

$$k = \sqrt{((\rho - \rho')n)^2 + 2(\rho - \rho')n} - (\rho - \rho')n \tag{7.12}$$

It is important to note that equations 7.11 and 7.12 assume that the compressive steel is yielding before the ultimate strain of the most compressed concrete fibre is reached. If this situation does not occur, the actual value of the steel stress should be substituted for f_y in these equations. Figure 7.6 gives a graphical representation of the curvature ductility of typical singly and doubly reinforced concrete sections.

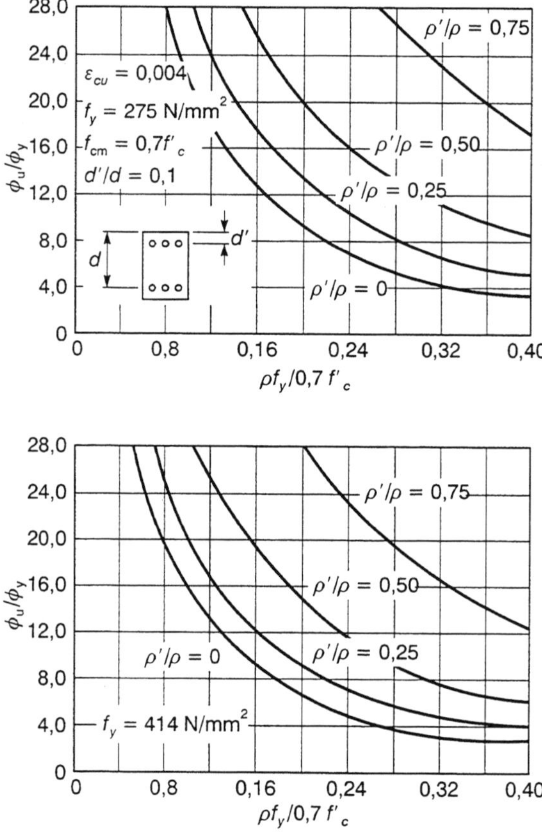

Figure 7.6 Available curvature ductility for singly and doubly reinforced concrete sections (from Blume et al., 1961).

We note that the available curvature ductility:

- decreases when the tensile steel ratio, ρ, increases;
- increases when the compression steel ratio, ρ', increases;
- decreases when the yield stress, f_y, increases.

7.3   INFLUENCE OF CONCRETE CONFINEMENT ON CURVATURE DUCTILITY

Many researchers have proven experimentally that the ductility and strength of concrete is greatly enhanced by confining the compression zone with closely spaced transverse reinforcement (Base and Read, 1985; Bertero and Felippa, 1964; Nawy et al., 1968). Park, Priestley, and Gill (1982) proposed a stress-strain model for compressed concrete confined by rectangular hoops (closed transverse reinforcement with 135° hooks). This model is a modification of the model originally proposed by Kent and Park (1971).

$$f_c = Kf'_c \left[\frac{2\,\epsilon_c}{0,002\,K} - \left(\frac{\epsilon_c}{0,002\,K} \right)^2 \right] \quad \text{for} \quad \epsilon_c \le 0,002\,K$$

$$f_c = Kf'_c \left[1 - Z_m \left(\epsilon_c - 0,002\,K \right) \right] \quad \text{for} \quad \epsilon_c > 0,002\,K$$

$$K = 1 + \frac{\rho_s\, f_{yh}}{f'_c} \tag{7.13}$$

$$Z_m = \frac{0,5}{\left(\dfrac{3 + 0,29\, f'_c}{145 f'_c - 1\,000} \right) + \dfrac{3}{4}\rho_s \sqrt{\dfrac{h''}{s_h}} - 0,002\,K}$$

where

ϵ_c	=	compressive strain of concrete
f_c	=	compressive stress of concrete (MPa)
f'_c	=	specified compressive strength of concrete (MPa)
ρ_s	=	ratio of the volume of transverse reinforcement to the volume of confined concrete (core volume)
f_{yh}	=	yield strength of the transverse reinforcement (MPa)
h''	=	width of the concrete core measured from the outside of the hoops
s_h	=	spacing of transverse reinforcement

Figure 7.7 shows typical compressive stress-strain curves for unconfined and confined concrete using different parameters K and Z_m in equation 7.13.

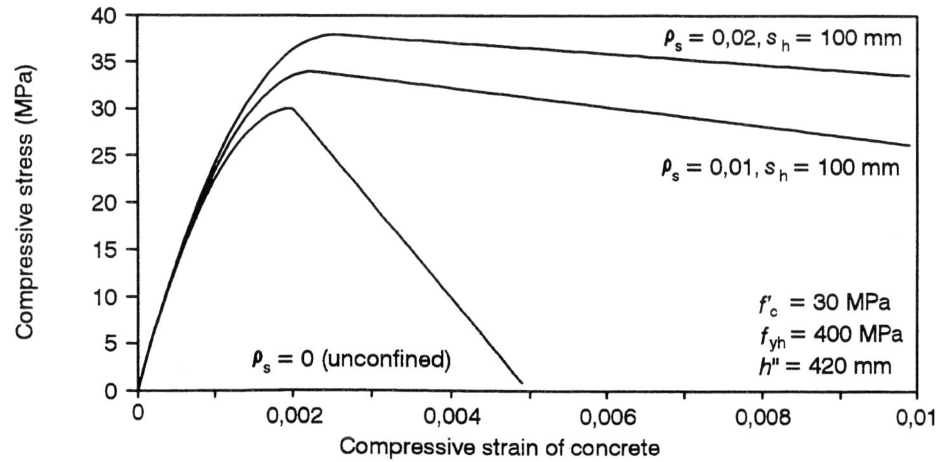

Figure 7.7 Typical compressive stress-strain curves for unconfined and confined concrete according to equation 7.13.

To calculate the available curvature ductility of a confined section, the equations presented above are used. However, the ultimate compressive strain of concrete, ε_{cu}, is modified to simulate the effect of confinement. The equation proposed by Corley (1966) is often used to estimate the value of ε_{cu} when the section is confined by rectangular hoops.

$$\varepsilon_{cu} = 0{,}003 + 0{,}02\frac{b}{l_c} + \left(\frac{\rho_{hc}\,f_{yh}}{138}\right)^2 \tag{7.14}$$

where f_{yh} = yield stress of the hoops (MPa)
  b = width of the section
  l_c = distance between the critical section (maximum moment) to the nearest inflexion point (zero moment)
  ρ_{hc} = ratio of the volume of confinement reinforcement (including the longitudinal reinforcement in compression) to the volume of the confined concrete (core volume)

Equation 7.14 does not take into account the increase in flexural strength of the confined concrete which will influence the shear demand in the section.

7.4   INFLUENCE OF AXIAL LOADS ON CURVATURE DUCTILITY

For columns, the curvature ductility decreases because of axial loads (fig. 7.8). Again, we see that the confinement of concrete increases the sectional curvature ductility.

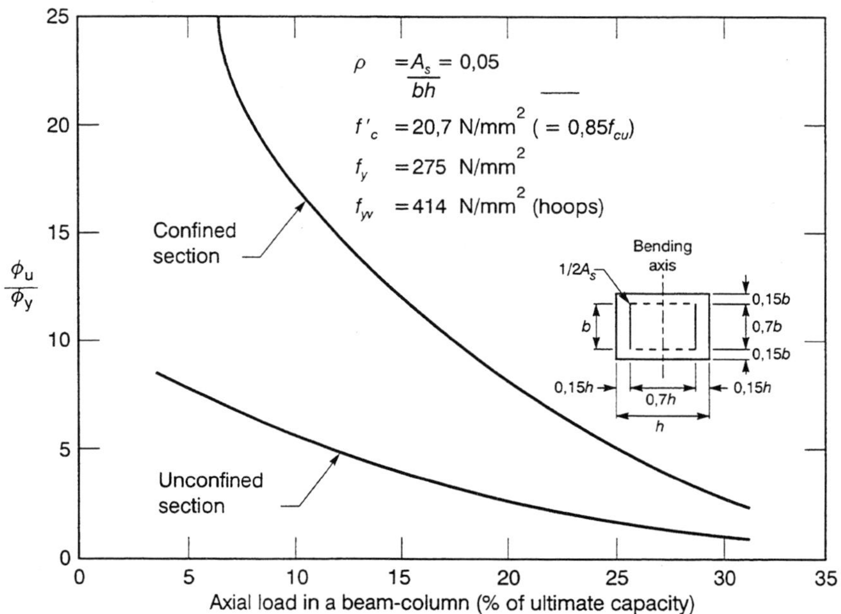

Figure 7.8 Influence of confinement on the curvature ductility of a column (from Blume et al., 1961).

For rectangular columns confined by closely spaced hoops, and in which the longitudinal reinforcement is mainly concentrated in two opposite faces, the curvature ductility can be estimated using the diagram in figure 7.9.

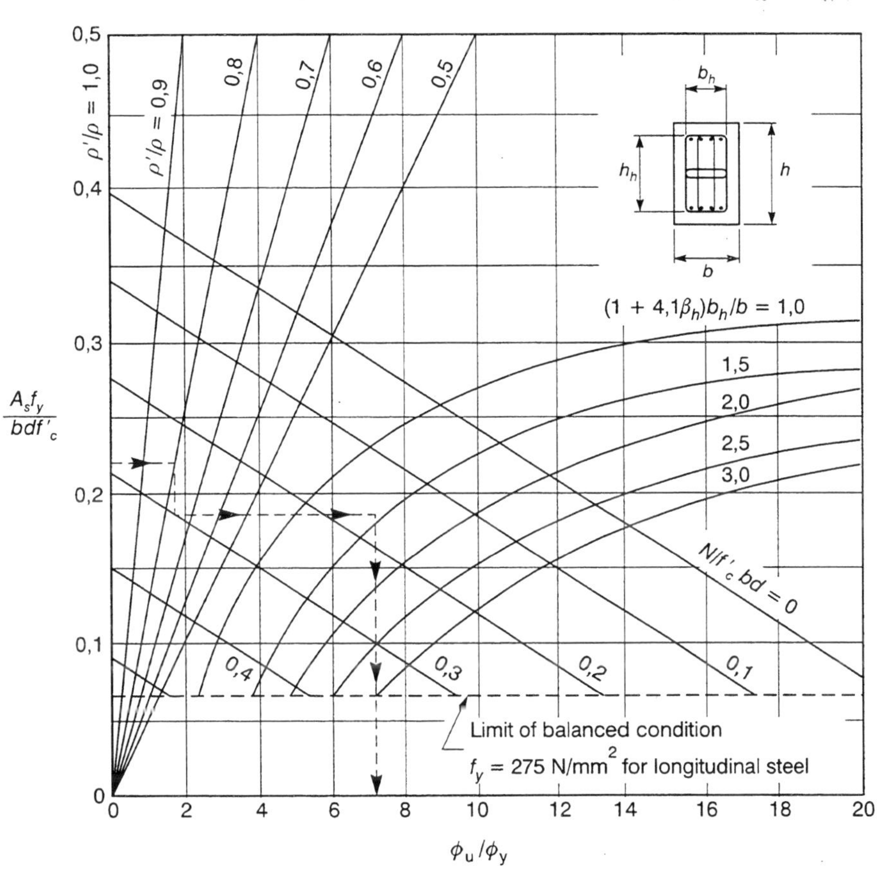

Figure 7.9 Available curvature ductility of confined concrete (from Blume et al., 1961).

In this diagram, A_s represents the area of the tensile reinforcement and

$$\beta_h = \frac{A_h f_{yh}}{s_h h_h f'_c} \tag{7.15}$$

where A_h = area of the hoops
  f_{yh} = yield stress of the hoops
  s_h = spacing of the hoops
  h_h = longest dimension of the confined concrete core

The available curvature ductility, ϕ_u/ϕ_y, is obtained by following the zigzag direction of the arrows.

7.5   EXAMPLE OF CALCULATION OF THE BENDING MOMENT-CURVATURE RELATIONSHIP FOR A BEAM SECTION AND A COLUMN SECTION

To understand better the difference between the curvature ductility of a beam and a column, let us consider the section shown in figure 7.10. Let us first make the following assumptions about the materials:

- f_y = yield strength of the reinforcement = 400 MPa
- E_s = elastic modulus of the reinforcement = 200 000 MPa
- f'_c = compressive strength of concrete = 30 MPa
- ε_{cu} = ultimate compressive strain of unconfined concrete = 0,003
- ε_{sy} = strain corresponding to the yield stress of the steel = 0,002

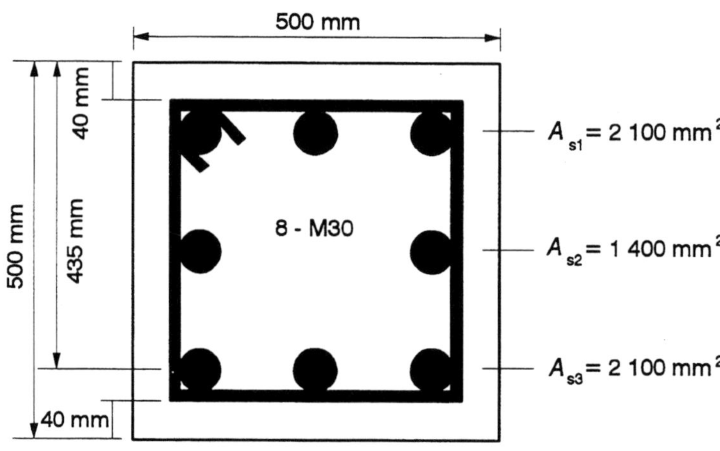

Figure 7.10   Cross-section of a beam or a column.

First, the ultimate compressive load of the section, P_u, is calculated (no bending moment is applied).

$$P_u = 0,85f'_c(bh - A_{st}) + f_y A_{st}$$

$$P_u = 8472\,kN; \quad \mu_c = 0$$

(7.16)

Then, the position of the neutral axis is determined for the balanced condition. The balanced condition is the condition for which the ultimate compressive strain of concrete is reached at the same time as the tensile steel yields (fig. 7.11).

$$\frac{\epsilon_{cu}}{c_b} = \frac{\epsilon_{sy}}{435 - c_b}$$

(7.17)

$$c_b = 261 \, mm$$

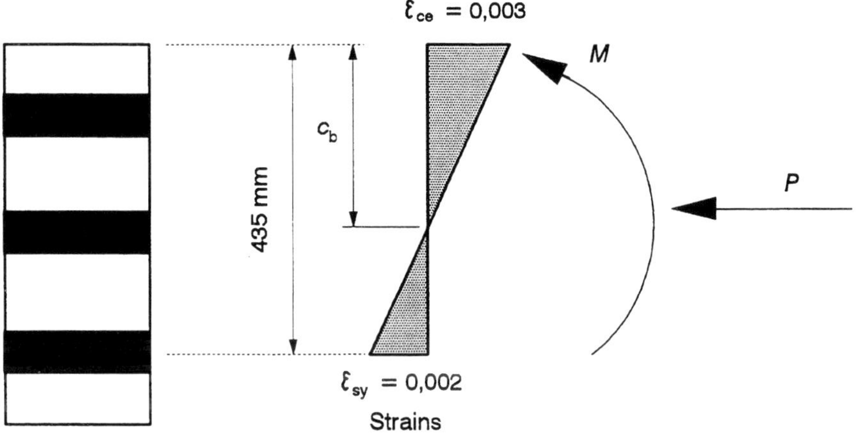

Figure 7.11 Balanced condition for a beam or a column section.

Table 7.1 Ultimate conditions for a beam or column.

c (mm)	500	400	261	200	100
ε_{s1}	0,000261(c)	0,00251(c)	0,00225(c)	0,00203(c)	0,00105(c)
ε_{s2}	0,00150 (c)	0,00113(c)	0,00013(c)	-0,00075(t)	-0,00450(t)
ε_{s3}	0,00039 (c)	-0,00263(t)	-0,0020(t)	-0,00353(t)	-0,01005(t)
f_{s1}(MPa)	400(c)	400(c)	400(c)	400(c)	210(c)
f_{s2}(MPa)	300(c)	225(c)	25(c)	-150 (t)	-400 (t)
f_{s3}(MPa)	78(c)	-53(t)	-400(t)	-400 (t)	-400 (t)
$a=0,85c$ (mm)	425	340	222	170	85
$C_{concrete}$(kN)	5 419	4 335	2 831	2 168	1 084
F_{s1}(kN)	840	840	840	840	441
F_{s2}(kN)	420	315	35	-210	-560
F_{s3}(kN)	164	-111	-840	-840	-840
P(kN)	6 843	5 379	2 866	1 958	125
M(kN-m)	328	523	704	668	468
$\phi_u=0,003/c$ (1/m)	0,0060	0,0075	0,0115	0,0150	0,0300

Now, let us consider the ultimate capacity of the section for different axial load and bending moment combinations, as shown in table 7.1. We assume a position for the neutral axis, c, and we calculate the *P-M* combination that causes the spalling of concrete in compression.

Then the *P-M* and $P\text{-}\phi_u$ diagrams are plotted, where ϕ_u is the curvature at the ultimate condition. These diagrams are shown in figure 7.12. When the neutral axis $c \leq 261$ mm, the tensile reinforcement yields before spalling of the concrete in compression. The *P-M* combinations that cause this yielding are found in table 7.2. As the compressed concrete has not yet reached the ultimate strain, we take the strain-stress curve in compression of the unconfined concrete, $K=1$, in equation 7.13.

$$f_c = f_c' \left[\left(\frac{2\,\epsilon_c}{0,002} \right) - \left(\frac{\epsilon_c}{0,002} \right)^2 \right] \quad \text{for} \quad \epsilon_c \leq 0,002 \qquad (7.18)$$

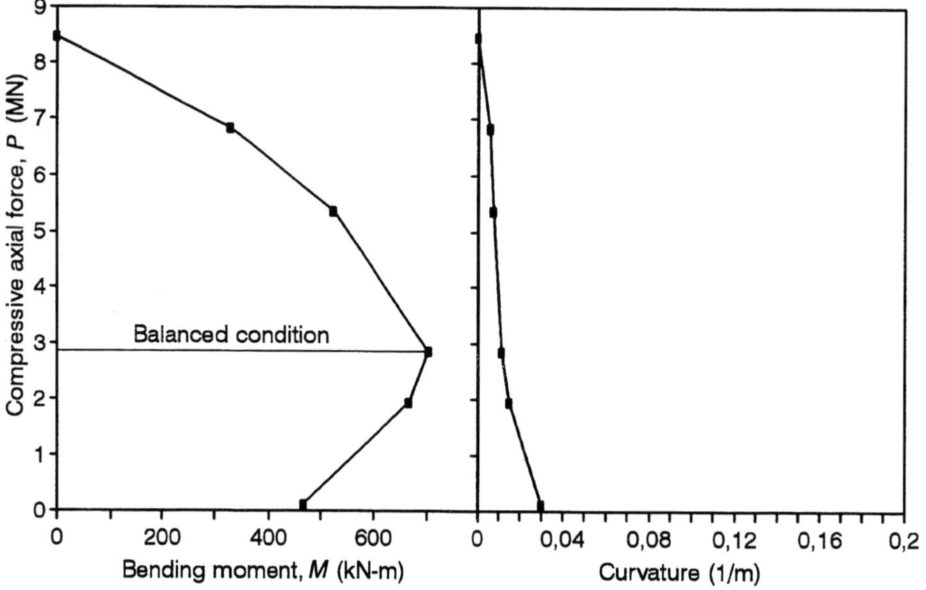

Figure 7.12   Ultimate interaction diagrams for an unconfined beam or column.

Table 7.2 Yield conditions for tensile reinforcement of a beam or column.

c **(mm)**	**200**	**150**	**100**
ε_c	0,00170	0,00105	0,00065
f_c(MPa)	29,30	23,20	16,43
f_{s1}(MPa)	230 (c)	119 (c)	42 (c)
f_{s2}(MPa)	- 85 (t)	-140 (t)	-179 (t)
f_{s3}(MPa)	-400 (t)	-400 (t)	-400 (t)
$C_{concrete}$(kN)	1 828	975	411
F_{s1}(kN)	483	250	88
F_{s2}(kN)	-119	-196	-251
F_{s3}(kN)	-840	-840	-840
P (kN)	1 352	189	-592
M (kN-m)	567	393	260
$\phi_u = \varepsilon_c/c$ (1/m)	0,0085	0,0070	0,0060

The corresponding P-M and P-ϕ diagrams are shown in figure 7.13.

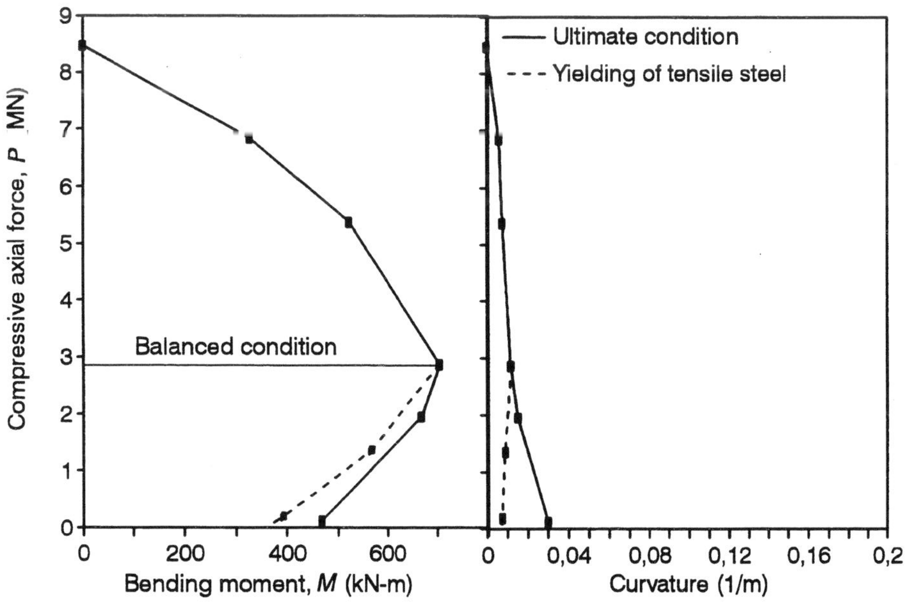

Figure 7.13 *P-M* and *P-ϕ* diagrams for a beam or a column.

After the spalling of concrete in compression, if the core of the concrete is confined by adequate transverse reinforcement, the section may continue to deform with a reduced cross-section, as shown in figure 7.14.

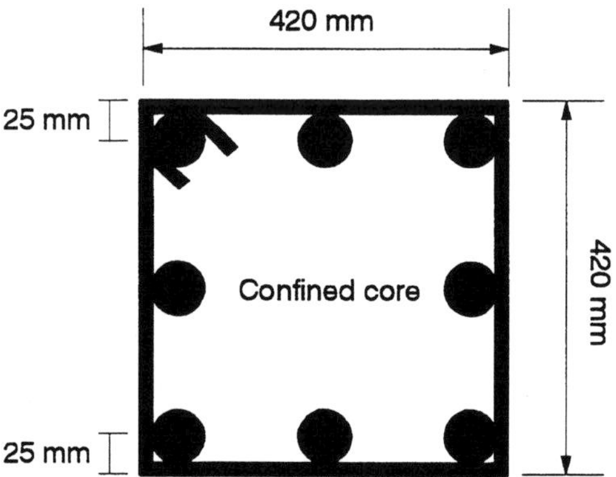

Figure 7.14 Effective section for a beam or a column after spalling of concrete in compression.

The effective section will continue to deform until the compressed concrete of the core reaches its ultimate confined strain. To calculate the *P-M* combinations that cause complete failure of the core, let us assume, to simplify calculation, the compressive stress-strain diagram for the confined concrete shown in figure 7.15. Computation of the *P-M* combinations are shown in table 7.3. Figure 7.16 presents the corresponding *P-M* and *P-ϕ* diagrams.

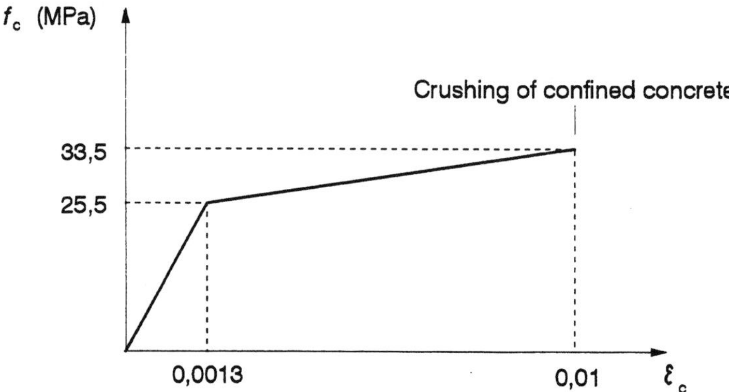

Figure 7.15 Stress-strain diagram for confined concrete in compression.

Table 7.3 Ultimate conditions for a confined beam or column.

c (mm)	400	300	250	200	150	50
f_{s1}(MPa)	400(c)	400(c)	400(c)	400(c)	400 (c)	400 (c)
f_{s2}(MPa)	400(c)	400(c)	320(c)	-100 (t)	-400 (t)	-400 (t)
f_{s3}(MPa)	25(c)	-400 (t)	-400 (t)	-400 (t)	-400 (t)	-400 (t)
$C_{concrete}$(kN)	4 590	3 442	2 869	2 295	1 721	574
F_{s1}(kN)	840	840	840	840	840	840
F_{s2}(kN)	560	560	448	-140	-560	-560
F_{s3}(kN)	53	-840	-840	-840	-840	-840
P (kN)	6 043	4 002	3 817	2 155	1 121	14
M (kN-m)	290	572	593	587	556	418
ϕ_u=0,01/c (1/m)	0,0250	0,0330	0,0400	0,0500	0,0670	0,2000

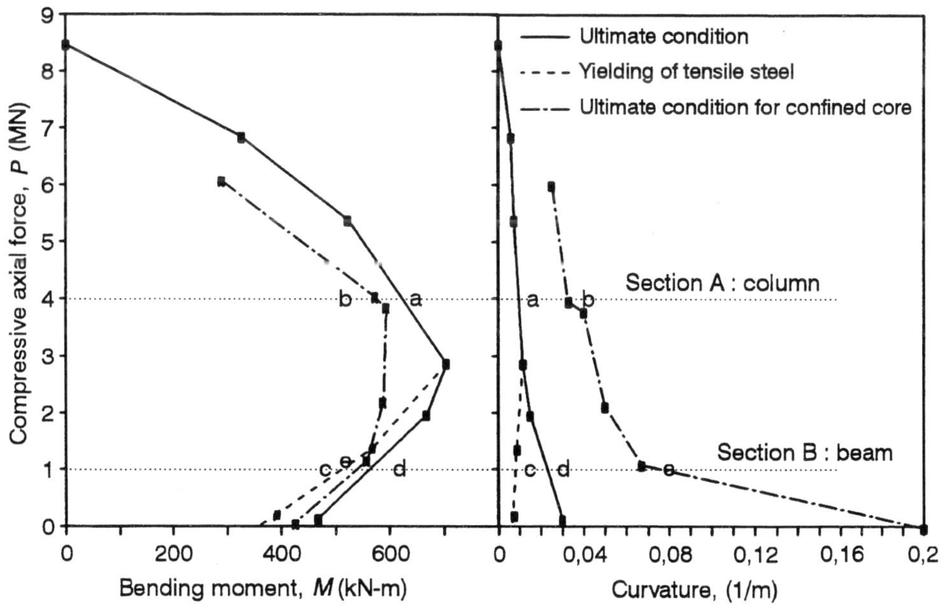

Figure 7.16 Ultimate and yield interaction diagrams for unconfined and confined beams or columns.

Let us now examine two sections that have different axial loads: section A, which has a high axial load noted as "column," and section B, which has a low axial load noted as "beam." For each section, the M-ϕ diagram is plotted by joining the corresponding points as shown in figure 7.17. We can clearly see that the available ductility for section A is much less than that for section B. This simple example demonstrates the influence of the axial load in drastically decreasing the available curvature ductility. This observation led to the concept of "weak beams and strong columns" in the earthquake-resistant design of reinforced-concrete buildings. In fact, the available ductility for a column is often referred to as the "last-resort ductility."

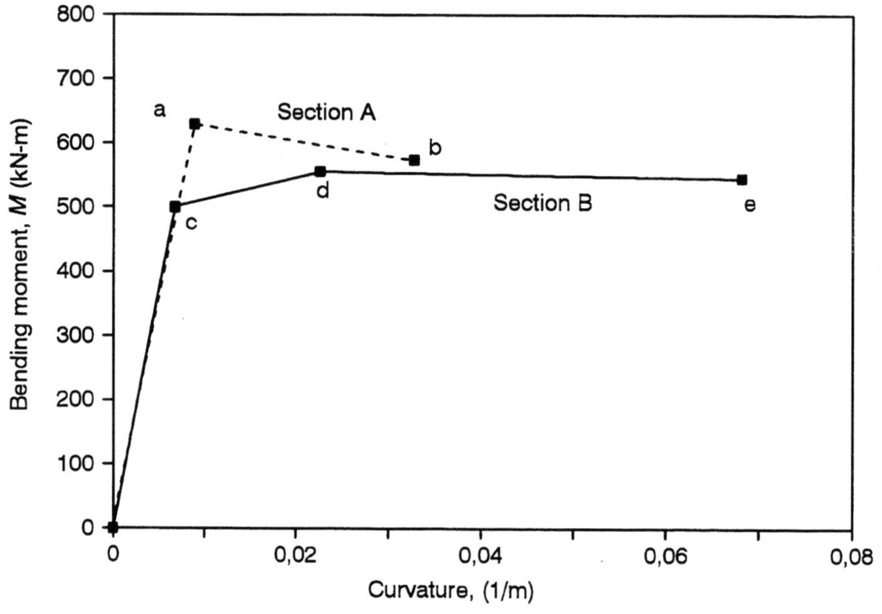

Figure 7.17   *M-ϕ* diagrams for different axial loads.

7.6   DETAILING REQUIREMENTS OF THE CANADIAN CONCRETE CODE FOR DUCTILE MOMENT-RESISTING FRAMES ($R = 4$)

7.6.1 Definition of Various Flexural Strengths

The Canadian concrete code (CSA, 1994) requires that special precautions be taken in the earthquake-resistant design of ductile moment-resisting frames. These frames are designed with a reduction factor $R = 4$, as we saw in section 6.6 (table 6.7). The main purpose of special detailing requirements for the frame is to minimize the possibility of shear failure and ensure that ductile flexural plastic hinges can occur in specified elements.

To fulfil this objective, the Canadian concrete code considers three different levels of flexural strengths:

- "factored" flexural strength, M_r;
- "nominal" flexural strength, M_n;
- "probable" flexural strength, M_p.

Factored flexural strength. The factored flexural strength of a frame element, M_r, is calculated with a resistance factor for concrete $\phi_c = 0,60$ and a resistance factor for the steel reinforcement $\phi_s = 0,85$. According to the limit-states design procedure, the factored flexural strength of each element must be greater or equal to the effects of the factored loads on this element.

Nominal flexural strength. The nominal flexural strength of a frame element, M_n, is determined with resistance factors for concrete and steel equal to one ($\phi_c = \phi_s = 1,0$). This strength level is used to ensure that in each beam-column subassembly, the columns are stronger than the beams. The following approximate relation is used between the nominal strength and the factored strength of a beam.

$$M_n \approx 1,20\, M_r \tag{7.19}$$

Probable flexural strength. The probable flexural strength of a frame element, M_p, is established with a resistance factor for concrete $\phi_c = 1,00$ and a resistance factor for steel reinforcement $\phi_s = 1,25$. The latter value corresponds to the maximum possible flexural strength that an element can reach taking into account the uncertainty on the yield strength, f_y, and the possibility of strain-hardening of the steel reinforcement. The probable strength of a frame element is used to compute the shear demand caused by the formation of flexural plastic hinges. The following approximate relation is used between the probable strength and the factored strength of a beam.

$$M_p \approx 1,47\, M_r \tag{7.20}$$

For a beam-column, a similar relation is used.

$$M_p \approx 1,57\, M_r \tag{7.21}$$

7.6.2 General Requirements

The Canadian concrete code requires that the concrete in all elements of a ductile moment-resisting frame have a compressive strength, f'_c, of between 20 and 55 MPa. As well, the steel reinforcement must be of weldable grade in conformity with CSA standard G30.18. It has been demonstrated that this type of steel shows behaviour that is more ductile than conventional steel and is therefore better suited to resist earthquakes.

7.6.3 Specific Requirements for Beams

The following specific requirements are aimed at lateral resisting elements that are designed mainly to sustain flexure. In this context, the factored compressive axial load on an element, P_f, must not exceed 10% of the gross compressive strength of the concrete section.

$$P_f \leq \frac{A_g f'_c}{10} \tag{7.22}$$

where A_g = gross area of the concrete cross section
       f'_c = specified compressive strength of concrete

Moreover, the beam must meet the following geometrical requirements:

- the clear span of the member should be no less than four times its effective depth;
- the width-to-depth ratio of the cross-section must be no less than 0,3;
- the width of the section must be no less than 250 mm.

At least two full-length bars must be provided for top and bottom reinforcement. The area of longitudinal reinforcement in the top and bottom reinforcements, A_l, must respect the following equation:

$$A_l \geq \frac{1,4\,b_w\,d}{f_y} \tag{7.23}$$

where b_w = width of the beam's web (mm)
       d = distance between the most compressed fibre and the centroid of the tensile reinforcement (mm)
       f_y = yield strength of the longitudinal reinforcement (MPa)

The code also requires an upper limit for the tensile reinforcement ratio, ρ.

$$\rho = \frac{A_s}{b\,d} \leq 0,025 \tag{7.24}$$

where A_s = area of the tensile reinforcement
       b = width of the compressed face of the beam

During a major earthquake, a moment-resisting frame sustains a very important lateral displacement, which may cause such rotation in the beams at the face of the joints that the negative moments, imposed by the gravity loads, may be overcome. To take this phenomenon into account, the code requires that the positive flexural resistance in a beam at the face of a joint must be no less than 50% of the negative flexural resistance at the same face. Furthermore, the positive and negative flexural resistance of any cross-section along the beam must be no less than 25% of the maximum flexural resistance provided at the face of either joint.

Lap splices of longitudinal reinforcement may be used only if hoops or spirals are provided over the lap length. The maximum spacing of the transverse reinforcement enclosing the lapped bars must be less than $d/4$ or 100 mm. Furthermore, lap splices cannot be used within the beam-column joints and within a distance of $2d$ from the face of the joint. Welded splices or mechanical connections can be used for splicing, provided that at any section, every second longitudinal bar is continuous. Furthermore, the centre-to-centre distance between adjacent splices must be more than 600 mm.

In a beam, closely spaced hoops are added near the face of the columns to allow the formation of plastic hinges. This reinforcement plays three important roles:

i)   it confines the concrete core;
ii)   it avoids buckling of the compressed longitudinal reinforcement;
iii)   it carries the maximum shear forces caused by the formation of "probable" moments at both ends of the beam and by the gravity loads (fig. 7.18).

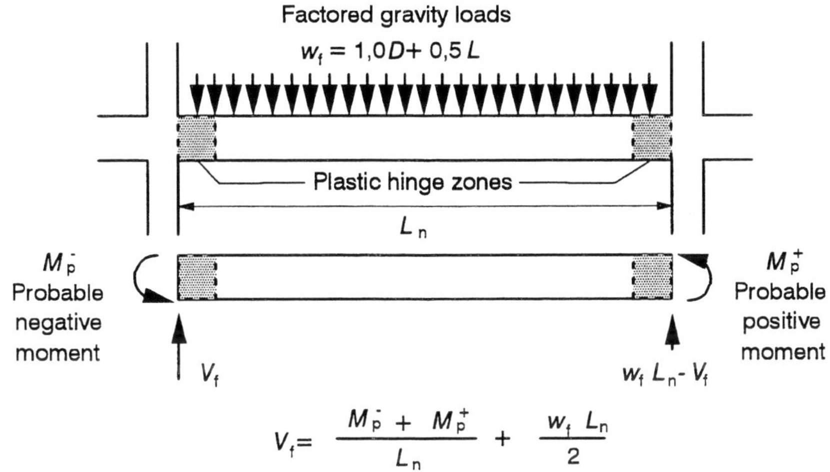

Figure 7.18 Design shear forces to ensure ductile flexural behaviour of a reinforced-concrete beam.

In a plastic hinge zone, the code makes the conservative assumption that concrete cracks too much to contribute effectively to the shear resistance, $V_c = 0$. In these zones, a 45° truss mechanism by the transversal reinforcement contributes to the factored shear strength, V_r.

$$V_r = V_s + V_c \geq V_f$$

$$V_s = \frac{\phi_s A_h f_{yh} d}{s_h} \qquad (7.25)$$

$$V_c = 0$$

where ϕ_s = resistance factor of the transverse reinforcement = 0,85
       A_h = cross sectional area of the transverse reinforcement
       f_{yh} = yield stress of the transverse reinforcement
       s_h = spacing of the transverse reinforcement

Hoops have to be installed over a distance equal to $2d$ measured from the face of the joint and over a distance d on each side of a zone where a plastic hinge may occur. The first hoop must be installed within 50 mm of the face of the joint. The maximum spacing of the hoops must not exceed $d/4$ to ensure adequate confinement of the concrete core, nor should it exceed eight times the diameter of the smallest longitudinal bars to avoid buckling, nor should it exceed 24 times the diameter of the hoops to avoid failure under the transverse pressure caused by the core, nor should it exceed 300 mm. For zones not requiring hoops, the stirrups must be spaced at a maximum distance of $d/2$ over the whole length of the member.

7.6.4 Specific Requirements for Columns

The following detailed requirements apply to elements of the lateral load-resisting system and are designed primarily to resist a bending moment and an axial load. For this purpose, the maximum factored axial compressive axial load on a member, P_f, must be more than 10% of the gross compressive strength of the column.

$$P_f > \frac{A_g f'_c}{10} \tag{7.26}$$

where $\quad A_g =$ gross cross-sectional area of the concrete
$\quad\quad\ \ f'_c =$ specified compressive strength of concrete

Furthermore, the column must satisfy the following geometric requirements:

- the smallest dimension of the cross-section cannot be less than 250 mm;
- the width-to-depth ratio of the cross-section cannot be less than 0,4.

To ensure the formation of plastic hinges in the beams (weak beams-strong columns philosophy), the code requires that the following equation be satisfied for all beam-column joints.

$$\sum M_{rc} \geq 1,1 \sum M_{nb} \tag{7.27}$$

where $\sum M_{rc} =$ the sum of moments at the centre of the joint, corresponding to the factored resistance of the columns framing into the joint. A consistent factored axial load must be used in the calculations
$\quad\ \ \ \sum M_{nb} =$ the sum of moments at the centre of the joint, corresponding to the nominal resistance of the beams framing into the joint

The cross-sectional area of the longitudinal reinforcement in a column, A_s, must be between 1% and 6% of the gross cross-sectional area, A_g.

$$0,01 \leq \frac{A_s}{A_g} \leq 0,06 \tag{7.28}$$

Lap splices must be designed as tensile splices and be placed at mid-height of a column. However, the code allows the use of lap splices anywhere except in the joint regions, if the development length is increased by 33% and if they are confined over the complete lap length by transverse reinforcement. This transverse reinforcement must be spaced at a distance of less than 200 mm, or 1/3 of the concrete core dimension measured perpendicularly to the longitudinal axis of the column. Welded splices and mechanical connections can be used for splicing the reinforcement at any section. The distance between splices must exceed 600 mm on the longitudinal axis of the column.

Closely spaced hoops must be installed in columns to provide "last resort" ductility. To avoid premature shear failure, the column's factored shear resistance must exceed the shear demand caused by the formation of simultaneous probable moments in the framing beams, as shown in figure 7.19.

$$V_f = \frac{M_{\text{top of column}} + M_{\text{bottom of column}}}{L}$$

$$M_{\text{top of column}} = \left(M_p^{(+)} + M_p^{(-)}\right)\frac{k_1}{k_1 + k_2}$$

$$M_{\text{bottom of column}} = \left(M_p^{(+)} + M_p^{(-)}\right)\frac{k_1}{k_1 + k_3}$$

(7.29)

where M_p = probable moment of the beams framing into the column ($\phi_c = 1{,}0$ and $\phi_s = 1{,}25$)

k_i = EI_i/L_i = representative flexural stiffness of the columns

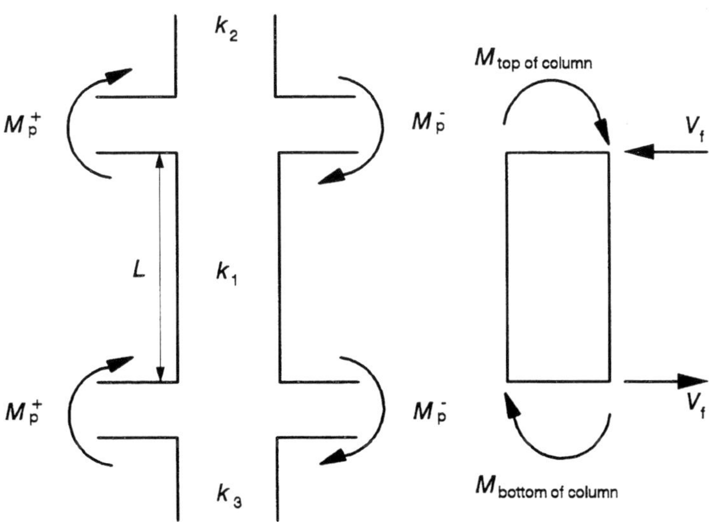

Figure 7.19 Design shear forces for columns.

Since beams must be the weakest links in flexure, the code assumes that the concrete in a column partly contributes to the shear strength.

$$V_r = V_s + V_c \geq V_f$$

$$V_s = \frac{\phi_s A_h f_{yh} d}{s_h} \tag{7.30}$$

$$V_c = 0,5 \left(0,2 \lambda \phi_c \sqrt{f'_c} \, b_w d\right)$$

These hoops must be installed within a distance l_o measured from the face of each joint. This distance cannot be less than the depth of the column at the face of the joint, nor than 1/6 of the clear height of the column, nor than 450 mm. Plastic hinges may occur in the first-floor columns, and these members must be confined over their complete height. Hoop spacing cannot be more than 1/4 of the smallest dimension of the column to ensure adequate confinement of the concrete core, nor 6 times the diameter of the smallest longitudinal bars to avoid buckling, nor 100 mm.

The code imposes two more restrictions regarding the transverse reinforcement in the columns. For spirals or circular hoops, the ratio of volume of transverse reinforcement to the volume of the confined concrete core, ρ_s, must be within the following limits.

$$\rho_s \geq \left(\frac{0,12 f'_c}{f_{yh}}\right) \tag{7.31}$$

If rectangular hoops are used, their total cross-sectional area, A_{sh}, must be within the following limits.

$$A_{sh} \geq 0,3 \, \frac{s_h h_c f'_c}{f_{yh}} \left(\frac{A_g}{A_{ch}} - 1\right)$$

$$A_{sh} \geq 0,09 \left(\frac{s_h h_c f'_c}{f_{yh}}\right) \tag{7.32}$$

where h_c = cross-sectional dimension of the column concrete core (mm)
$\,A_{ch}$ = cross-sectional area of the column concrete core (mm)

7.6.5 Specific Requirements for Beam-Column Joints

To avoid premature shear failure in a beam-column joint, the Canadian concrete code requires that the shear force demand in a joint be based on tensile stress in the longitudinal reinforcement of the adjacent beams equal to $1,25 f_y$. In other words, the joint shear strength must be at least equal to the shear force imposed by probable moments in adjacent framing beams. Figure 7.20 shows the free-body diagram from which the factored shear force in an interior beam-column joint, V_{fj}, can be obtained.

$$V_{fj} = 1{,}25 A_s' f_y + 1{,}25 A_s f_y - V_{col} \tag{7.33}$$

where

$$V_{col} = \frac{M_{cb(i+1)} + M_{ch(i+1)}}{L_{i+1}} \tag{7.34}$$

where the top and bottom bending moments, $M_{ch(i+1)}$ et $M_{cb(i+1)}$, can be obtained by equation 7.29.

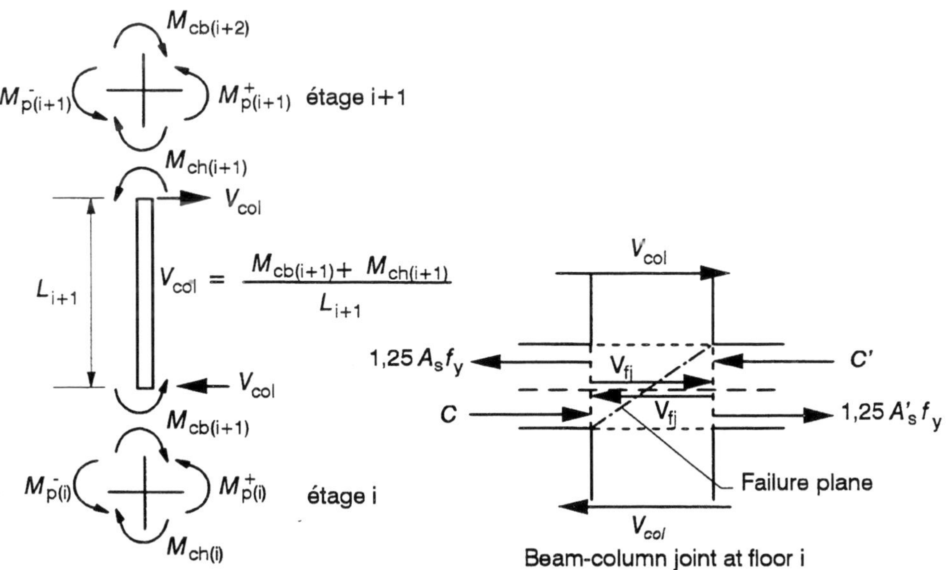

Figure 7.20   Factored shear force in a beam-column joint.

This shear force, V_{fj}, must be less than the factored shear strength of the joint, V_{rj}:

$$V_{rj} = K_j \lambda \phi_c \sqrt{f_c'} A_j \tag{7.35}$$

where A_j = cross-sectional area of the joint (mm^2), which must be less than the gross cross-sectional area of the column and twice the width of the web of the adjacent beam multiplied by the dimension of the column in the direction of the applied shear force on the joint

K_j = confinement factor equal to 2,4 if the joint is confined on four sides by framing beams or equal to 1,8 if the joint is confined on three or two sides by framing beams or equal to 1,5 in all other cases

The code specifies that the same transverse reinforcement installed in the column must be placed in the joint region unless it is confined on four sides by framing beams. In this case, only half of the transverse reinforcement in the column can be placed in the joint. Furthermore, the code imposes a limit on the longitudinal bar diameter d_b, which can go through the joint region.

$$d_b \leq \frac{l_j}{24} \tag{7.36}$$

where l_j is the joint dimension (mm) parallel to the longitudinal bar crossing the joint.

7.6.6 Example of Earthquake-Resistant Design of a Beam-Column Assembly

Building Description. Let us study a four-storey building with four spans located in Quebec City (fig. 7.21). This building is designed for a reduction factor $R = 4$, and we focus specifically on the seismic design of the first-floor beam-column assembly of an interior frame that resists lateral loads in the east-west direction. The five lateral spans of the frame each have a width of 6 m. The floor system is a unidirectional concrete slab, 110 mm thick, supported by floor joists. These joists are then supported by the main beams and by columns. To simplify the analysis, we neglect torsional effect on the building. To meet the inter-storey drift requirements of the NBCC (1995), the following dimensions are assigned for the members:

- interior and exterior columns: 400 × 400 mm;
- main beams: 400 × 500 mm;
- floor joists: 350 × 450 mm.

Material properties. In the calculations, a normal-density concrete is used with a compressive strength $f'_c = 30$ MPa; the yield strength of the longitudinal reinforcement is taken as $f_y = 400$ MPa.

Design loads. The design gravity and lateral loads are obtained according to the requirements of the NBCC (1995).
- live loads at the floors: 2,4 kN/m^2
- snow load on the roof: 3,14 kN/m^2
- dead loads: - partition walls: 1,0 kN/m^2
  - mechanical and electrical equipment: 0,5 kN/m^2
  - roof insulation: 0,5 kN/m^2
  - weight of frame and concrete slab: 24 kN/m^3
- wind loads: in this example, the wind loads were omitted.

- seismic loads:
  - $v = 0{,}15$ (Quebec City)
  - $S = 2{,}94$ (table 6.4 for $T = 0{,}4\ s$, $Z_a/Z_v > 1$)
  - $I = F = 1{,}0$
  - $W = 937$ kN/floor for levels 1 to 3
    and 1 053 kN for level 4 = 3 864 kN/bay
  - $R = 4$
  - $V = 256$ kN/slab $= 0{,}0663\ W$

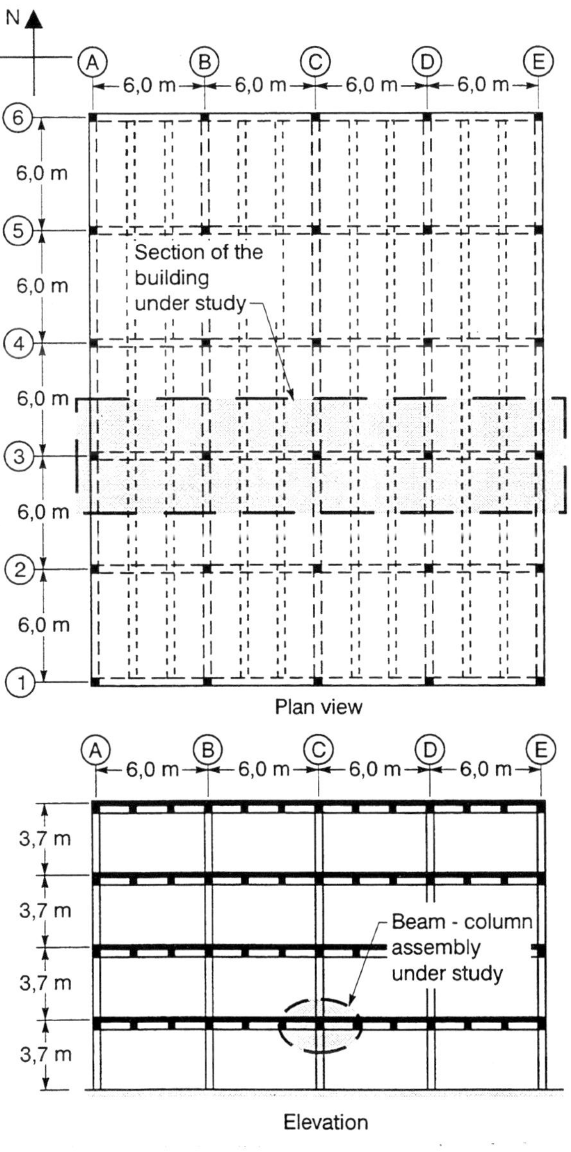

Figure 7.21 Elevation of a building designed for Quebec City.

Figure 7.22 shows the distribution of the unfactored design loads on the frame. The gravity loads are taken as concentrated loads where the floor joists are supported. This simplification is acceptable for this floor system. Tributary areas were used to calculate the concentrated loads and the distribution of the lateral seismic loads was obtained using equation 6.27.

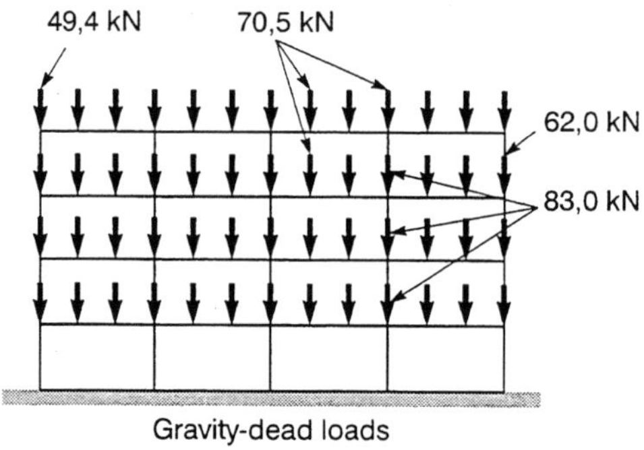

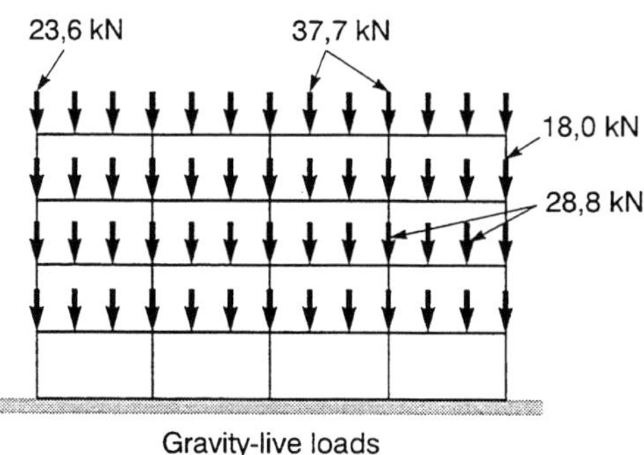

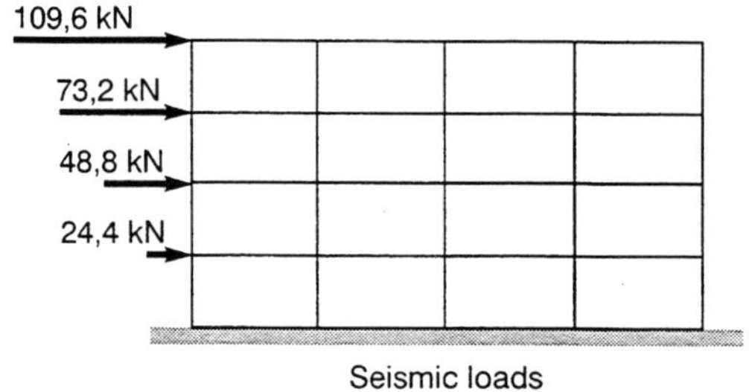

Figure 7.22   Unfactored design loads on the frame.

Internal forces. A linear-elastic static analysis program was used to obtain the internal forces in the members. The floor slab was not taken into account in the computation of geometric properties. In reality, the contribution of the floor slab must be included. Rigid links were used to compute internal forces but were omitted to compute lateral displacements. Cracking was considered by using 70% of the gross moment of inertia of the columns and 40% of the gross moment of inertia of the beams. The gross cross-sectional area of the columns was also taken into account.

To compute the design factored forces in the members, the limit states procedure was used (equation 6.29). Table 7.4 shows the factored bending moments in the beams of the interior first-floor beam-column assembly. Table 7.5 shows the bending moments and the axial forces in the columns of this assembly.

Table 7.4 Factored bending moments in the beams of the first-floor interior beam-column assembly.

Combination	Bending moments (kN-m)			
	Face of the B axis column	One third of the span between B and C axes	Two thirds of the span between B and C axes	Face of the C axis column
D	-87	+40	+41	-86
L	-36	+16	+17	-35
E →	+87	+31	-31	-87
(1) 1,25D + 1,5L	-163	+74	+77	-160
(2) 1,00D + 1,0E →	0	+71	+10	-173
(3) 1,00D + 1,0E ←	-174	+9	+71	+1
(4) 1,00D + 0,5L + 1,0E →	-18	+79	+19	-191
(5) 1,00D + 0,5L + 1,0E ←	-192	+17	+81	-17

Table 7.5 Factored bending moments and axial forces in the columns of the first-floor interior beam-column assembly.

Combination	Bending moments (kN-m)		Axial forces (kN)	
	Base of the C axis column	Two thirds of the height of C axis column	Base of the C axis column	Two thirds of the height of C axis column
D	0	0	-882	-659
L	0	0	-372	-285
E →	-136	+70	0	0
(1) 1,25D + 1,5L	0	0	-1 661	-1 251
(2) 1,00D + 1,0E →	-136	+70	-882	-659
(3) 1,00D + 1,0E ←	+136	-70	-882	-659
(4) 1,00D + 0,5L + 1,0E →	-136	+70	-1 068	-802
(5) 1,00D + 0,5L + 1,0E ←	+136	-70	-1 068	-802

To decrease the negative gravity bending moments in the beams, the Canadian concrete code (CSA, 1994) allows for redistribution of up to 20% of these moments. However, the moments caused by seismic loads cannot be redistributed. Table 7.5 shows the factored bending moments, after redistribution, in the beams of the first-floor interior beam-column assembly.

Table 7.6 Factored bending moments after redistribution in the beams of the first-floor interior beam-column assembly.

Combination	Bending moments (kN-m)			
	Face of the B axis column	One third of the span between B and C axes	Two thirds of the span between B and C axes	Face of the C axis column
D	-70	+57	+58	-69
L	-29	+23	+24	-28
E →	+87	+31	-31	-87
(1) 1,25D + 1,5L	-131	+106	+109	-128
(2) 1,00D + 1,0E →	+17	+88	+27	-156
(3) 1,00D + 1,0E ←	-157	+24	+89	+18
(4) 1,00D + 0,5L + 1,0E →	+3	+100	+39	-170
(5) 1,00D + 0,5L + 1,0E ←	-172	+38	+101	+4

Interstorey drifts. Table 7.7 shows the lateral displacements and interstorey drifts caused by lateral seismic loads. The *P-Delta* effects were taken into account in this analysis by using the unfactored dead loads at the floor levels. As seen in section 6.6.7, the interstorey drift must not exceed 2% of the storey height: 3 700 mm × 0,02 = 74 mm. Table 7.7 demonstrates that these displacements are within this limit.

Table 7.7 Lateral displacements and interstorey drifts caused by seismic lateral loads.

Storey	Lateral displacements (mm)	Lateral displacements $\times R$ (mm)	Interstorey drifts (mm)
4 (roof)	46,1	184,4	31,6
3	38,2	152,8	50,8
2	25,5	102,0	60,8
1	10,3	41,2	41,2

Preliminary design of beams and columns. According to equation 7.36, the maximum longitudinal bar diameter through the joint is calculated as follows:

$$d_b \leq \frac{l_j}{24} = \frac{400 \text{ mm}}{24} = 17,7 \text{ mm} \tag{7.37}$$

Longitudinal steel bars in the beam with a diameter of 15 mm (15M) meet this requirement.

According to table 7.6, the maximum negative factored moment at the face of the C axis column on the first floor is equal to -170 kN-m. The maximum positive factored moment where the first joist is supported is +101 kN-m. To sustain both maximum factored moments, it is estimated that eight longitudinal 15 M bars ($A_s = 1600$ mm^2) should be used at the top of the beam and four bars of 15 M should be used at the bottom of the beam ($A'_s = 800$ mm^2). It is also estimated that the effective depth reaches $d = 400$ mm when using a concrete cover of 40 mm. Table 7.8 shows the different flexural strengths of the chosen beam section.

Table 7.8 Flexural strengths (kN-m) of the chosen beam section.

M_f^+	M_f^-	M_r^+	M_r^-	M_n^+	M_n^-	M_p^+	M_p^-
+101	-170	+120	-196	+144	-235	+176	-288

For preliminary design of the columns, equation 7.36 determines once again the maximum longitudinal bar diameter through the joint.

$$d_b \leq \frac{l_j}{24} = \frac{500 \text{ mm}}{24} = 20,8 \text{ mm} \tag{7.38}$$

A diameter of 20 mm (20M) is then chosen for the longitudinal bars of the column.

In preliminary design of the longitudinal reinforcement, the different load combinations shown in table 7.6 must be considered. Three different arrangements of longitudinal reinforcement are possible.

1. 16 15M bars: $\rho_s = 0,02$;
2. 12 15M bars + 4 20M bars: $\rho_s = 0,0225$;
3. 16 20M bars: $\rho_s = 0,03$.

The factored interaction curves of the column for these arrangements are shown in figure 7.23. The effect of each factored combination of table 7.5 is also shown. We can see that combinations (4) and (5) are the most critical.

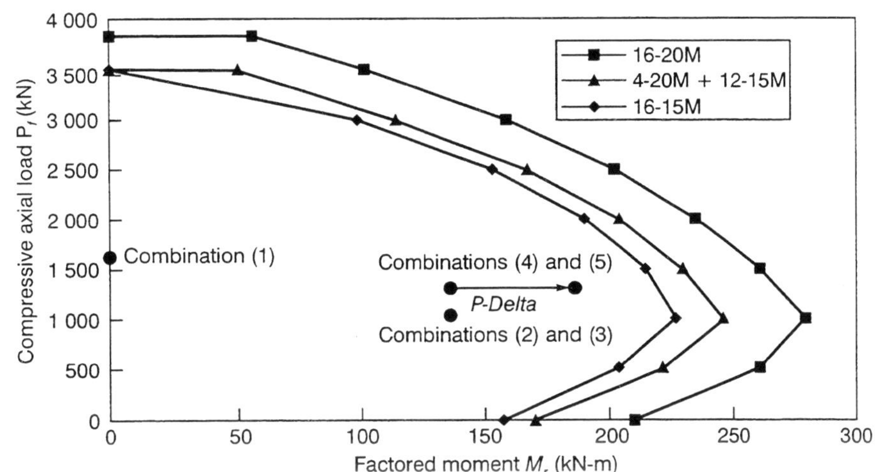

Figure 7.23   Factored strength of a 400 × 400 mm column.

The increase in the bending moment for combinations (4) and (5) caused by the *P-Delta* effects can be estimated by multiplying the displacement of the first floor (41,2 mm in table 7.2) by the axial load of 1 068 kN to yield an increase in the bending moment of 44 kN-m. As we can see, all reinforcing arrangements show adequate strength. Arrangement (3) is chosen to ensure that the columns will be strong enough to comply with the "weak beams-strong columns" criterion. Figure 7.24 shows the preliminary design of the longitudinal reinforcement in the beams and columns of the assembly.

Seismic design of beams. First, let us verify that the geometric properties are respected. The free span of each beam, 5,6 m, corresponds to 14 times its depth (400 mm) and therefore exceeds the minimal limit of 4. The width-to-depth ratio of the section is 400/500 = 0,8, greater than the limit of 0,3. Finally, the width of the section is 400 mm, which is greater than the limit of 250 mm.

Now, the restrictions concerning the minimum (equation 7.23) and maximum (equation 7.24) reinforcement ratio must be verified for the top and bottom steel.

Minimum steel = $1,4\ b_w\ d/f_y = 1,4 \times 400 \times 400/400 = 560$ mm^2 < 800 mm^2 O.K.
Maximum steel = $0,025\ b_w\ d = 0,025 \times 400 \times 400 = 4\ 000$ mm^2 > 1 600 mm^2 O.K.

Then, we must make sure that the positive bending strength at the face of the joint is at least 50% of the negative bending strength at the same face. According to the results shown in table 7.8, this requirement is respected. To simplify this example, we are not taking into account lap splice requirements.

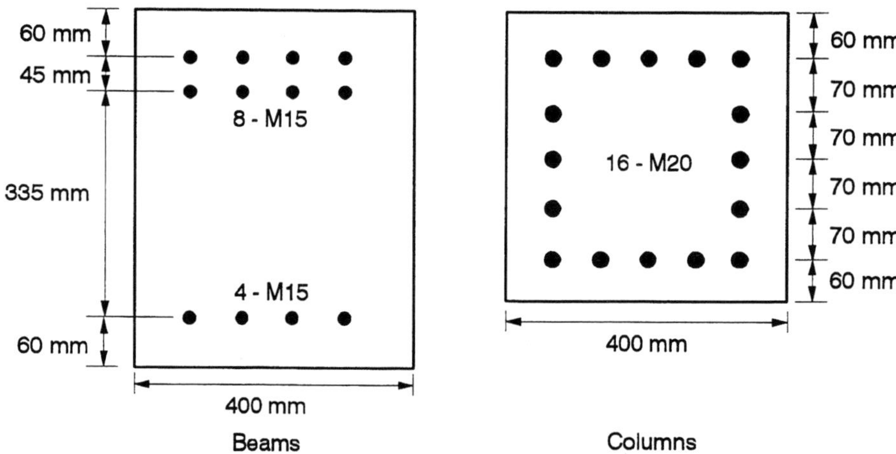

Figure 7.24 Preliminary design of the beams and columns of the assembly.

Now, we calculate the transversal reinforcement that must sustain the shear forces caused by the formation of positive and negative probable moments at the ends of each beam when factored gravity loads are applied on the beam. The span between the B and C axes is used in the computation, as shown in figure 7.26. The 85 kN load corresponds to loading combinations (4) and (5). It is important to note that the free-body diagram and the shear force diagram of figure 7.25 can be inverted according to the direction of the earthquake. It is therefore necessary to apply the peak forces symmetrically. The factored shear force, V_f, is calculated according to the free-body diagram shown in figure 7.25.

$$V_f = \frac{288\,\text{kN}-\text{m} + 176\,\text{kN}-\text{m}}{5,6\,\text{m}} + 85\,\text{kN} = 168\,\text{kN} \tag{7.39}$$

To sustain the factored shear force, only the contribution of the transverse steel reinforcement, V_s, is used.

$$\frac{A_h}{s_h} \geq \frac{168\,\text{kN}}{\phi_s\,f_{yh}\,\text{d}} = \frac{168\,\text{kN}}{0,8 \times 0,4\,\dfrac{\text{kN}}{\text{m}^2} \times 400\,\text{mm}} = 1,24\,\text{mm} \tag{7.40}$$

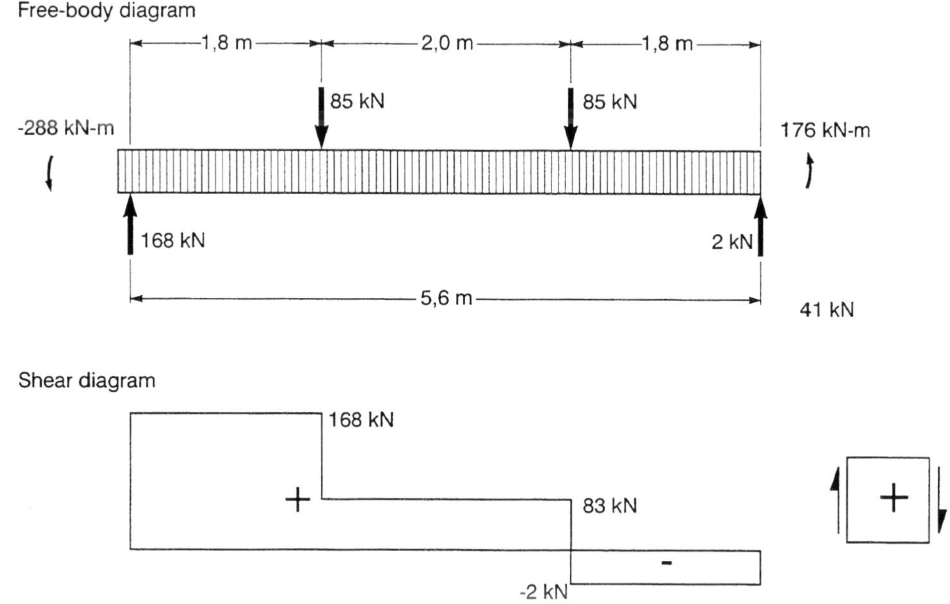

Figure 7.25 Design shear for given beams.

If double hoops of 10M bars are used as transverse reinforcement, it yields $A_h = 400$ mm², which will require a spacing of $s_h = 392$ mm. The seismic confinement requirements must also be respected over a distance $2d$ of the face of the column.

i) $d/4 = 400$ mm$/4 = 100$ mm
ii) $8d_b = 8 \times 15$ mm $= 120$ mm
iii) $24d_h = 24 \times 10$ mm $= 240$ mm
iv)   300 mm

The first criterion governs the spacing of the hoops in the plastic hinge regions ($2d$ from the face of the column). The first hoop must be installed 50 mm from the face of the column. Outside of these regions, the spacing cannot exceed $d/2$ or 200 mm. In the central region of the beam, single hoops spaced at 200 mm are installed. Figure 7.26 illustrates the transverse reinforcement distribution in the first-floor beam between axes B and C. Figure 7.27 shows the longitudinal and transverse steel configuration of the beam sections for the regions with single and double hoops.

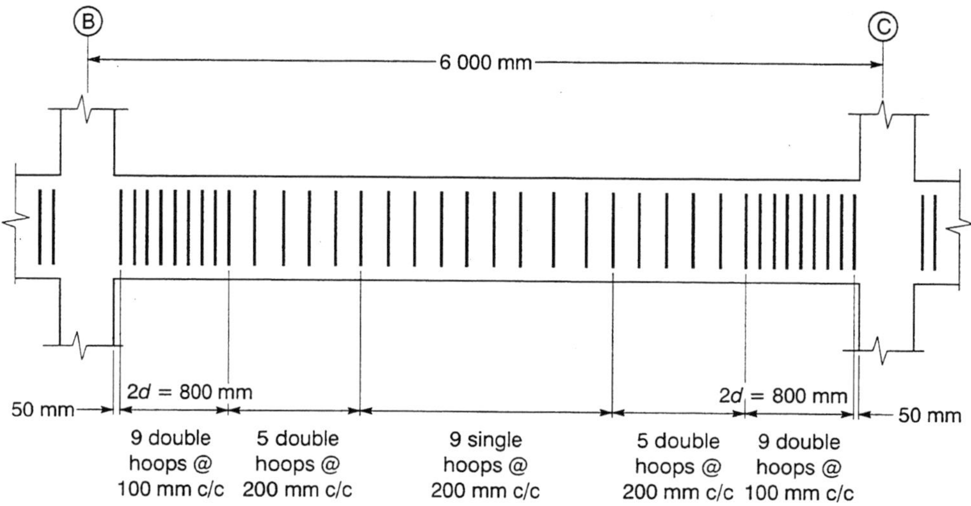

Figure 7.26   Hoop distribution in the first-floor beam between axes B and C.

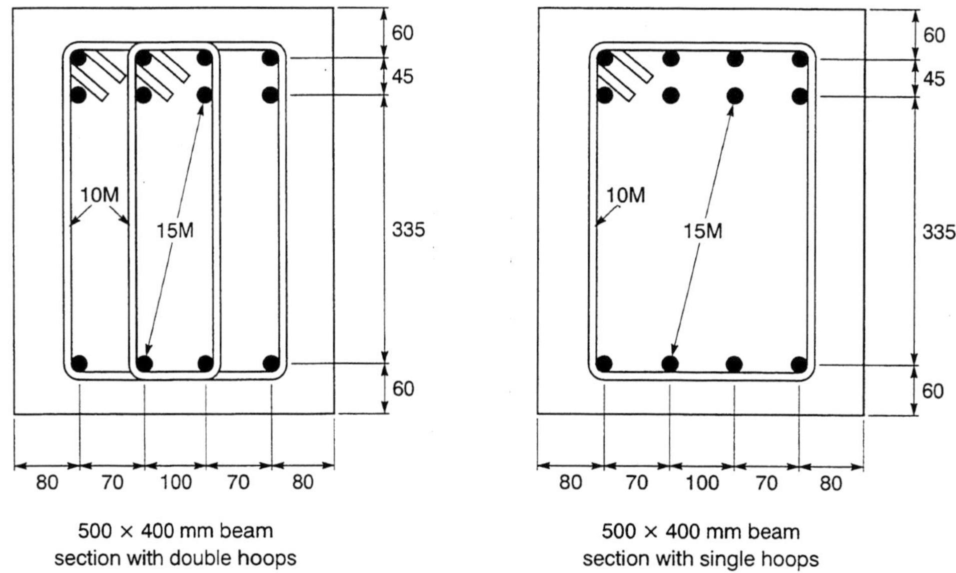

Figure 7.27   First-floor beam sections.

Seismic design of columns. The maximum compressive axial force that meets the criterion of the Canadian concrete code for columns is $A_g f'_c /10 = 400$ mm $\times$ 400 mm $\times$ 0,03 kN/mm^2 /10 = 480 kN. Table 7.5 shows that all load combinations produce a factored compressive axial load that exceeds this limit.

The seismic design of columns is strongly influenced by the "weak beams-strong columns" concept. Equation 7.27 represents this concept. According to this equation and table 7.8, the contribution of the nominal flexural strength of the beams adjacent to the column on axis C can immediately be calculated.

$$1{,}1 \sum M_{nb} = 1{,}1 \times 235\,\text{kN-m} + 1{,}1 \times 144\,\text{kN-m} = 417\,\text{kN-m} \tag{7.41}$$

To calculate the sum of the factored flexural strength of the columns in equation 7.27, an axial load in the columns is assumed and the interaction diagram of figure 7.24 is used. To be conservative, the minimum factored flexural strength of the columns corresponding to a zero axial load can be used.

$$\sum M_{rc} = 2 \times 210\,\text{kN-m} = 420\,\text{kN-m} > 417\,\text{kN-m} \tag{7.42}$$

The factored bending strength of the columns is therefore sufficient to meet the Canadian code requirements to ensure the formation of plastic hinges in the beams.

The minimum and maximum longitudinal reinforcement criteria for columns must also be verified according to equation 7.28.

$$0{,}01 < \frac{A_s}{A_g} = \frac{16 \times 300\,\text{mm}^2}{400\,\text{mm} \times 400\,\text{mm}} = 0{,}03 < 0{,}06 \tag{7.43}$$

The first-floor column must now sustain the shear force produced by the formation of plastic hinges occurring in the beams. The moment on top of the first-floor column can be calculated using equation 7.29.

$$M_{\text{top of column}} = (288\,\text{kN-m} + 176\,\text{kN-m})\frac{\dfrac{1}{3{,}45\,\text{m}}}{\dfrac{1}{3{,}45\,\text{m}} + \dfrac{1}{3{,}2\,\text{m}}} = 225\,\text{kN-m} \tag{7.44}$$

For the moment at the bottom of the first floor column, it is considered that the probable moment can be reached. To be conservative, an axial load of 1000 kN corresponding to the balanced condition shown in figure 7.23 is used.

$$M_{\text{bottom of column}} = 1{,}57 \times 280\,\text{kN-m} = 440\,\text{kN-m} \tag{7.45}$$

Figure 7.28 illustrates the design shear force, V_f, in the first-floor column caused by a plastic hinge mechanism in the adjacent beams.

The design shear force is obtained by the moment equilibrium in the first-floor column.

$$V_f = \frac{225\,\text{kN-m} + 440\,\text{kN-m}}{3{,}4\,\text{m}} = 196\,\text{kN} \tag{7.46}$$

The concrete contribution, V_c, and the transverse steel contribution, V_s, to resist the shear force are calculated by equation 7.30. The concrete contribution is calculated as follows:

$$V_c = 0.5 \times 0.2 \times 1.0 \times 0.6 \times \sqrt{30} \times 400 \times 340 = 45\,\text{kN} \qquad (7.47)$$

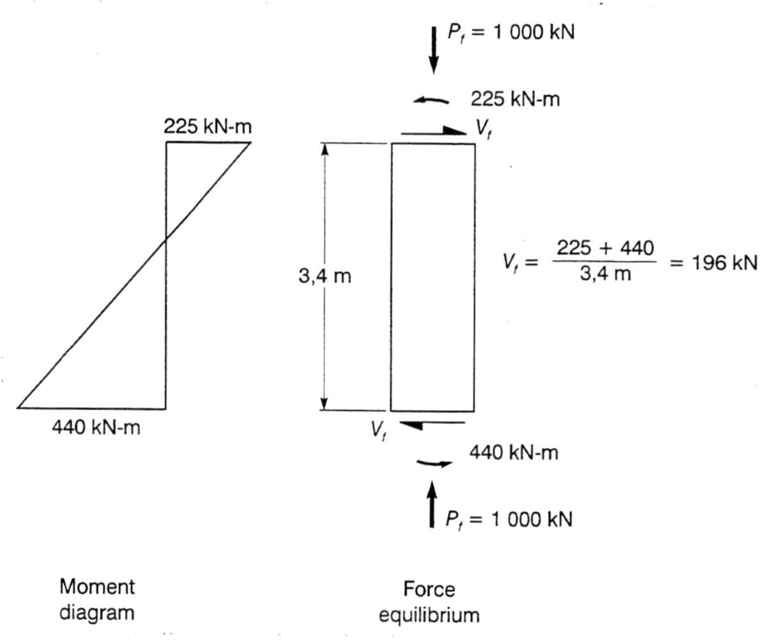

Figure 7.28 Shear forces in the first-floor column on axis C caused by formation of plastic hinges in the adjacent beams.

The transverse steel must then sustain 196 kN - 45 kN = 151 kN.

$$\frac{A_h}{s_h} \geq \frac{151\,\text{kN}}{\phi_s f_{yh} d} = \frac{151\,\text{kN}}{0.85 \times 0.4\,\dfrac{\text{kN}}{\text{m}^2} \times 340\,\text{mm}} = 1.31\,\text{mm} \qquad (7.48)$$

If 10M bars are used as rectangular hoops with 45° frames ($A_h = 341$ mm^2), the required spacing is $s_h = 261$ mm. Additional requirements must be observed, however, to ensure the confinement of the column in regions where plastic hinges may potentially develop. The spacing cannot exceed the following limits:

1. $h/4 = 400$ mm $/ 4 = 100$ mm
2. $6\,d_b = 6 \times 20$ mm $= 120$ mm
3. 100 mm

For the first-floor column, the maximum spacing cannot exceed 100 mm. The criteria given by equation 7.32 must also be verified.

$$\frac{A_{sh}}{s_h} \geq 0{,}3 \frac{h_c f_c}{f_{yh}} \left(\frac{A_g}{A_{ch}} - 1 \right)$$

$$\frac{A_{sh}}{s_h} \geq 0{,}09 \left(\frac{h_c f_c}{f_{yh}} \right)$$

$$\frac{A_{sh}}{s_h} \geq 0{,}3 \frac{320 \times 30}{400} \left(\frac{400 \times 400}{320 \times 320} - 1 \right) = 4{,}05 \text{ mm} \tag{7.49}$$

$$\frac{A_{sh}}{s_h} \geq 0{,}09 \left(\frac{320 \times 30}{400} \right) = 2{,}16 \text{ mm}$$

Using the same transverse reinforcement configuration ($A_{sh} = 341$ mm^2), the maximum spacing allowed is $s_h = 85$ mm. The first-floor column must be confined over its entire height. For the column located above the joint, these requirements are applied over a distance l_o at least equal to:

1. the depth of the column at the face of the joint $= 400$ mm
2. 1/6 of the column's free height $= 3{,}2$ m $/ 6 = 540$ mm
3. 450 mm

Outside of the confined regions, the transverse reinforcement must sustain the factored shear forces. In these regions, simple square hoops are installed and are spaced at 150 mm. Figure 7.29 illustrates the distribution of the hoops in the second-floor column and figure 7.30 shows the details of the steel reinforcement in the column.

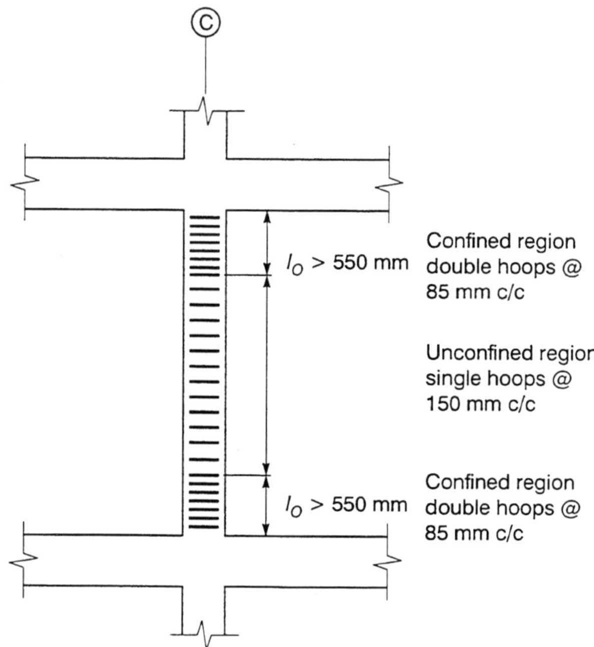

Figure 7.29 Hoop distribution in the second-floor column.

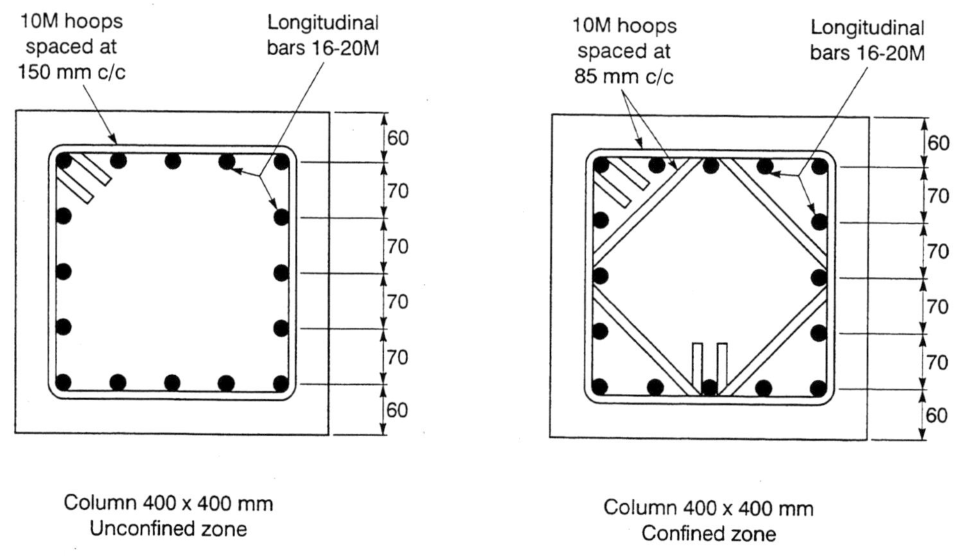

Figure 7.30 Column cross-sections.

Seismic design of the beam-column joint. The design shear force in the joint is obtained from equation 7.33. For the shear in the column, V_{col}, we assume that the design of the second-floor columns is identical to the design of the first floor (fig. 7.31).

$$V_{fj} = 1{,}25 \times 1600 \times 400 + 1{,}25 \times 800 \times 400 - 147{,}2 = 1\,053\,\text{kN} \tag{7.50}$$

We can assume that the joint of an interior frame is confined by framing beams on all four sides. The shear strength of the joint is found using equation 7.35.

$$V_{rj} = 2{,}4 \times 1{,}0 \times 0{,}6 \times \sqrt{30} \times 400 \times 400 = 1\,262\,\text{kN} > 1\,053\,\text{kN} \tag{7.51}$$

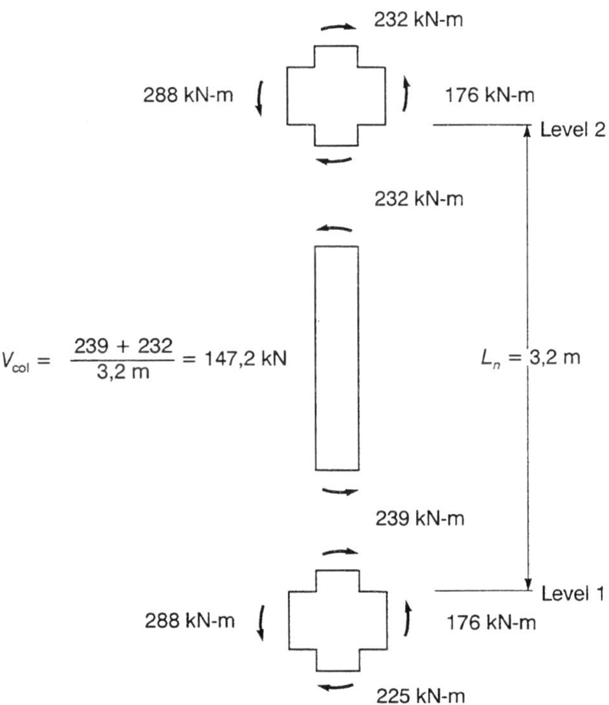

Figure 7.31   Shear in the second-floor column used to calculate the design shear force in the joint.

Even though confinement of all four faces of the joint would allow installation in the joint of only half of the transverse reinforcement specified for the column, the same specified lateral reinforcement for the column is installed in the joint to be conservative. The installation thus comprises five sets of double hoops between the longitudinal bars of the beams and a set of hoops over and under the bars, as shown in figure 7.32.

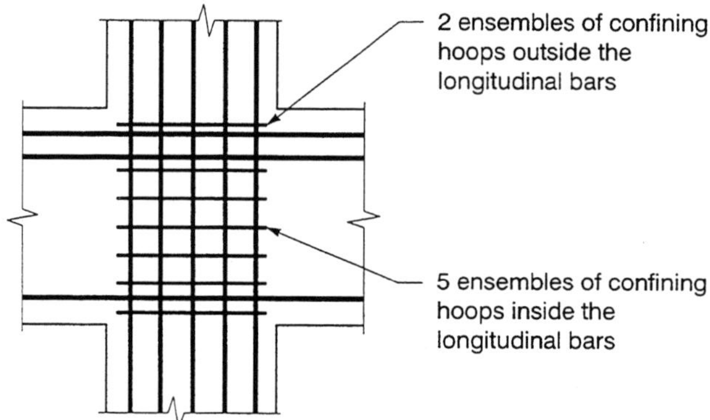

Figure 7.32 Transverse reinforcement in the joint.

Figure 7.33 illustrates the final seismic detailing configuration of the beam-column assembly, between the mid-height of the columns and the mid-span of the beams.

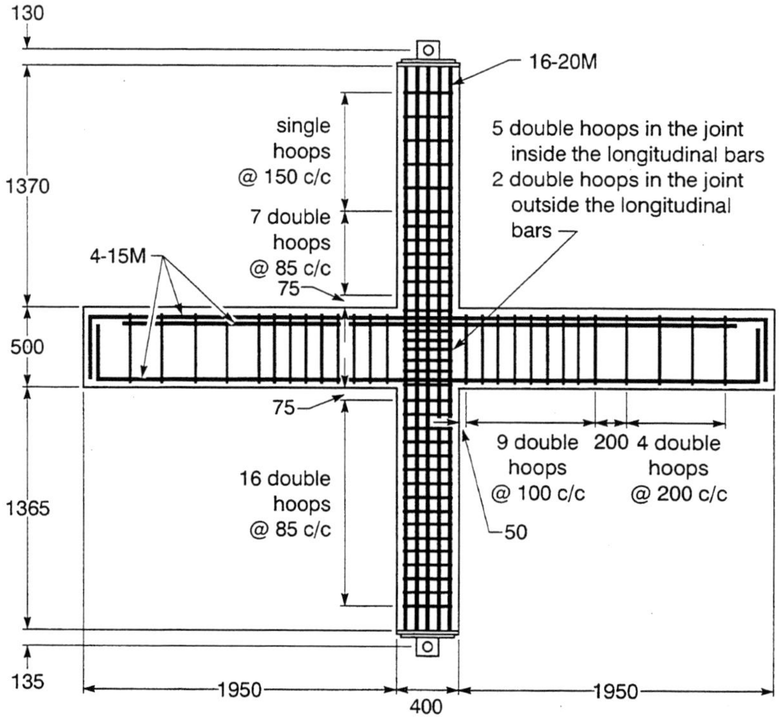

Figure 7.33 Seismic detailing of the beam-column assembly.

7.6.7 Real Cyclic Behaviour of the Designed Assembly

The assembly designed in the previous section was the subject of quasi-static testing in the structures laboratory at École Polytechnique in Montreal (Pineau, 1994). Figure 7.34 shows the experimental set-up used for these tests. The loading system was made of four hydraulic actuators, which operated in pairs in order to test the specimen in reverse cyclic loading. A constant axial load was also applied at the top of the column. Hinges, installed at both ends of the column, simulated the inflexion points at mid-height of each column.

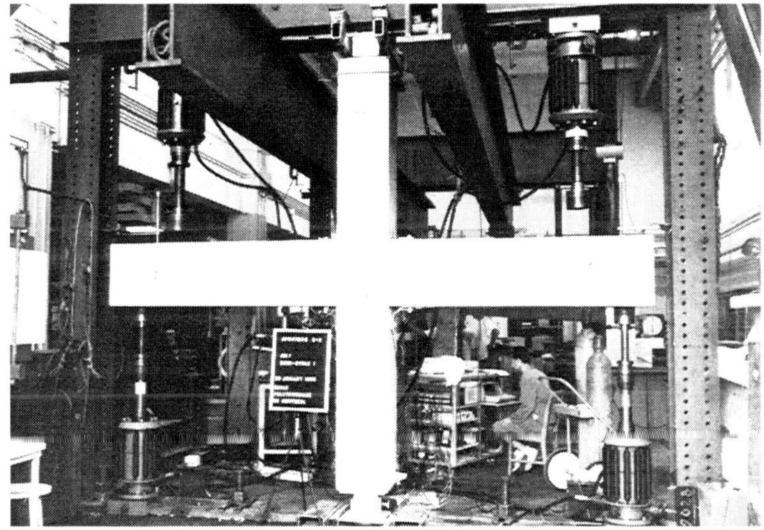

Figure 7.34 Experimental set-up for quasi-static tests of the beam-column assembly.

Figure 7.35 shows the shear-interstorey drift hysteresis loops obtained from these tests. A photograph of the assembly at the end of the test is also shown. The hysteresis loops are very stable. Each loop is large, which indicates good energy dissipation. As well, the lateral shear corresponding to the probable moments of the beams, V_{pr+} and V_{pr-}, is reached and maintained throughout the duration of the test. The plastic hinge mechanism in the beams is complete and the joint sustains the shear force within the elastic range of the material.

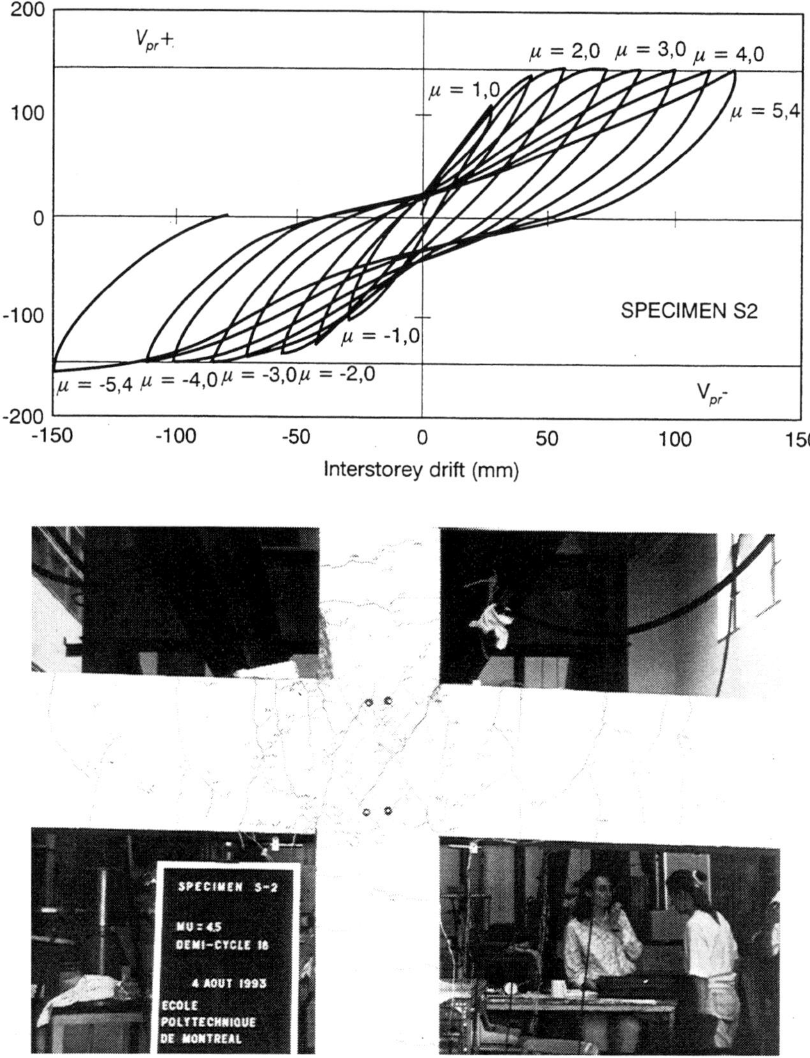

Figure 7.35 Experimental results from quasi-static tests of the assembly.

7.7   PROBLEMS

7.1   Consider a singly reinforced concrete beam section, as shown in figure 7.36. The compressive strength of concrete is $f'_c = 25$ MPa and the yield strength of the steel reinforcement is $f_y = 400$ MPa.

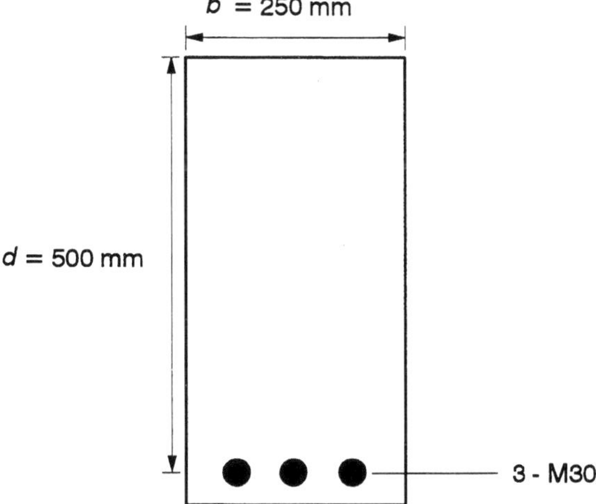

Figure 7.36   Singly reinforced beam section.

a)   Estimate the available curvature ductility for this section.
b)   Find the increase in the curvature ductility when the section is confined (figure 7.37) by rectangular hoops made of 10M bars (f_{yn} = 400 MPa) spaced at 75 mm. Let us assume that b/l_c = 1/8. Neglect the temperature top steel.

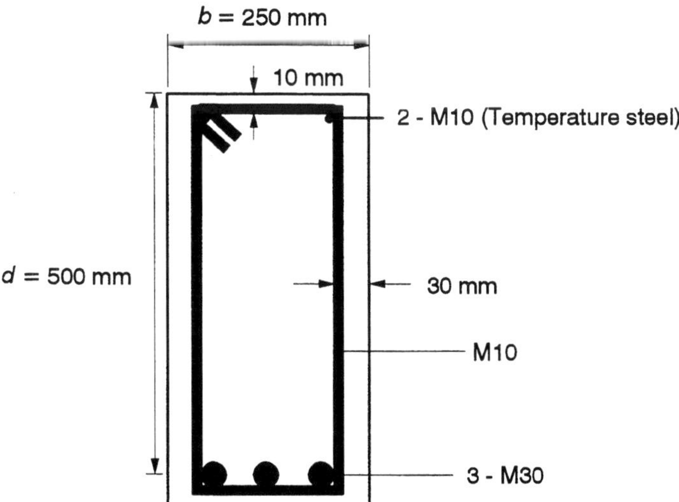

Figure 7.37   Confined beam section.

7.2 Figure 7.38 shows plan and elevation views of a portion of a small office building. The east-west lateral resisting system is made of reinforced concrete ductile frames. The floor is made of a reinforced concrete slab (reinforced in one direction only) and of joists supported by the main beams.

The seismic design of the structure must be made according to the NBCC (1995) and must meet the CSA A23.3-94 code with a load reduction factor of $R = 4$. The dead load on the floor is equal to 5,5 kN/m^2 and the live load is 2,4 kN/m^2.

A preliminary design of the structure, based on the factored loads, has allowed us to find the dimensions of beams B_1 and B_2 and of columns C_1, C_2 ,and C_3:

Beams B_1 and B_2:   350 mm × 450 mm;
Columns C_1, C_2 and C_3:   350 mm × 350 mm.

The flexural strengths at each end of these elements, also based on factored loads, are shown in table 7.9.

Table 7.9 Flexural strengths of beams B_1 and B_2 and columns C_1, C_2 and C_3.

Element	Factored moment (kN-m)		Nominal moment (kN-m)		Probable moment (kN-m)	
	Positive	Negative	Positive	Negative	Positive	Negative
Beams B_1, B_2	110	-180	130	-215	160	-265
Columns C_1, C_2, C_3*	135	-135	160	-160	210	-210

* The flexural strengths of the columns are the same for all floors and are calculated for the factored axial load.

a) Find the minimal shear strength for beam B_2 in order to ensure its ductile behaviour in flexure.

b) Check that column C_2 is adequate to ensure that plastic hinges first occur in beams B_1 and B_2.

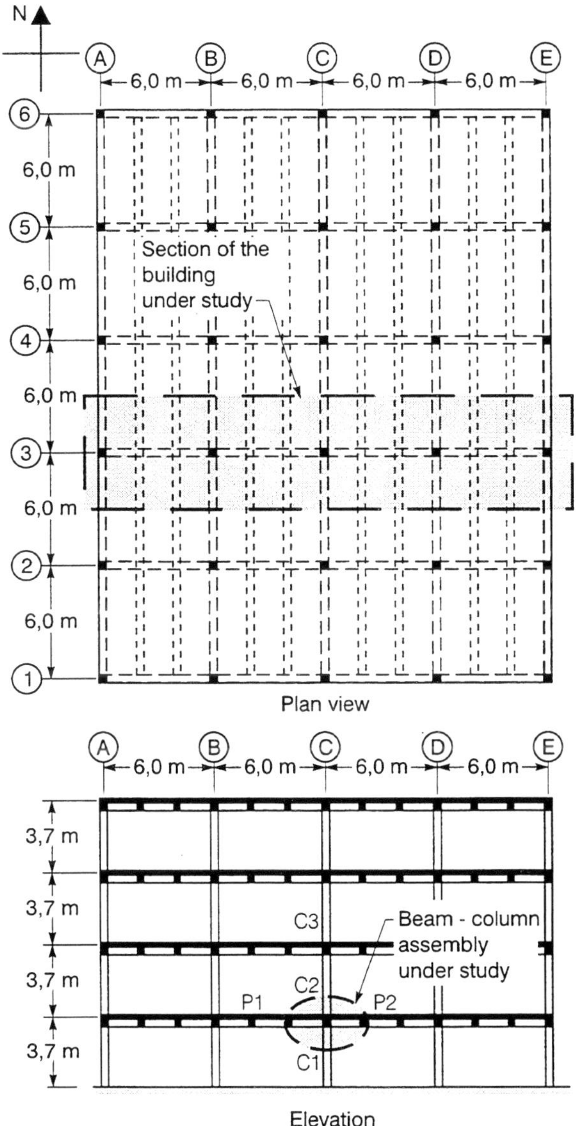

Figure 7.38 Portion of a small office building.

7.8   REFERENCES

Base, G.D., and Read, J.B. (1965). "Effectiveness of Helical Binding in the Compression Zone of Concrete Beams." *ACI Journal,* 62(7): 763–81.

Bertero, V.V., and Felippa, C. (1964). "Discussion of a paper by Roy, H., and Sozen, M., Ductility of Concrete," *Proceedings of International Symposium on Flexural Mechanics for Reinforced Concrete*, ASCE-ACI, Miami, Vol. 4, pp. 213–35.

Blume, J.A., Newmark, N.M., and Corning, L.H. (1961). *Design of Multi-Storey Reinforced Concrete Buildings for Earthquake Motions*. Chicago: Portland Cement Association.

Corley, W.G. (1966). "Rotational Capacity of Reinforced Concrete Beams." ASCE, *Journal of the Structural Division,* 92(ST5): 121–46.

CSA (1994). *Design of Concrete Structures*. CAN A23.3-M94. Rexdale, ON: Canadian Standards Association.

Kent, D.C., and Park, R. (1971). "Flexural Members with Confined Concrete." ASCE, *Journal of the Structural Division,* 91(ST7): 1969–90.

National Building Code of Canada (1995). 18th ed. Ottawa: National Research Council of Canada.

Nawy, E.G., Danesi, R.F., and Grosko, J.J. (1968). "Rectangular Spiral Binders Effect on Plastic Rotation Capacity in Reinforced Concrete Beams." *ACI Journal,* 65(12): 1001–16.

Park, R., Priestly, N.J.M., and Gill, W.D. (1982). "Ductility of Square Confined Concrete Columns." ASCE, *Journal of the Structural Division* 108(ST4): 929–50.

Pineau, S. (1994). *Étude expérimentale du comportement sismique d'assemblages poutre-colonne en béton renforcé de fibres*. Master's thesis, Department of Civil Engineering, École Polytechnique de Montréal, Montreal.

Tanaka, H., Park, R., and McNamee, B. (1985). "Anchorage of Transverse Reinforcement in Rectangular Reinforced Concrete Columns in Seismic Design." *Bulletin of the New Zealand Society for Earthquake Engineering,* 18(2): 165–90.

CHAPTER 8

Elements of Seismic Detailing for Steel Structures in Canada

8.1 INTRODUCTION

For many years, steel-framed structures were considered to have the best overall behaviour during earthquakes, mainly because of the available ductility of the steel. However, the collapse of the Pino-Suarez complex during the Mexico City earthquake in 1985 (Hansen and Martin, 1987) and the numerous weld fractures that occurred during the 1994 Northridge earthquake in California (Thiel, 1994) have demonstrated that steel structures can also be vulnerable during major earthquakes. In Canada, until the 1989 edition of the Canadian steel code was published, no specific requirements governed structural detailing to ensure adequate ductility of steel structures. The most recent edition of the Canadian steel code (CSA, 1994) incorporates specific requirements for seismic detailing of steel structures relevant to the load-reduction factor, R, used in design. These requirements are concerned with the seismic design of three types of lateral-load-resisting systems:

- moment-resisting frames;
- concentric braced frames;
- eccentric braced frames.

8.2 CANADIAN STEEL CODE DETAILING REQUIREMENTS FOR DUCTILE MOMENT-RESISTING FRAMES ($R = 4$)

8.2.1 General Requirements

In active seismic zones such as California, ductile moment-resisting frames are the most often used. Their ability to sustain lateral seismic forces relies entirely on the behaviour of the beam-column connections. If the system incorporates appropriate structural detailing, high ductility may be achieved. The Canadian steel code requires definition of the critical element of each beam-column connection, as shown in figure 8.1. Although the code does not specifically list an order of preference for the critical elements, it should be as follows:

1. beams;
2. panel zone;
3. columns.

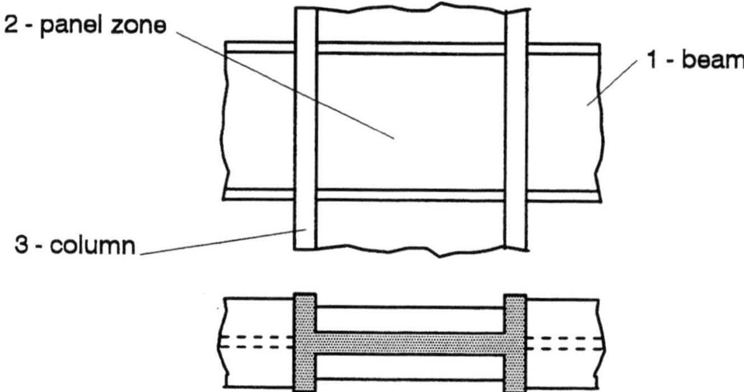

Figure 8.1 Possible critical elements of a beam-column connection.

The first step in seismic design of a steel frame is to identify the critical element of each joint. Thus, each element is considered separately as if it had completely yielded. As a result, loads corresponding to the nominal strength of the element $(\phi = 1)$ are applied to the joint and a yield stress multiplied by a factor of 1,20 is used to take into account the strain hardening of the steel. Then, the forces produced in the other two elements are compared with their factored resistance. Figure 8.2 shows an example of a beam-column connection for which the beams are critical.

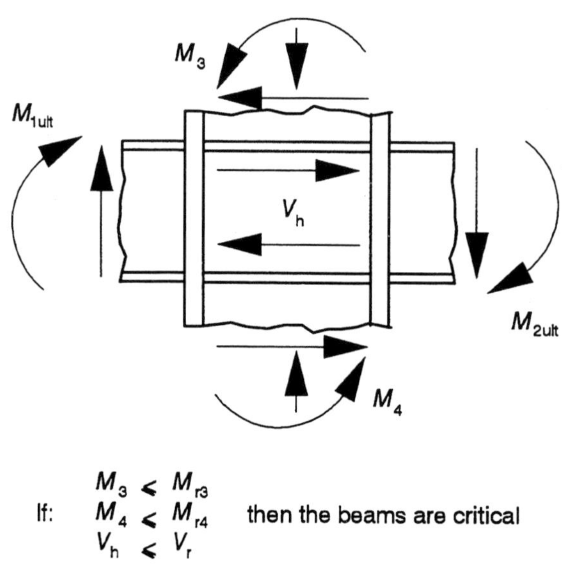

$$\text{If:}\quad \begin{array}{l} M_3 \leqslant M_{r3} \\ M_4 \leqslant M_{r4} \\ V_h \leqslant V_r \end{array} \quad \text{then the beams are critical}$$

Note: $M_{ult} = 1{,}2\,M_p$

Figure 8.2 Critical beams in a beam-column connection.

8.2.2 Requirements for Beams

The flexural strength of a beam must take the influence of the concrete slab into account in the case of a composite floor system. In this situation, the positive plastic moment must be based on a compressive stress for concrete of $1,3f'_c$ to account for the fact that the concrete slab may be bearing against the flange of the column, causing a confinement of the concrete.

If the beam is critical in the connection, the flanges and web must meet the requirements of a class 1 section (plastic section). Furthermore, the beam must be retained laterally. If the beam is not critical in the connection, the flanges must meet the requirements of a class 1 section (plastic section), and the web must meet the requirements of a class 2 section (compact section).

8.2.3 Requirements for Columns

If the column is critical in a connection, it must meet the requirements of a class 1 section (plastic section). The column must also be retained laterally. The plastic moment of the column, M_{pc}, must be reduced because of the presence of the axial load.

$$M_{pc} = 1,18 M_p \left(1 - \frac{C_s}{C_y} \right) \leq M_p \tag{8.1}$$

where C_s = representative gravity axial load
C_y = yield compressive load
M_p = plastic moment (unfactored) of the column for a zero axial load

As for reinforced concrete columns (section 7.4), the presence of a compressive axial load greatly reduces the available curvature ductility of a steel column. If the column is critical in a connection, and if the velocity zone is $Z_v \geq 4$, seismic requirements dictate a limit on the factored axial load, C_f, originating from any loading combination.

$$C_f \leq 0,30 \, \phi A F_y \tag{8.2}$$

The welded splices of the columns must be located at a minimum of 1/4 of the free height between the beams and at least at one metre from the beam-column joint.

If the column is not critical in a connection, it must meet the requirements of a class 2 section (compact section).

8.2.4 Requirements for Panel Zones

As it is subjected to lateral and gravity loads, the column web (panel zone) must sustain very high shear force. During a major earthquake, the negative gravity bending moment at one end of the beam increases, while the moment at the other end decreases. The bending moments of the two beams at the face of the joint will thus be in the same direction and cause a high shear force in the column web. Krawinkler and Popov (1982) demonstrated that the column web can dissipate cyclic shear energy as long as local buckling can be avoided.

In fact, laboratory tests have shown that the shear yield stress of the web is exceeded because of the strain hardening of the steel and the contribution of the column flanges. For this reason, the Canadian code requires the use of the following formula when calculating the factored shear strength, V_r, of the panel zone.

$$V_r = 0.55 \, \phi d_c w_c F_{yc} \left[1 + \frac{3 b_c t_c^2}{d_c d_b w_c} \right] \tag{8.3}$$

where
$$\phi = 0.9$$
$$d_c = \text{depth of the column}$$
$$w_c = \text{thickness of the column web}$$
$$F_{yc} = \text{yield stress of the column steel}$$
$$b_c = \text{width of the column flanges}$$
$$t_c = \text{thickness of the column flanges}$$
$$d_b = \text{depth of the beam}$$

The second term in the brackets represents the flexural contribution of the column flanges (5–20% of the strength). To avoid local buckling of the panel zone in the zones where Z_a or $Z_v \geq 2$, the Canadian code limits its slenderness with the following equation:

$$\frac{T_p + H_p}{w_c} \leq 90 \tag{8.4}$$

where
$$T_p = \text{width of the panel zone}$$
$$H_p = \text{height of the panel zone}$$
$$w_c = \text{thickness of the panel zone}$$

The code recommends that doubler plates with plug welds (fig. 8.3) be used to ensure that equation 8.3 is satisfied. If the panel zone is critical, the code requires that horizontal stiffeners be added at the levels of the column flanges.

8.2.5 Requirements for Beam-Column Joints

The connection between the beam flanges and the column must be able to sustain the probable plastic moment of the beam or the moment in the beam caused by the probable shear strength of the panel zone. Although the Canadian code does not disallow use of bolted connections, use of welded connections with full penetration welds is recommended.

A connection can be bolted between the beam web and the column flange. Figure 8.4 illustrates the typical connection recommended by the Canadian code. It is important to note that following the damage sustained during the 1994 Northridge earthquake in the Los Angeles region, this type of welded connection is currently being re-evaluated in the United States (Thiel, 1994).

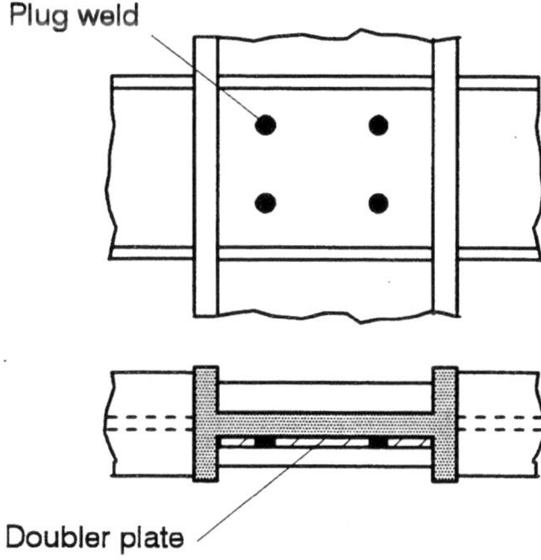

Figure 8.3 Doubler plate for the panel zone.

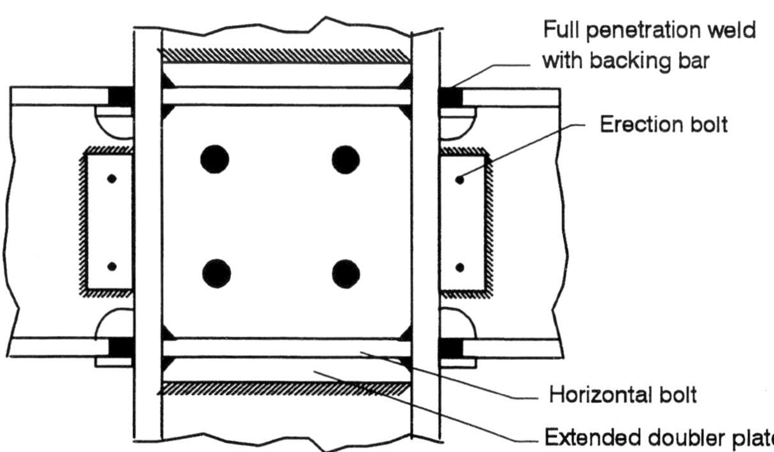

Figure 8.4 Beam-column connection recommended by the Canadian code.

8.2.6 Example of Earthquake Resistant-Design of a Beam-Column Connection

Table 8.1 shows the different load combinations acting on the interior beam-column connection shown in figure 8.5. We assume a positive sign convention according to the figure. The connection is part of a ductile moment-resisting frame. We assume that the bay span adjacent to the connection is 7,4 m and that the storey height above and below the connection is 3,7 m. The steel grade is $F_y = 300$ MPa and the gravity loads on the adjacent beams are $w_D = 18$ kN-m for the dead load and $w_L = 22$ kN-m for the live load. In this example, the wind load is neglected. In this connection, we wish to design the beams, columns, and panel zone in order for the beams to be the critical elements.

Table 8.1 Load combinations for the interior beam-column connection shown in figure 8.5.

Combination	V_1 kN	V_2 kN	V_3 kN	V_4 kN	M_1 kN-m	M_2 kN-m	M_3 kN-m	M_4 kN-m	C_3 kN	C_4 kN
D	-67	+67	0	0	-88	+88	0	0	+1 112	+1 246
L	-81	+81	0	0	-108	+108	0	0	+1 334	+1 496
E→	+40	+40	+74	+88	+163	+136	+136	+163	+445	+445
(1) 1,25D + 1,50L	-205	+205	0	0	-272	+272	0	0	+3 391	+3 801
(2) 1,00D + 1,0E→	-27	+107	+74	+88	+75	+224	+136	+163	+1 557	+1 691
(3) 1,00D + 1,0E←	-107	+27	-74	-88	-251	-48	-136	-163	+667	+801
(4) 1,00D + 0,5L + 1,0E→	-68	+147	+74	+88	+21	+278	+136	+163	+2 224	+2 438
(5) 1,00D + 0,5L + 1,0E←	-148	+68	-74	-88	-305	+6	-136	-163	+1 334	+1 548

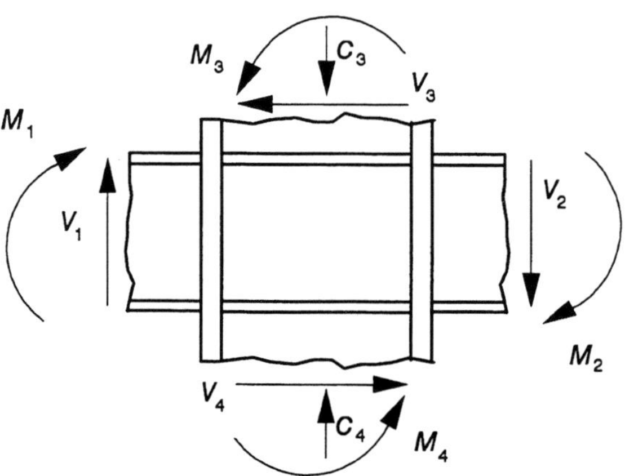

Figure 8.5 Interior beam-column connection that is part of a ductile moment-resisting frame.

The critical factored moment in the beams at the face of the column originates from combination (5) and is equal to $M_f = 305$ kN-m. To sustain this factored moment, a class 1 W360 × 72 section is chosen whose factored bending strength is equal to $M_r = 346$ kN-m. In the design of the columns and panel zone, the factored forces originate from the factored moment, M_p in each beam assumed to be equal to 1,2 times the unfactored plastic moment. We assume a 310 mm deep section for these columns.

$$M_f = 1,2 Z_x F_y = (1,2)(1,28 \times 10^{-3} \text{m}^3)(300 \times 10^3 \text{kN/m}^2) = 461 \text{kN-m} \qquad (8.5)$$

First, the shear force is calculated at the end of the beams, V_1 and V_2, when plastic hinges are formed in the beams and when gravity-factored loads are applied. The free-body diagram shown in figure 8.6 is used to calculate the shear force.

$$V_1 = 27\text{kN}$$
$$V_2 = 233\text{kN} \qquad (8.6)$$

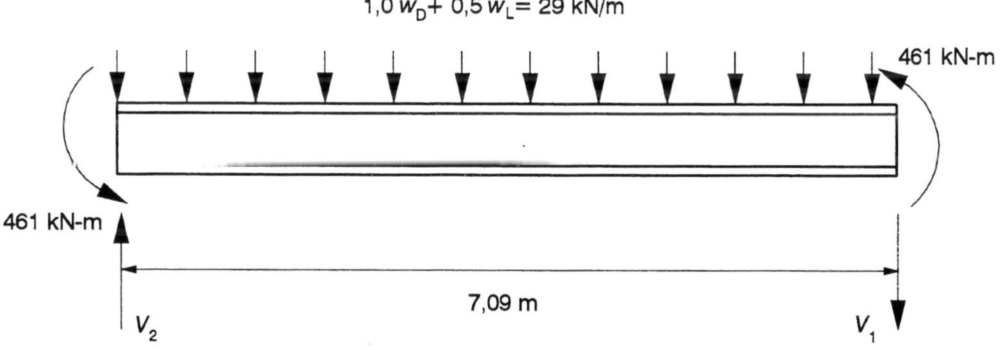

Figure 8.6   Shear force when plastic hinges form in the beams.

We then obtain the free-body diagram of the beam-column connection shown in figure 8.7.

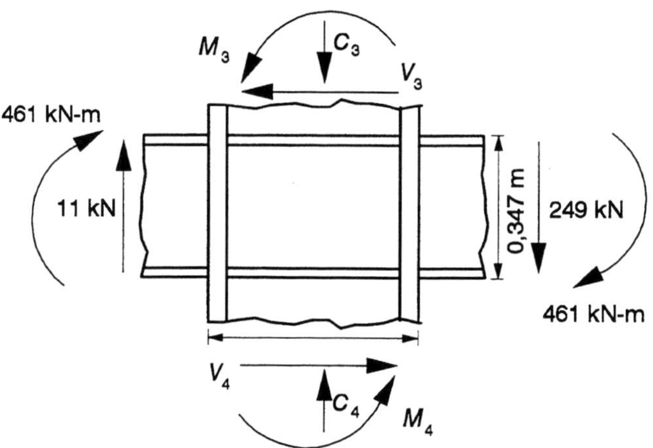

Figure 8.7   Free-body diagram of the beam-column connection corresponding to the formation of plastic hinges in the beams.

Taking the sum of the moments with respect to the centre of the joint, it yields:

$$M_3 + M_4 + 0{,}17\,V_3 + 0{,}17\,V_4 = 962\,\text{kN}-\text{m} \tag{8.7}$$

The shear forces in the columns, V_3 and V_4, can be written as a function of the bending moments in the columns, M_3 and M_4, assuming inflexion points at the mid-height of the columns.

$$V_3 = \frac{M_3}{1{,}68\,\text{m}}$$

$$\tag{8.8}$$

$$V_4 = \frac{M_4}{1{,}68\,\text{m}}$$

Finally, we can assume that the ratio of the moments in the columns is the same as the ratio of the moments caused by the seismic loads in table 8.1.

$$\frac{M_3}{M_4} = \frac{136}{163} = 0{,}83 \tag{8.9}$$

Substituting equations 8.8 and 8.9 into equation 8.7, we obtain the moments in the columns corresponding to the formation of plastic hinges in the beams.

$$M_3 = 396\,\text{kN-m}$$
$$M_4 = 477\,\text{kN-m} \tag{8.10}$$

To design the column, we must assume a factored axial load level. Combination (4) is chosen from table 8.1, which corresponds to the same combination used to calculate shear forces at the ends of the beams. So, for the column below the connection we have: $M_f = 477$ kN-m and $C_f = 2\,438$ kN. We assume that the stability criteria are met. Then, the column must meet the following strength criterion:

$$\frac{C_f}{C_r} + \frac{0,85\,M_f}{M_r} \leq 1,0 \tag{8.11}$$

A class 1, W310 × 179 section is chosen with $M_r = 823,5$ kN-m and $C_r = 6156$ kN.

$$\frac{2\,438}{6\,156} + \frac{(0,85)(477)}{823,5} = 0,89 < 1 \quad \text{o.k.} \tag{8.12}$$

To design the panel zone, we find the factored shear, V_f, in the column web, which corresponds to the formation of plastic hinges in the beams. We use the free-body diagram shown in figure 8.8.

$$V_f = T_2 + C_1 - V_3 \tag{8.13}$$

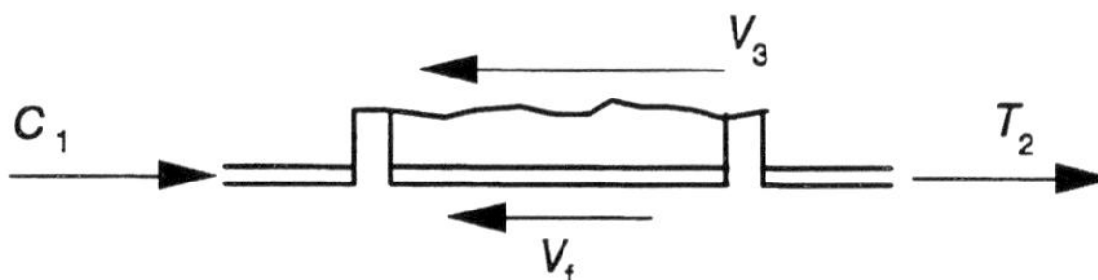

Figure 8.8 Shear force in the panel zone corresponding to the formation of plastic hinges in the beams.

The shear force in the column, V_3, is defined by equation 8.8. The tension and compressive force, T_2 and C_1, are obtained by assuming that the amplified plastic moment is taken entirely by the flanges of the beams.

$$T_2 = C_1 = \frac{461\,\text{kN-m}}{0,347\,\text{m}} = 1\,329\,\text{kN} \tag{8.14}$$

$$V_f = \frac{461}{0,347} + \frac{461}{0,347} - \frac{396}{1,68} = 1\ 329 + 1\ 329 - 236 = 2\ 422\,\text{kN} \qquad (8.15)$$

The factored shear strength of the panel zone, V_r, is obtained by equation 8.3.

$$V_r = (0,55)(0,9)(333)(18)(0,3)\left[1 + \left(\frac{(3)(313)(28,1)^2}{(333)(347)(18)}\right)\right] \qquad (8.16)$$

$$V_r = 1\ 207\,\text{kN} \le 2\ 422\,\text{kN} \quad \text{thus a reinforcement is needed}$$

To reinforce the panel zone, a 25 mm-thick doubler plate is welded to the column web by plug welds. With this reinforcement, $w_c = 18 + 25 = 43$ mm.

$$V_r = (1\ 207)\left(\frac{30}{18}\right)\left[1 + \left(\frac{(3)(313)(28,1)^2}{(333)(347)(30)}\right)\right] \qquad (8.17)$$

$$V_r = 2\ 442\,\text{kN} > 2\ 422\,\text{kN} \quad \text{o.k.}$$

Finally, the connection between the beams and columns must be designed for the forces corresponding to the formation of plastic hinges in the beams: $V_f = 249$ kN and $M_f = 461$ kN-m.

8.3   DETAILING REQUIREMENTS OF THE CANADIAN STEEL CODE FOR MOMENT-RESISTING FRAMES WITH NOMINAL DUCTILITY ($R = 3$)

Nominally ductile moment-resisting frames should sustain limited inelastic deformations. If all elements of a beam-column joint (beams, columns, and panel zone) are designed to reach their ultimate capacity, it is not necessary to identify the critical elements. All elements must be class 1 or 2. The connection between the beam flange and the column must be able to sustain the probable plastic moment of the beam or the moment in the beam caused by the probable shear resistance of the panel zone. However, the code sets a limit on the strength of the connection equal to 1,25 times the force produced by design seismic loads.

8.4   DETAILING REQUIREMENTS OF THE CANADIAN STEEL CODE FOR DUCTILE BRACED FRAMES ($R = 3$)

8.4.1 Degradation of Hysteresis Loops

For many years, the use of diagonal steel bracing was considered to be an efficient, versatile, and economical system, able to sustain lateral loads caused by wind or earthquakes. Figure 8.9 illustrates typical arrangements of bracing systems.

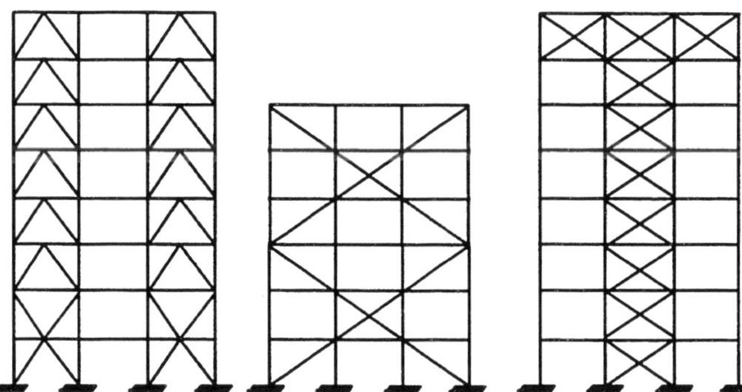

Figure 8.9 Typical bracing systems.

Normally, the bracing is positioned so that its line of action coincides with the centre line of the beam-column connection. This is called concentric bracing (or X bracing). It is often used along the peripheral walls of a building where it does not interfere with the architectural features. In the last 15 years, however, it was found that this system is not as efficient against earthquakes as it was first believed to be, especially when slender braces are used. In fact, there are two major problems associated with slender concentric bracing systems:

- a braced structure is stiffer than others (at least initially) and generally attracts high seismic loads;
- a braced structure may produce hysteresis loops that deteriorate with the number of cycles.

In theory, greater stiffness is preferred to reduce lateral deformations and prevent structural and architectural damage. However, the mechanical problems created by this bracing system cancel out these advantages. This is why many engineers, especially in California, prefer to use bracing systems with a second line of defence such as a moment-resisting frame.

The degradation of hysteresis loops occurs mostly on very slender, tension-only bracing systems. The behaviour of this system is shown in figure 8.10. The degradation of the hysteresis loops can be explained as follows:

- A lateral load on the right-hand side stretches the right diagonal member beyond its yield stress and causes buckling in the left diagonal member.
- When the load drops back to zero in the right diagonal member, the frame is not in its initial position and the left diagonal member is still buckled. From this point, until the frame returns to its initial position, the lateral stiffness of the frame is basically equal to zero as the two diagonal members are buckled.

- When the frame returns to its initial position, the right diagonal member is now buckled and the left diagonal member suddenly starts to stretch, causing an impact on the connection. This diagonal member also stretches beyond its yield stress.
- When the load drops back to zero in the left diagonal member, the frame is not in its initial position and the right diagonal member is still buckled. Again, the lateral stiffness of the frame is basically equal to zero as both diagonal members are buckled. However, the zero stiffness lasts longer than in the previous cycle as the frame must return to the previous deformed position before the right diagonal member starts to stretch, because this diagonal member has been permanently elongated.
- The same phenomenon is reproduced during the subsequent cycles, causing increasing degradation of the hysteresis loops.

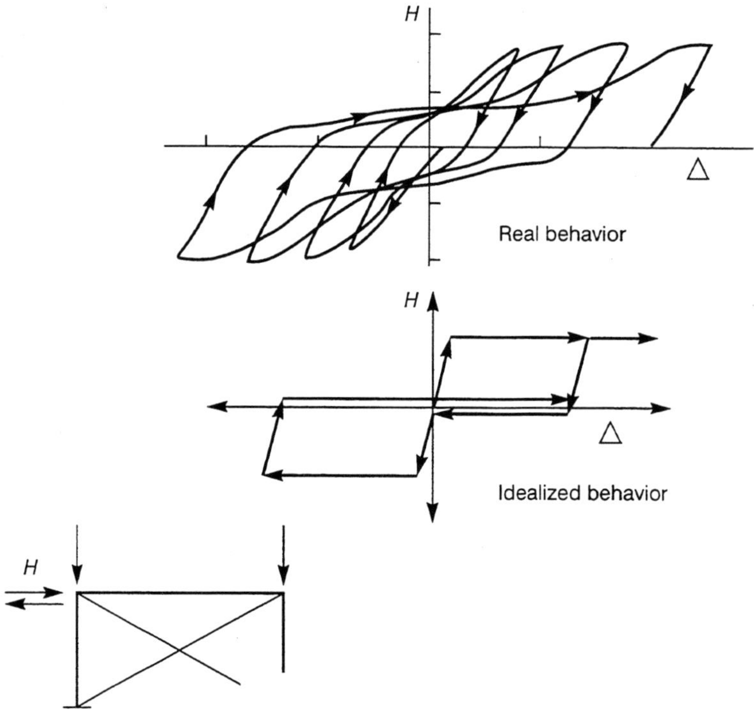

Figure 8.10   Degradation of the hysteresis loops of a portal frame with a tension-only bracing system.

The result of this phenomenon is mediocre seismic energy dissipation. Recognizing this problem, some building codes (in California, for example) reject the use of tension-only braced frames. In Canada, this system is permitted but the structure must be designed for higher lateral loads ($R = 1,5$).

The objective of the Canadian steel code concerning ductile braced frames is to ensure ductile behaviour through yielding of the braces in tension and inelastic buckling of the same braces in compression. The other elements (beams and columns) remain in the elastic range of the steel. To accomplish this, the code incorporates requirements for the global strength of the system, the slenderness of the diagonal members, the strength of the connection, and the strength of other members (beams and columns). Figure 8.11 illustrates the most important requirements of this clause.

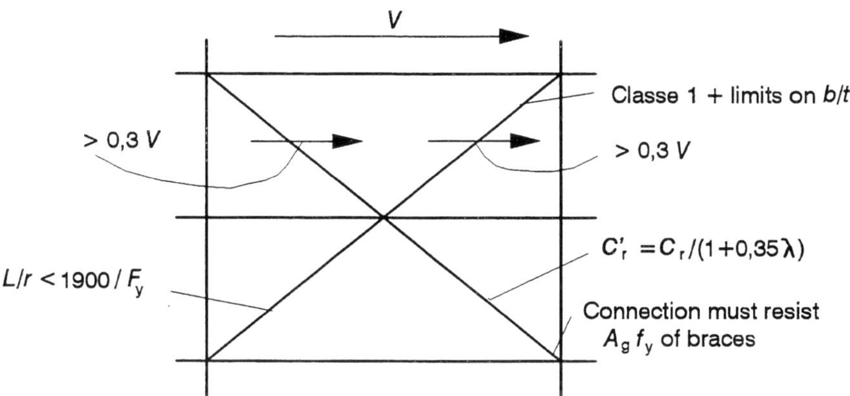

Figure 8.11 Summary of the most important requirements of the Canadian steel code for ductile bracing systems.

8.4.2 Global Strength Requirements

The diagonal members, first in tension then in compression, must be able to sustain at least 30% of the horizontal shear force prescribed by the NBCC (1995) to ensure adequate redundancy of the system. This requirement completely eliminates the use of tension-only braced frames. Figure 8.12 shows the bracing systems that can and cannot be used as ductile bracing. Chevron (*V* or *K)* braces cannot be used as ductile bracing.

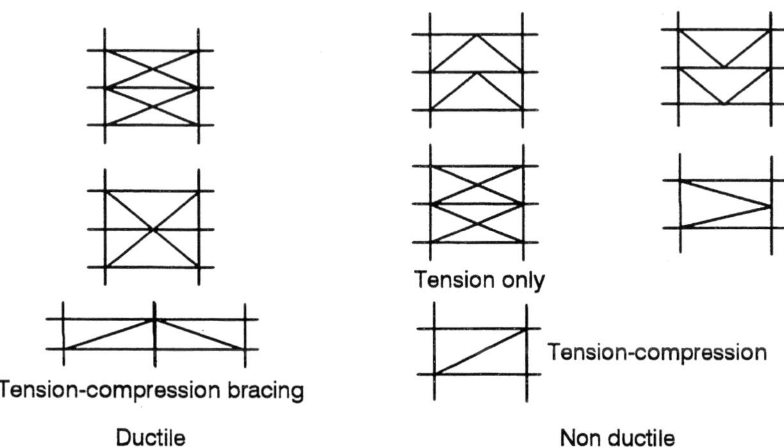

Figure 8.12 Ductile and non-ductile bracing according to the Canadian steel code.

8.4.3 Requirements for Bracing Members

To ensure adequate energy dissipation in compression, the global slenderness of a diagonal member, L/r, must be limited according to the following equation:

$$\frac{L}{r} \leq \frac{1\ 900}{\sqrt{F_y}} \qquad (8.18)$$

The code also limits the local slenderness of a diagonal member in order to eliminate all possibility of local buckling. All diagonal members must be class 1 and respect the following limits:

$$\frac{b}{t} \leq \frac{145}{\sqrt{F_y}}$$

for angles, T-sections and flanges of C-sections

$$\frac{b}{t} \leq \frac{330}{\sqrt{F_y}} \qquad (8.19)$$

for rectangular or square HSS

$$\frac{b}{t} \leq \frac{13\ 000}{F_y}$$

for circular HSS

In a seismic zone in which Z_a or $Z_v \leq 1$, the diagonal member must meet the limits of only class 1 or 2. If the diagonal member is made of a built-up section, the global slenderness of any element must be less than 50% of the global slenderness of the built-up section.

The ultimate compressive load that a diagonal member can sustain decreases with the number of cycles (figure 8.13). To take this phenomenon into account, the Canadian code requires use of a reduced compressive factored strength, C'_r, which is a function of the global slenderness of the diagonal member.

$$C'_r = \frac{C_r}{1 + 0{,}35\lambda}$$

$$\qquad (8.20)$$

$$\lambda = \frac{L}{r}\sqrt{\frac{F_y}{\pi^2 E}}$$

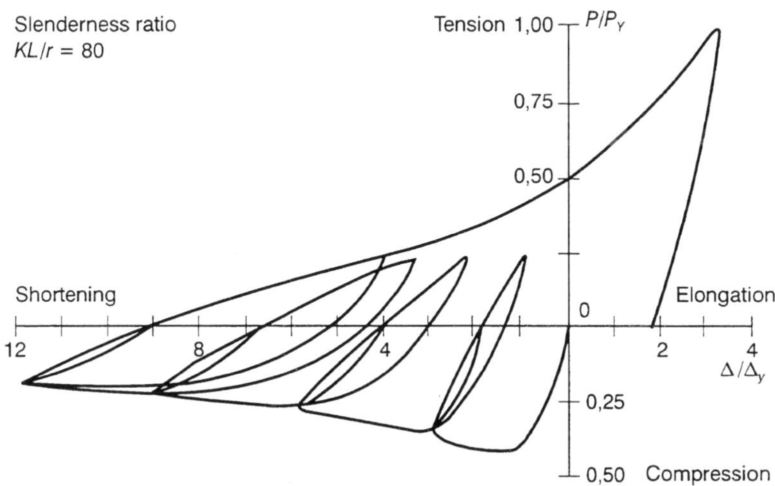

Figure 8.13   Hysteresis curve of a diagonal member (from Jain et al., 1980).

8.4.4 Requirements for Bracing Connections

The connections must be as strong as the diagonal members for seismic zones where Z_a or $Z_v \geq 3$. In these zones, the connections must have a factored strength at least equal to the nominal tensile yield load of the diagonal members, $A_g F_y$. In seismic zones 1 and 2, the code allows an upper limit to the connection strength equal to the forces caused by twice the factored loads plus the gravity loads. In fact, the diagonal members, mostly on higher floors, will be over-designed because of the slenderness limits. In this case, it is unlikely that the diagonal members will yield in tension because of the damping increase and the natural period shift that occurs when the diagonal members of lower floors yield. However, this clause must be used with caution, as the connections are weaker than the diagonal members. Some authors (Tremblay et al., 1991) demonstrated that shear forces on higher floors may be much higher than the limit set by the code.

8.4.5 Requirements for Other Members

Beams and columns must be able to sustain gravity loads plus the maximum loads in the diagonal members. The loads may be redistributed to take into account yielding and buckling of the diagonal members. As well, the reduction factor given by equation 8.20 must not be used to calculate the factored strength of the diagonal members in compression, unless it causes a worst critical situation for the beams or columns.

8.5   DETAILING REQUIREMENTS OF THE CANADIAN STEEL CODE FOR BRACED FRAMES WITH NOMINAL DUCTILITY ($R = 2$)

Braced frames with nominal ductility must be able to absorb a limited amount of energy through plastic deformations of the diagonal members in tension. The Canadian steel code requires that the diagonal members must be at least class 2. In seismic zones where Z_a or $Z_v \geq 2$, the connections must at least sustain the yield tensile load of the diagonal members, $A_g F_y$. In a seismic zone 1, the code sets an upper limit for the strength of connections that is equal to the seismic loads times a factor of 1,33 plus the gravitational loads. The beams and columns must also be able to sustain these forces. If the diagonal members are used in tension only, the forces acting on the connections must be increased by 10% to take possible impacts into account. Chevron bracings are permitted here for nominally ductile frames. The beam joining the diagonal members must be able to support the gravity loads on its own, without the help of the diagonal members.

8.6   INTRODUCTION TO ECCENTRIC BRACED FRAMES

Since the late 1970s, the use of eccentric bracing has spread in active seismic zones such as California. It was soon clear that when well designed, an eccentric braced frame can develop very high ductility during a major earthquake (Roader and Popov, 1978). The eccentric braced frame is illustrated in figure 8.14.

Unlike concentric bracing (section 8.4.1), the diagonal members of eccentric bracing do not coincide with the centre line of the beam-column connection. An eccentricity, e, is deliberately introduced. Then, the axial load in the diagonal member is transmitted to the column through moments and shear forces in the beam. If the system is designed correctly, the beam section between the diagonal member and the column, called the link beam, can dissipate the seismic energy by flexural and shear yielding. This system has the advantages of great ductility and stiffness compared to moment-resisting frames. The Canadian steel code also contains requirements for this type of system.

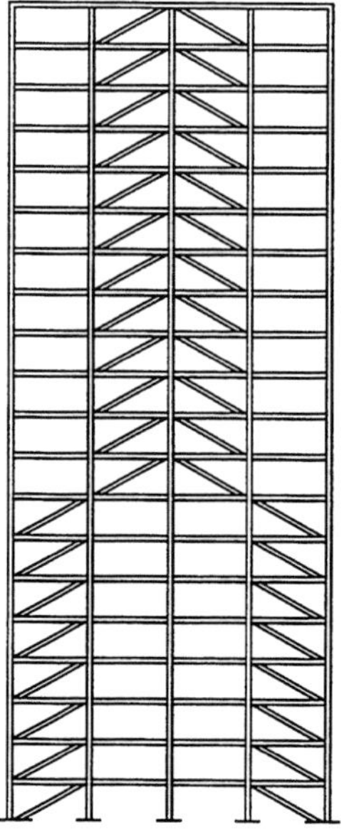

Figure 8.14   Eccentric braced frame (from Roader and Popov, 1978).

8.7 REFERENCES

CAN/CSA-S16.1094. (1994). *Limit States Design of Steel Structures*, Rexdale, ON: Canadian Standard Association.

Hanson, R.D., and Martin, H.W. (1987). "Performance of Steel Structures in the September 19 and 20, 1985 Mexico Earthquakes." *Earthquake Spectra* 3(2): 329–46.

Jain, A.K., Subhash, C.G., and Hanson, R.D. (1980). "Hysteretic Cycles of Axially Loaded Steel Members." ASCE, *Journal of the Structural Division*, 106(S-8): 1777–95.

Krawinkler, H., and Popov, E.P. (1982). "Seismic Behaviour of Moment Connections and Joints." ASCE, *Journal of the Structural Division*, 108(S2): 373–91.

National Building Code of Canada. (1995). 18th ed. Ottawa: National Research Council of Canada.

Popov, E.P., and Bertero, V.V. (1980). "Seismic Analysis of Some Steel Building Frames." ASCE, *Journal of the Engineering Mechanics Division,* 106(EM1): 75–92.

Popov, E.P., and Stephen, R.M. (1972). *Cyclic Loading of Full-Size Steel Connections*, Bulletin no. 21, American Iron and Steel Institute.

Roader, C.W., and Popov, E.P. (1978). "Eccentrically Braced Steel Frames for Earthquake." ASCE, *Journal of the Structural Division,* 104(S3): 391–412.

Thiel, C.C., ed. (1994). *Proceedings of the International Workshop on Steel, Seismic Issues, September 8-9, 1994.* Sacramento, CA: SAC Joint Venture.

Tremblay, R., Stiemer, S.F., and Filiatrault, A. (1991). "Seismic Behaviour of Concentrically Braced Frames." *Proceedings of the Annual Conference of Canadian Society of Civil Engineering*, Vancouver, BC, pp. 315–24.

CHAPTER 9

Elements of Energy Balance and Introduction to Innovative Earthquake-Resistant Systems

9.1 INTRODUCTION

As seen in section 6.4, the basic principle of conventional earthquake-resistant design is to ensure an acceptable safety level while avoiding catastrophic failures and loss of life. This principle is illustrated in table 9.1.

Table 9.1 Basic principle of conventional earthquake design.

Earthquake	To avoid	Required factor
Minor	Architectural damage	Stiffness
Moderate	Structural damage	Strength
Major	Collapse	Ductility

When a structure does not totally collapse during a major earthquake, we consider that this structure has fulfilled its function even though it may never be used again. Generally, this approach is used for most types of buildings. However, for important buildings, safer methods are required, while keeping economic factors in mind. Avoiding collapse is not sufficient for buildings that must remain functional immediately after an earthquake: hospitals, police stations, communication centres, and so on.

Over the last 20 years, a great deal of research has been conducted into developing new earthquake-resistant systems in order to raise the safety level while keeping construction costs reasonable. The purpose of this chapter is to discuss two of these systems that have demonstrated considerable potential:

- the base isolation system (elastomeric isolator);
- the friction-damped bracing system (Pall's system).

However, to understand better the contribution of each system during ground shaking, we first develop the energy balance equation of a single-degree-of-freedom system.

9.2   ENERGY BALANCE EQUATION

Let us consider a simple portal frame with only one degree of freedom, as shown in figure 9.1.

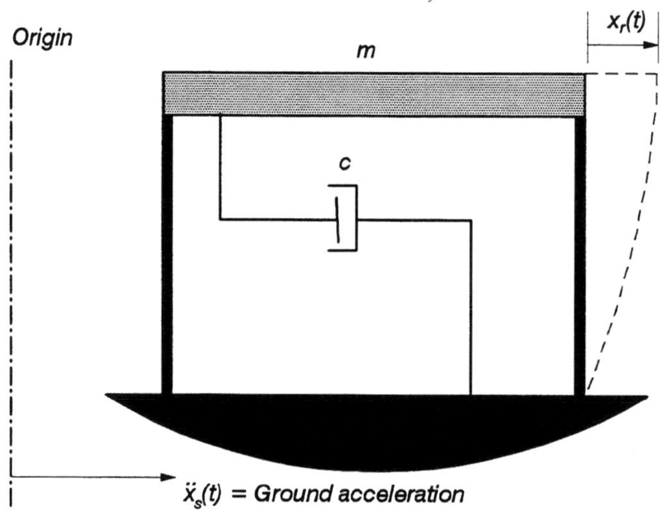

Figure 9.1   SDOF simple portal frame.

The equation of motion for this system subjected to a base motion is:

$$m\ddot{x}_r(t) + c\dot{x}_r(t) + F_s(t) = -m\ddot{x}_s(t) \tag{9.1}$$

where m =   mass of the system

c =   viscous damping coefficient of the system

$\dot{x}_r(t), \ddot{x}_r(t)$ = relative velocity and acceleration with respect to the ground

$\ddot{x}_s(t)$ =   ground acceleration (accelerogram)

$F_s(t)$ =   nonlinear restoring force (hysteretic) produced by the stiffness of the structure (figure 9.2)

Equation 9.1 is multiplied by the relative velocity $\dot{x}_r(t)$, and we integrate with respect to time. This yields:

$$m\int_0^t \ddot{x}_r\dot{x}_r\,dt + c\int_0^t \dot{x}_r\dot{x}_r\,dt + \int_0^t F_s(t)\dot{x}_r\,dt = -m\int_0^t \ddot{x}_s\dot{x}_r\,dt \tag{9.2}$$

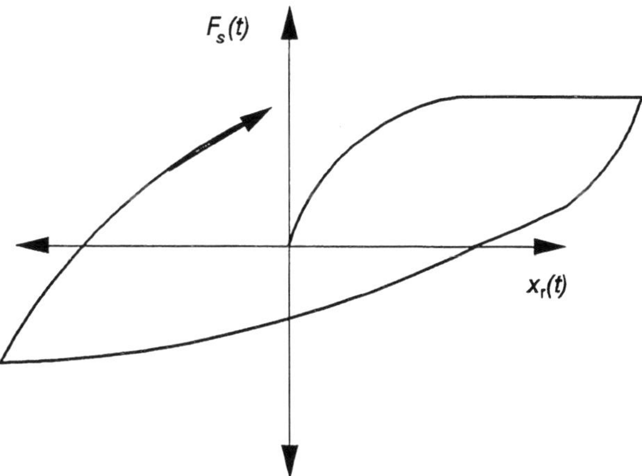

Figure 9.2 Nonlinear restoring force.

Differential relations are now used between the displacement, velocity, and acceleration.

$$\dot{x}_r = \frac{dx_r}{dt}$$

$$\ddot{x}_r = \frac{d\dot{x}_r}{dt} \tag{9.3}$$

Substituting equation 9.3 into equation 9.2 yields:

$$m \int_0^{\dot{x}_r} x_r\, dx_r + c \int_0^{x_r} x_r\, dx_r + \int_0^{x_r} F_s(t)\, dx_r = -m \int_0^{x_r} \ddot{x}_s\, dx_r \tag{9.4}$$

The first term on the left-hand side of equation 9.4 can be integrated directly to yield:

$$\frac{1}{2}\, m\dot{x}_r^2 + c \int_0^{x_r} \dot{x}_r\, dx_r + \int_0^{x_r} F_s(t)\, dx_r = -m \int_0^{x_r} \ddot{x}_s\, dx_r \tag{9.5}$$

or:

$$T_r(t) + D(t) + U(t) = I_r(t) \tag{9.6}$$

where $T_r(t)$ = kinetic energy at time t caused by the relative motion of the mass with respect to the base

$D(t)$ = energy dissipated by viscous damping up to time t

$U(t)$ = strain energy of the system at time t. Part of this energy can be restored (elastic strain energy) when the other part is dissipated by the nonlinear behaviour of the structure (hysteretic energy)

$I_r(t)$ = relative input energy introduced into the system

Notice that the input energy introduced into the system is equal to the equivalent seismic force, $-m\ddot{x}_s$, integrated through the relative displacement at the roof level, dx_r. In other words, the input energy depends on the characteristics not only of the earthquake (accelerogram) but also of the structure. In fact, the structure acts as a filter attracting the input energy at specific frequencies; it is the resonance phenomenon seen from an energy perspective.

It is imperative to understand that for a time-step nonlinear dynamic analysis, we can calculate each energy component of equation 9.6 individually at the end of each time-step. The accuracy of the analysis can be quantified by first setting a tolerance level on the energy error obtained, then comparing the sum of the energy components (left-hand-side terms in equation 9.6) with the input energy introduced into the system (right-hand-side term of equation 9.6). There are many techniques available to evaluate numerically the different energy components at the end of each time increment. For example, if we use the trapezoidal rule, it yields:

$$T(t) = \frac{1}{2}m\dot{x}_r^2$$

$$D(t) = D(t-\Delta t) + \frac{1}{2}c[\dot{x}_r(t-\Delta t) + \dot{x}_r(t)][x_r(t) - x_r(t-\Delta t)]$$

$$U(t) = U(t-\Delta t) + \frac{1}{2}[F_s(t-\Delta t) + F_s(t)][x_r(t) - x_r(t-\Delta t)] \tag{9.7}$$

$$I(t) = I(t-\Delta t) - \frac{1}{2}m[\ddot{x}_s(t-\Delta t) + \ddot{x}_s(t)][x_r(t) - x_r(t-\Delta t)]$$

where Δt represents the time-step used in the analysis.

Figure 9.3 illustrates an example of energy calculation for a simple single-degree-of-freedom elasto-plastic system.

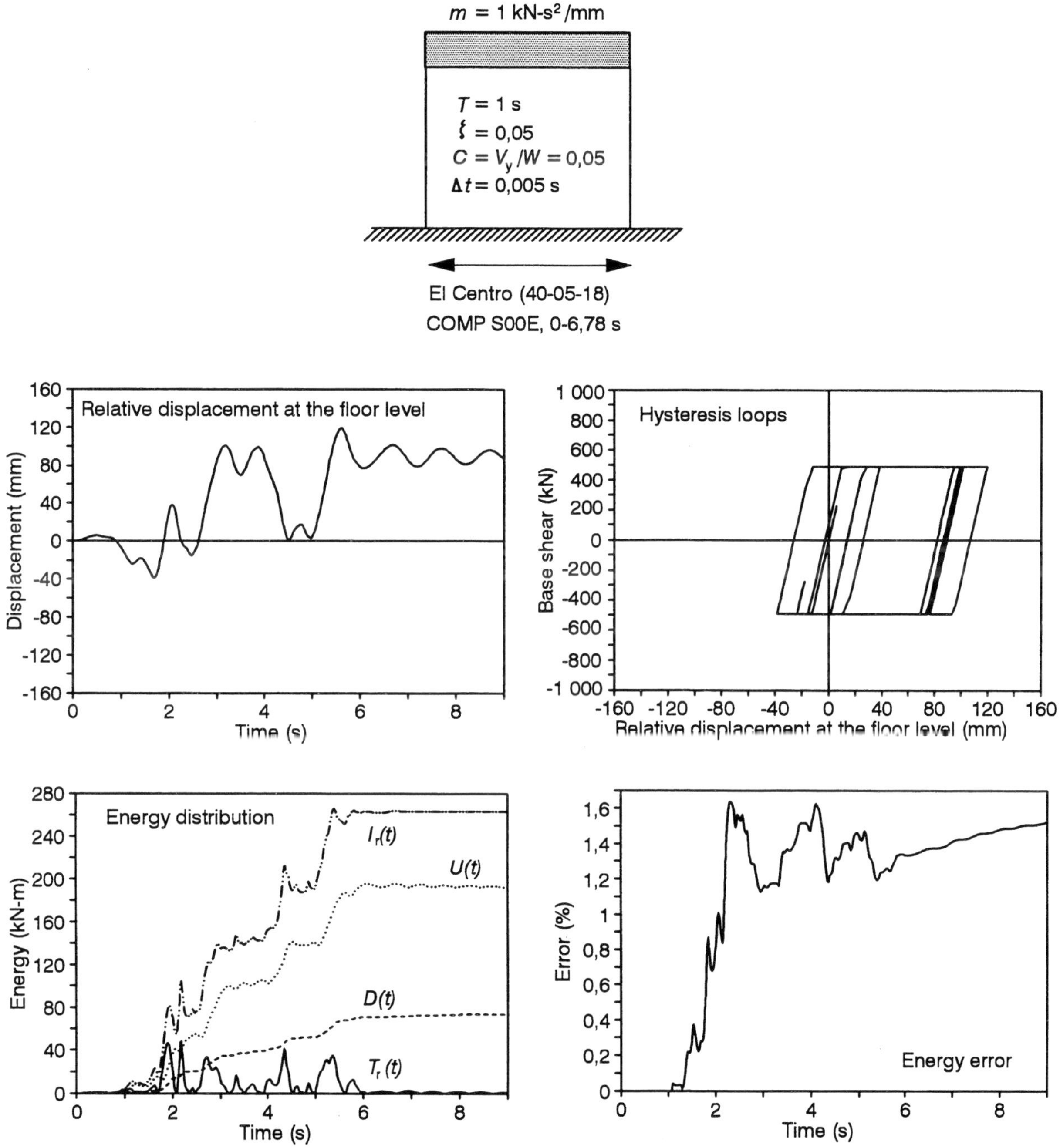

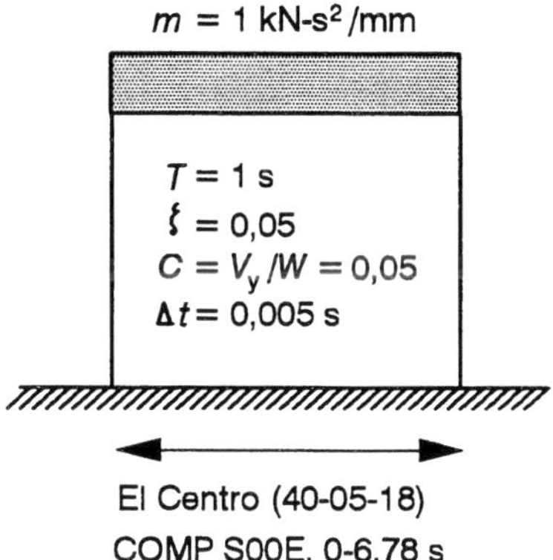

Figure 9.3   Energy calculation for a single-degree-of-freedom elasto-plastic system.

9.3   RELATIVE INPUT ENERGY AND ABSOLUTE INPUT ENERGY

The energy balance given by equation 9.6 is for an equivalent seismic load applied to a rigid base system (fig. 9.4).

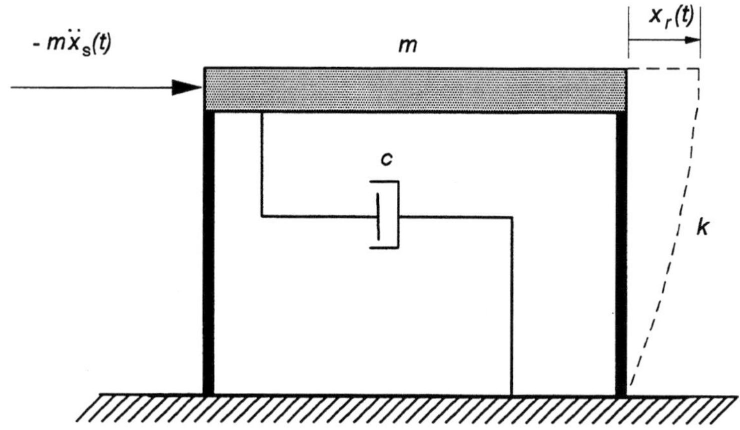

Figure 9.4   Equivalent seismic problem.

This relative approach does not directly consider the energy associated with the rigid body translation of the structure. In other words, the total kinetic energy is not considered explicitly. An absolute energy formulation is obtained by integrating by parts the right-hand-side term in equation 9.6.

$$I_r(t) = -m \int_0^t \ddot{x}_s \dot{x}_r \, dt = -m \dot{x}_s \dot{x}_r + m \int_0^t \dot{x}_s \ddot{x}_r \, dt \tag{9.8}$$

The relative acceleration, $\ddot{x}_r$, is written as a function of the absolute acceleration, $\ddot{x}_a$.

$$\ddot{x}_r = \ddot{x}_a - \ddot{x}_s \tag{9.9}$$

Substituting equation 9.9 in the right-hand-side term of equation 9.8 yields:

$$I_r(t) = -m \dot{x}_s \dot{x}_r + m \int_0^t \dot{x}_s \ddot{x}_a \, dt - m \int_0^t \dot{x}_s \ddot{x}_s \, dt \tag{9.10}$$

where

$$\ddot{x}_s = \frac{d\dot{x}_s}{dt}$$

$$\dot{x}_s \, dt = dx_s \tag{9.11}$$

Substituting equation 9.11 into equation 9.10 yields:

$$I_r(t) = -m\dot{x}_s\dot{x}_r + m\int_0^{x_g}\ddot{x}_a\,dx_s - m\int_0^{\dot{x}_g}\dot{x}_s\,d\dot{x}_s \tag{9.12}$$

Integrating the last term of the right-hand side, it yields a new expression for the relative input energy.

$$I_r(t) = -m\dot{x}_s\dot{x}_r + m\int_0^{x_g}\ddot{x}_a\,dx_s - \frac{1}{2}m\dot{x}_s^2 \tag{9.13}$$

Substituting equation 9.13 into the energy balance equation:

$$\frac{1}{2}m\dot{x}_r^2 + c\int_0^{x_r}\dot{x}_r\,dx_r + \int_0^{x_r}F_s(t)\,dx_r = -m\dot{x}_s\dot{x}_r - \frac{1}{2}m\dot{x}_s^2 + m\int_0^{x_g}\ddot{x}_a\,dx_s \tag{9.14}$$

Rearranging the right-hand-side terms:

$$\frac{1}{2}m\dot{x}_r^2 + m\dot{x}_s\dot{x}_r + \frac{1}{2}m\dot{x}_s^2 + c\int_0^{x_r}\dot{x}_r\,dx_r + \int_0^{x_r}F_s(t)\,dx_r = m\int_0^{x_g}\ddot{x}_a\,dx_s \tag{9.15}$$

or:

$$\frac{1}{2}m(\dot{x}_r+\dot{x}_s)^2 + c\int_0^{x_r}\dot{x}_r\,dx_r + \int_0^{x_r}F_s(t)\,dx_r = m\int_0^{x_g}\ddot{x}_a\,dx_s \tag{9.16}$$

or:

$$\frac{1}{2}m\dot{x}_a^2 + c\int_0^{x_r}\dot{x}_r\,dx_r + \int_0^{x_r}F_s(t)\,dx_r = m\int_0^{x_g}\ddot{x}_a\,dx_s \tag{9.17}$$

Equation 9.17 can be rewritten as follows:

$$T_a(t) + D(t) + U(t) + = I_a(t) \tag{9.18}$$

where $T_a(t)$ = kinetic energy at time t caused by the absolute motion of the mass
$\quad\quad\quad D(t)$ = energy dissipated by viscous damping up to time t
$\quad\quad\quad U(t)$ = strain energy at time t. Part of this energy can be restored (elastic strain energy) but the other part is dissipated by the nonlinear behaviour of the structure (hysteretic energy)
$\quad\quad\quad I_a(t)$ = absolute input energy introduced at time t

The absolute input energy introduced in the system is equal to the base shear, $m\ddot{x}_a$, integrated through the ground displacement, dx_s.

This absolute energy expression requires the ground displacement for the computations, while the relative approach requires only the acceleration, which must nevertheless be used for the dynamic analysis.

Uang and Bertero (1990) demonstrated that absolute and relative input energy are essentially identical for a natural period range of 0,1 to 5 s.

9.4 ENERGY BALANCE AT THE END OF AN EARTHQUAKE

From an energy point of view, the seismic problem can be interpreted as a finite quantity of energy being filtered by a structure. Figure 9.5 illustrates the seismic energy distribution at the end of an earthquake when the structure is at rest (kinetic energy and elastic strain energy = 0). Let us now see how the two previously discussed energy-dissipation systems (base isolation and friction-damped bracing) influence energy distribution.

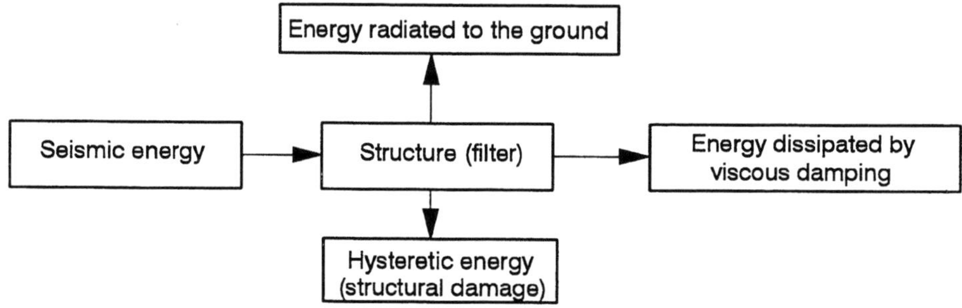

Figure 9.5 Energy distribution at the end of an earthquake.

9.5 BASE ISOLATION SYSTEM

A base isolation system requires the installation of isolators beneath every supporting point of the structure. The isolators, which have a much lower lateral stiffness than the structure, separate it from the ground. From an energy point of view, the base isolation system limits seismic energy transfer to the structure. Theoretically, if seismic energy is zero, it is not necessary to design the structure to sustain lateral loads. Although there are many different systems, the principle of all base isolation systems remains the same. The system combines two main components:

- An isolator, such as ball bearings or rubber pads, placed between the foundation and the columns of the structure (Kelly, 1979). Because it has a lateral stiffness much lower than the stiffness of the structure, the isolator shifts the natural frequency of the structure below the most prevailing frequency band of typical earthquakes.
- An energy-dissipation mechanism (damper) that dissipates residual input energy and limits the forces transmitted to the structure.

The base isolation system that is most highly developed is the lead-plug elastomeric bearing (Robinson, 1982). This system is composed of elastomeric layers (rubber) alternating with steel plates solidly joined together under high pressure and high temperature (vulcanized), and a central lead plug. This system is illustrated in figure 9.6.

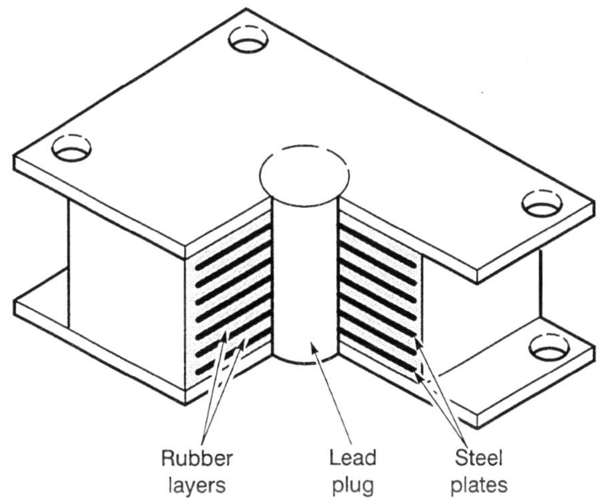

Figure 9.6 Lead plug elastomeric bearing.

Experimental studies (Robinson, 1982) have demonstrated that an elastomeric isolator can simply be modelled as a solid with a bi-linear shear behaviour (figure 9.7). The parameters of this model represent initial stiffness, $K_1 = 10\,K_r$, post-elastic stiffness, $K_2 = K_r$, and the yielding load, F_y, which is a function of the yielding force of the lead plug (about 10 MPa).

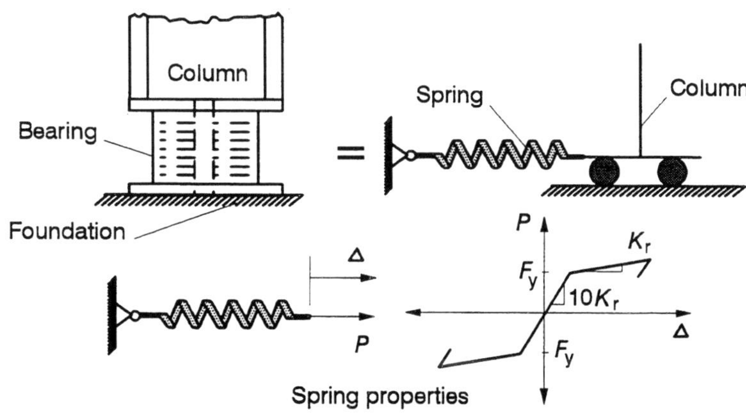

Figure 9.7 Modelling of a lead plug elastomeric bearing.

The isolator's vertical stiffness is much greater than its lateral stiffness (on the order of 300 to 400 times). We can thus assume that the isolator is perfectly rigid vertically.

Experimental and theoretical studies (Blakeley, 1980; Eisenberg and Rutenberg, 1986; Lee and Medland, 1978; Megget, 1978) have shown reasonable values for lateral stiffness and yield load for lead-plug elastomeric bearings.

$$K_r = (1 \text{ to } 2) \, W$$
$$F_y = (0{,}05 \text{ to } 0{,}10) \, W \tag{9.19}$$

where W = total weight of the structure in kN
K_r = total lateral stiffness in kN/m
F_y = total yield load in kN

9.6   FRICTION-DAMPED BRACING SYSTEM

The friction-damped bracing system was first proposed by Canadian researchers (Pall and Marsh, 1982). Its goal is to counter problems caused by the hysteresis loop degradation of conventional steel bracing, as studied in section 8.4.1. This system is a simple mechanism containing friction devices installed at the intersection of the diagonal members (figure 9.8). These devices are designed not to slip during the building's normal operations or minor earthquakes. During major earthquakes, the devices slip at a predetermined load before yielding occurs in the structural elements. The slipping dissipates input energy mechanically without yielding in the diagonal or other structural members.

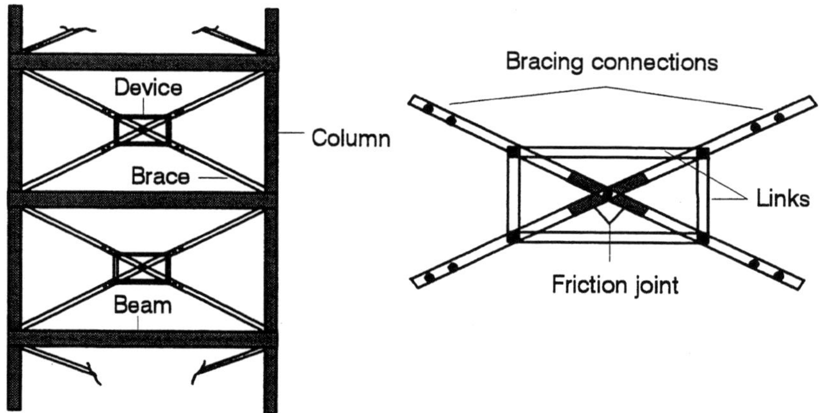

Figure 9.8   Friction-damped bracing system.

Moreover, deformation of the device allows larger and more stable hysteresis loops compared to traditional bracing. To explain this behaviour, let us consider a simple portal frame with a friction damped bracing system, as shown in figure 9.9.

Let us define:

V = lateral load at floor level representing the seismic load

Δ_1 = relative displacement at node C with respect to node A (defined as positive in figure 9.9)

Δ_2 = relative displacement at node B with respect to node D (defined as positive in figure 9.9)

P_1 = axial load in diagonal member 1 (positive in tension)

P_2 = axial load in diagonal member 2 (positive in tension)

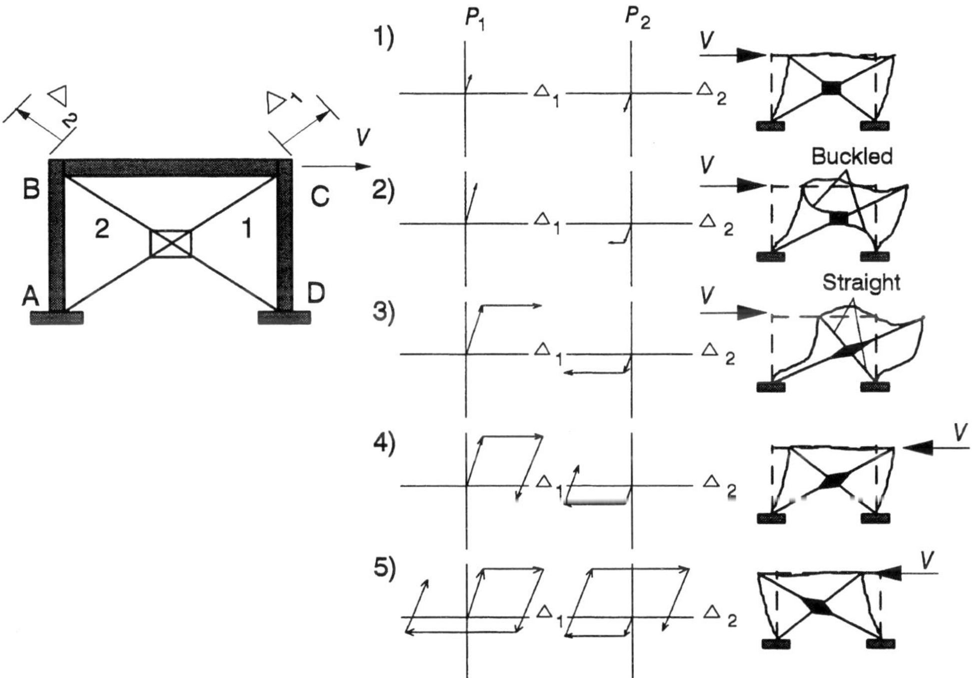

Figure 9.9 Hysteretic behaviour of a simple portal frame with friction bracing.

Figure 9.9 illustrates five different steps during a typical loading cycle. For each step, we also show the modelled hysteresis loop for each diagonal member and the frame's deformation. These steps are:

1. Under low lateral load, the two diagonal members remain elastic in tension and compression.
2. The buckling load is reached in diagonal member 2 while diagonal member 1 is still stretching elastically.
3. The device slips before yielding occurs in diagonal member 1. During slipping, the four hinges deform the rectangular mechanism into a diamond shape. This deformation compensates for the buckling deformation occurring in diagonal member 2. At the end

of slipping, P_2 is still equal to the critical buckling load, but diagonal member 2 is almost straight and taut.

4. When the load is reversed, diagonal member 2 becomes ready to stretch in tension.
5. At the end of the cycle, the loops are identical for both diagonal members. Degradation is virtually eliminated because the diagonal members do not yield in tension. Moreover, the deformation of the device compensates for the post-buckling deformation.

To develop the hysteresis loops shown in figure 9.9, the devices must themselves show load-slipping diagrams that are very stable, without deterioration. Figure 9.10 shows a hysteresis loop obtained on a typical friction device (Filiatrault and Cherry, 1987).

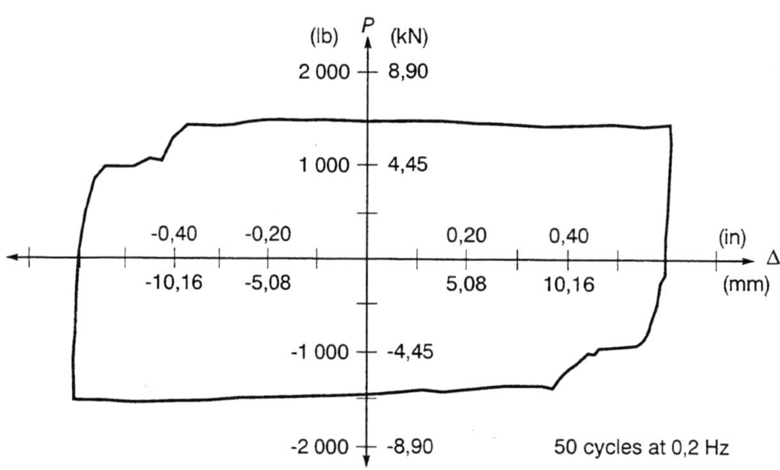

Figure 9.10   Hysteresis loop of a friction device (from Filiatrault and Cherry, 1987).

We note that even after 50 cycles, the loop resembles a rectangle and shows very little deterioration. Shake-table tests were also performed on a three-storey frame equipped with a friction-damped bracing system (Filiatrault and Cherry, 1987). It was observed that the system allowed for a great reduction in seismic response.

9.7   COMPARATIVE STUDY BETWEEN BASE ISOLATION AND FRICTION-DAMPED BRACING SYSTEMS

A comparison has been made (Filiatrault and Cherry, 1988) between the seismic performance of a steel frame equipped with (i) a tension-only braced moment resisting frame with a stiff base (BMRF), (ii) a tension-only base isolated braced moment-resisting frame (BIBMRF), and (iii) a friction-damped braced frame (FDBF). The frame studied had 10 floors with alternating bracing, as shown in figure 9.11. Time-step nonlinear analyses were done with three different accelerograms.

- The 1940 El Centro earthquake, S00E component, 0-6 s
- The 1977 Romania earthquake, Bucharest, N-S component, 0-16 s
- The 1985 Mexico earthquake, SCT, E-W component, 0-75 s

The structural damage was simply evaluated by setting a damage ratio, DR, defined as follows:

$$\text{DR} = \frac{\text{Number of yielded members}}{\text{Total number of members}} \tag{9.20}$$

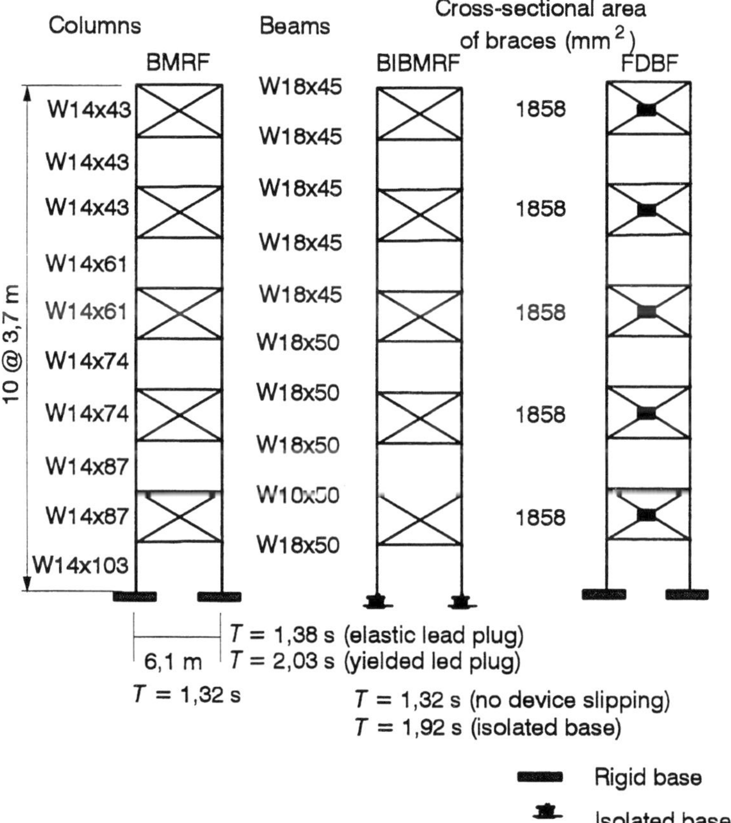

Figure 9.11   Ten-storey steel frame.

Some of the results of this comparative study are presented in figure 9.12. We note that the two systems (base isolation and friction-damped bracing) are very efficient during the El Centro earthquake. However, the friction-damped system is better than the base isolation system for the Romania and Mexico earthquakes. This can be explained by the fact that the fundamental period of the base isolated structure is almost identical to the fundamental period of the ground motion, causing a quasi-resonance phenomenon. Induced resonance is difficult to achieve with the friction-damped bracing system.

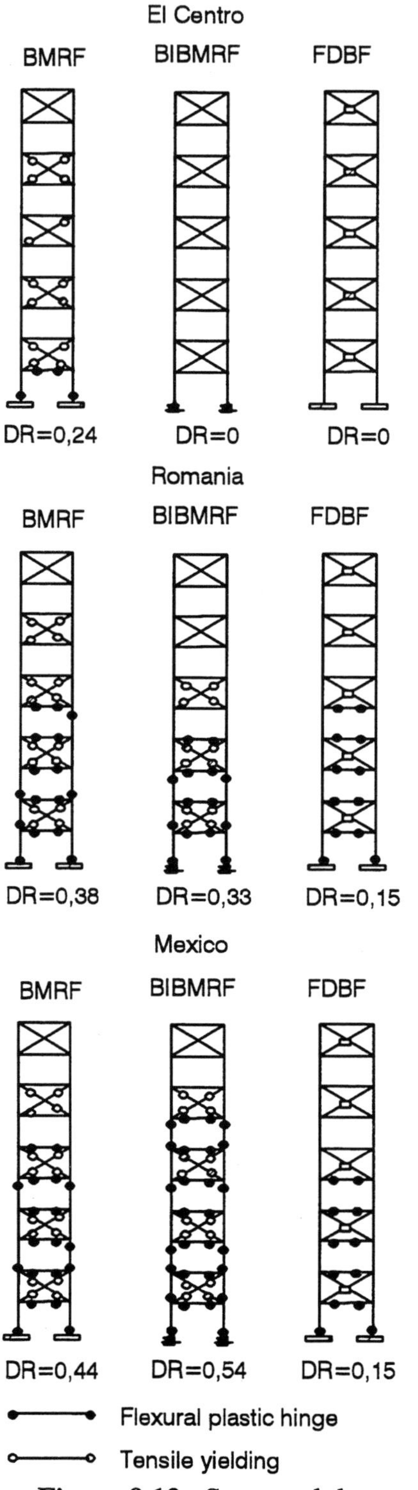

Figure 9.12 Structural damage.

9.8 REFERENCES

Blakeley, R.W.G., Cormack, L.G., and Stockwell, M.J. (1980). "Mechanical Energy Dissipation Devices." *Bulletin of the New Zealand National Society for Earthquake Engineering*, 13(3): 264–68.

Eisenberger, M., and Rutenberg, A. (1986). "Seismic Base Isolation of Asymmetric Shear Buildings." *Engineering Structures*, 8(1): 2–8.

Filiatrault, A., and Cherry, S. (1987). "Performance Evaluation of Friction Damped Braced Steel Frames Under Simulated Earthquake Loads." *Earthquake Spectra*, 3(1): 57–78.

— (1988). "Comparative Performance of Friction Damped Systems and Base Isolation Systems for Earthquake Retrofit and Aseismic Design." *Earthquake Engineering and Structural Dynamics*, 16(3): 389–416.

Kelly, J.M. (1979. "Aseismic Base Isolation: A Review." *Proceedings of the 2nd US National Conference on Earthquake Engineering*, Stanford, CA, pp. 823–37.

Lee, D.M., and Medland, M. (1978). "Estimation of Base Isolated Structure Responses." *Bulletin of the New Zealand Society for Earthquake Engineering*, 11(4): 234–44.

Megget, L.M. (1978). "Analysis and Design of a Base-Isolated Reinforced Concrete Frame Building." *Bulletin of the New Zealand National Society for Earthquake Engineering*, 11(4): 245–54.

Pall, A.S., and Marsh, C. (1982. "Response of Friction Damped Braced Frames." *ASCE, Journal of the Structural Division*, 108(ST6): 1313–23.

Robinson, W. (1982). "Lead-Rubber Hysteretic Bearings Suitable for Protection of Structures During Earthquake." *Earthquake Engineering and Structural Dynamics*, 10(4): 593.

Uang, C.M., and Bertero, V.V. (1990). "Evaluation of Seismic Energy in Structures." *Earthquake Engineering and Structural Dynamics*, 19(1): 77–90.

User's Manual for the RESAS Computer Program

A.1 INTRODUCTION

The RESAS program allows the computation of exact response spectra, *SD*, *SV* and *SA* (section 4.3.3). The program is interactive and has a graphics module that allows viewing of the response spectra. The available number of vibration periods and critical damping ratios is limited only by the memory available in the personal computer used.

A.2 HARDWARE REQUIREMENTS

The RESAS program was written in Turbo Pascal 5.1 language. The RESAS version available on the web site http://www.polymtl.ca/pub requires the following computer hardware:

- an IBM or compatible computer with 640k RAM and a math co-processor;
- MS-DOS operating system, version 4.0 or later;
- a graphics module with VGA screen.

A.3 INSTALLATION AND START-UP

The RESAS program has already been compiled and is ready to run. To install on a hard disk, the same sub-directories as on the web page must be created on the hard drive and all files must be copied to the appropriate sub-directories.

To run RESAS, choose drive A: (or other if the program is installed on hard drive) and type in: "CD A:\RESAS←" and "RESAS←". The main menu will then appear on screen.

A.4 MAIN KEYSTROKES

The following keys provice access to the program's menus.

→	moves the cursor to the right of the screen
←	moves the cursor to the left of the screen;
↑	moves the cursor to the top of the screen;
↓	moves the cursor to the bottom of the screen;
←	return, selects the field or function or enters information;
Esc	quits a window, cancels the choice of a field or interrupts computations.

A.5 MENUS AND WINDOWS

When running the program, the main menu is displayed on screen. This menu allows the user to:

- get a file containing information from the data entry window;
- save information from the data entry window;
- open the data entry window;
- quit the RESAS program.

The data entry window gives the necessary information for computation of response spectra. The three fields in the "TITLE" section allow the user to define a project title. These fields will be displayed when numerical and graphical information is displayed and when saving. By default, the information in the "DEFAUT.RES" file will be displayed in this window.

Field descriptions for the "Accelerograms" section are found in table A.1.

Table A.1   Field descriptions for the "Accelerograms" section.

Field	Description
File	Name of the accelerogram file, including the extensions
Number of lines to ignore	Number of lines containing the title that precede the acceleration numerical values in the accelerogram file
Number of points	Number of acceleration data points
dt	Time-step between two acceleration values
Units	Acceleration units. The units can be modified by typing ↵ and choosing the desired units.

Field descriptions for the "Period(s)" and "Damping" sections are shown in table A.2.

Table A.2   Field description for the "Periods(s)" and "Damping" sections.

Field	Description
Start	First natural period (or critical damping ratio) by which the response spectra are calculated. The first period cannot be equal to zero.
End	Last natural period (or critical damping ratio) by which the response spectra are determined.
Step	Increment between the first and last natural period (or critical damping ratio).

The "Output units" field allows the user to define the units for the response spectra. These units can be modified by typing ← and typing in the desired units.

The "Analysis" field allows calculation of the spectral values.

The "Data viewing" field allows viewing of the numerical results or plotting of response spectra graphs. To print these graphs, the "GRAPHICS" command of the DOS system must be activated before the RESAS program is run.

The "Data saving" field allows the user to save the calculated response spectra into a file.

The "Quit" field closes the "Data entry" window and returns to the main menu.

A.6 EXAMPLE

An example is given in the data file under the name "DEMO.RES". This file is loaded using the "Loading a file" field in the main menu. The full name with extension is: A:\RESAS\DEMO.RES. The accelerogram associated with this example is the longitudinal component of the 1988 Saguenay earthquake, recorded at site no. 20, "Les Éboulements" (QC). It presents an absolute acceleration spectrum of 10% critical damping. After loading this data file, activate the "Analysis" field in the "Data entry" window.